T0192432

Computational Fluid Dynamics for Incompressible Flows

Computational Fluid Dynamics for Incompressible Flows

D. G. Roychowdhury

CRC Press is an imprint of the
Taylor & Francis Group, an **Informa** business

First edition published 2020
by CRC Press
6000 Broken Sound Parkway NW, Suite 300, Boca Raton, FL 33487-2742

and by CRC Press
2 Park Square, Milton Park, Abingdon, Oxon, OX14 4RN

CRC Press is an imprint of Taylor & Francis Group, LLC

ISBN: 978-0-367-40806-0 (Hardback)
ISBN: 978-0-367-52432-6 (Paperback)
ISBN: 978-0-367-80917-1 (eBook)

Visit the eResource page: www.routledge.com/9780367408060

Dedication

To my Parents, Wife and Daughters

Contents

Preface

Computational Fluid Dynamics (CFD) has become a powerful and essential tool for fluid-flow analysis with the increase of computational power, better solution algorithms and reduced cost of the computer. CFD can provide insight into detailed flow behavior in a system. Today, engineers are increasingly using it for the cost-effective, optimal design of a component/system based on performance analysis. CFD has been successfully applied to various fields of engineering and even medicine; the range of application is wide and encompasses many different fluid phenomena.

Many commercial general-purpose CFD codes and very specialized CFD programs are available. All the commercial codes are user-friendly and have a pre-processing, solver and post-processing modules. However, while using the commercial CFD codes, the user has no idea what is going on inside the solver. It will be very difficult for the user to effectively use these codes unless he/she is exposed to basic knowledge of the CFD process.

This book is written mainly for incompressible flow. It is prepared in simple language with enough background for practicing engineers to find the book useful for strengthening their fundamental knowledge in CFD. During my study and teaching, I felt the need for a book that deals in detail with both finite difference and finite volume methods. This is necessary as the finite difference method provides the basic foundations in CFD and strengthens the fundamentals of CFD for the students/researchers engaged in the field.

I hope that this book will be very useful for both the undergraduate and graduate students. Wherever necessary, some advanced topics are added, which may be useful for research students to gain insight and appreciate the CFD procedure. At the end of the book, a chapter on best practice guidelines helps students and practicing engineers.

Most chapters provide workout examples and questions to allow students to check their understanding of the subject manually as well as with the help of computer either by writing their own program or using commercial codes. Undergraduate and graduate students undergoing CFD coursework are advised to undertake CFD problems as their project work for a better understanding of the essentials of the CFD process.

SCOPE

Chapter 1 is devoted to an overview of CFD and a basic introduction to various CFD processes. The structure of commercial codes, basic governing equations and different types of discretization procedures with desirable properties are discussed.

Chapter 2 deals with the governing equations used in CFD and their derivations, even though students have already had an introduction to these equations in their earlier courses. The basic difference between conservative and non-conservative forms of equations and their merit are discussed, as are the physical and mathematical significance of partial difference equations and various types of boundary conditions.

Chapter 3 explores the fundamentals of the finite difference method, along with the errors associated with various numerical schemes. The consistency and stability of the method is discussed, and Von Neumann's method of assessing the stability of the finite difference scheme is narrated.

Chapter 4 looks into the application of various finite difference methods. This chapter discusses in detail, various common numerical methods of diffusion equation, wave equation, complete transport equation and inviscid and viscid Burgers' equations with associated error, consistency and stability consideration using explicit and implicit methods.

Chapter 5 deals with the finite volume method applied to one-dimensional steady state diffusion and convection-diffusion problems. Discretization schemes for convection fluxes and their merits are discussed in detail, and high-resolution and bounded convective schemes are provided, especially for researchers. The implementation of convective schemes in the code is discussed, so that readers can appreciate them. The extension of these schemes to two and three dimension is described. This chapter also deals with time dependent methods, including various discretization schemes for the transient term, and the implementation of boundary conditions in code.

Chapter 6 addresses the solution of incompressible N-S equations. The merit and demerits of co-located and staggered grid arrangements are highlighted. Both vorticity-stream function and primitive variable methods are explored. The SIMPLE algorithm and its variants used to solve the incompressible N-S equations are described in detail along with their associated merits and demerits.

Chapter 7 is devoted to the application of finite volume method in complex geometries. The advantage of using the non-orthogonal co-located grid is highlighted, and the detailed discretization procedure for non-orthogonal structured grid, Cartesian structured grid and unstructured grid with co-located meshes are described. Finally, the implementation of the SIMPLE algorithm for these cases is introduced.

Chapter 8 discussed the solution of algebraic equations arising from finite difference/finite volume discretization. Various direct and iterative methods of solving the system of equations and their merits are highlighted. Advanced methods like conjugate gradient methods, preconditioning and the multigrid method are discussed.

Chapter 9 deals with an introduction to turbulence and the need for turbulence modeling. Since most real-life flow is turbulent in nature, characteristics of turbulence and the task of turbulent modeling is discussed in detail. Popularly used RANS turbulence models like k-ε, k-ω, SST k-ω are discussed. The RSTM models and LES are presented.

Chapter 10 explores various types of grids and their classification. This chapter briefly discusses body-fitted grids, adaptive grid and mesh quality.

A short introduction to grid-generation methods for structured and unstructured grids is provided.

Chapter 11 involves sources of errors and the best practice guidelines to be followed to reduce/avoid such types of error. An analysis of results, sensitivity studies and uncertainties is provided to help CFD practitioners. The chapter also discusses the need for and methods of verification and validation.

In Appendix 1, the method for calculating the area and volume for non-orthogonal structured geometry is discussed, while in Appendix 2, the transformation of equation from physical space to computation space is addressed. In Appendix 3, basic vector calculus is discussed. Appendix 4 provides problems for readers to solving with commercial codes or by writing their own codes and gain knowledge in CFD.

Acknowledgements

I would like to put on record the role played by my teachers, Prof. S.M. Deshpande of the Indian Institute of Science, Bangalore, and Prof. T. Sundararajan and Prof. Jayanti Srinivas of the Indian Institute of Technology, Madras, for exposing me to the exciting field of CFD and providing me a basic background. I would like to acknowledge the Indira Gandhi Centre for Atomic Research, Kalpakkam, for providing the opportunities and training required during my entire career. I would also like to put on record the contributions made by my undergraduate and graduate students of the Hindustan Institute of Technology and Science, Chennai, in the form of useful discussions that provided me insight while writing this book.

Finally, I would like to thank my wife, Subhasree Roychowdhury, and daughters, Somasree Roychowdhury and Suryasree Roychowdhury, for their affectionate support, encouragement, tolerance of all my tantrums, and sacrifices to their personal lives that they have endured while I was writing this book.

D. G. Roychowdhury

1 Overview of CFD

1.1 INTRODUCTION

'Computational fluid dynamics' (CFD) is the science or art of solving problems involving fluid flow using computers and numerical techniques.

Industrial flow is often complex and involves intricate geometries in heat exchangers, turbomachinery, aircraft, electrical and electronic components, meteorology, biomedical engineering, nuclear reactor, etc. In practical situations, the flow is generally three dimensional and turbulent. Also, the governing equations involving fluid flow are non-linear. Hence, obtaining analytical solutions is impossible, although analytical solutions are available for simplified cases (like pipe flow, etc.) that are obtained after various degrees of assumptions. These very useful techniques lack universal applications; we shall see that turbulent flows are not amenable to analytical solutions. In earlier days, designers used various correlations derived from experiments for designing components. However, experiments normally provide global parameters like drag, lift, pressure drop and heat transfer coefficients. A basic understanding of the local details of flow and transport phenomena is essential for optimal design performance of equipment. If we want to obtain any local flow parameters, various sophisticated instruments are necessary, making the method very costly and, in some cases, altering the flow phenomena. Also, not all flow parameters can be simulated in the same setup.

CFD simulations can provide insight into a system's detailed flow behavior and help the designer arrive at an optimum design based on the "virtual" performance analysis. With the increase of computational power, better solution algorithm and reduced cost of the computer, the cost of "virtual prototyping" has decreased with the use of CFD. Now, various aspects of flow field and transport phenomena, which are not amenable to direct experimental technique, can be studied in greater detail by CFD. However, one should remember that CFD cannot replace the experimental technique completely. Only through CFD analysis we can find the parameter important for a flow situation in a component such as pump and design the experiment accordingly. Hence, both CFD and experiments are complementary, and whatever we are designing by CFD analysis must be validated through experiment.

As mentioned earlier, with the increase of computational power and reduced cost of computers, CFD has become very popular and powerful in the design stage for fluid flow analysis of the components. CFD has already been successfully applied to various fields of engineering and medicine. The range of applications is wide and encompasses many different fluid phenomena.

1.2 BASIC PRINCIPLES OF CFD

Fluid dynamics is governed by the conservation of mass, momentum, energy and any constituents normally expressed in terms of partial differential equations (PDEs), which are continuously varying functions. Equations can be expressed in *differential* or *integral form* (explained henceforth). We normally approximate PDEs at a finite number of points, a process called *discretization.* In this process, we get a set of equations known as *difference equations (DEs)*. By discretization, we convert the PDE to difference equations. This is shown in Figure 1.1.

The basic process of any CFD simulation

- The *Governing Equation*, normally represented by PDE and related to field variables (u, v, w, p, ...) that are continuous varying functions, is discretized or approximated by their values at a finite number of points called *nodes.*
- By this process, we get discrete equation, which is known as difference equation (DE), for each node. Differential or integral equations, which are continuous functions, are converted into a set of algebraic equations consisting of each node by the discretization process.
- The system of algebraic equations thus obtained is solved to obtain values at the nodes.

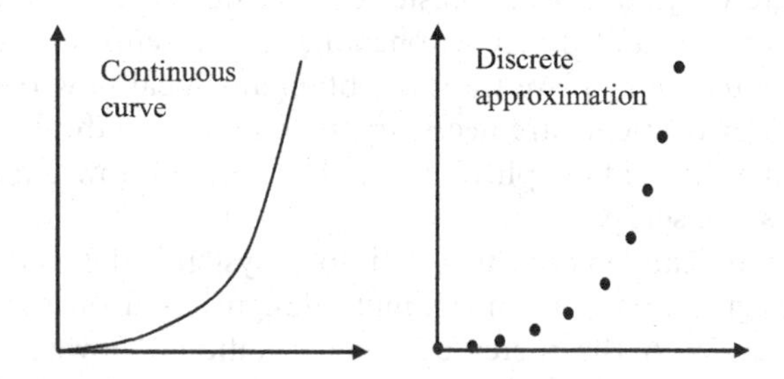

FIGURE 1.1 Continuous vs discrete curve.

1.3 WHAT DOES A CFD ALGORITHM DO?

Many commercial general-purpose CFD programs are available (e.g. Fluent, CFX, Star-CD, FLOW-3D and Phoenics). Some very specialized programs are also available (e.g. simulating combustion in engines, electronic cooling systems, etc.). All the commercial codes are user friendly and have pre-processing, solver and post-processing modules. OpenFoam is an open-source program and can handle most CFD problems. However, this code is not as user friendly as commercial codes. In solving a problem using CFD, many steps must be defined, as illustrated in Figure 1.2.

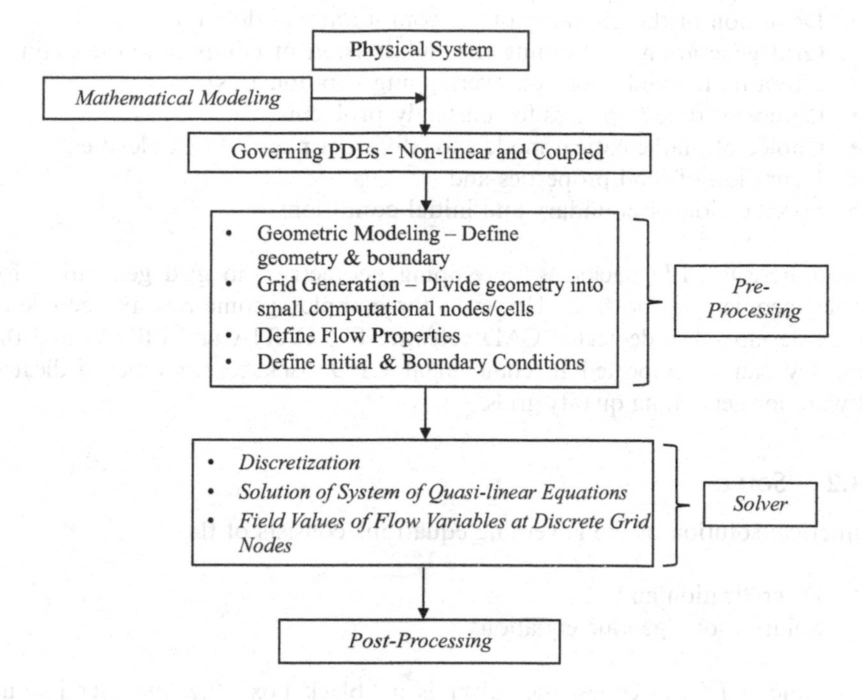

FIGURE 1.2 CFD Process.

1.4 STAGES OF A CFD ANALYSIS

A complete CFD analysis consists of:

- *pre-processing*;
- *solving*;
- *post-processing*.

In Figure 1.2, we saw that the solver plays the most important part in the CFD process; in this course, we shall be focusing our attention on the solving process. However, we shall see that pre-processing and post-processing play important roles in CFD. In commercial CFD codes, pre-processing and post-processing are integral, and its user-friendly GUI helps the user to operate the codes effortlessly.

1.4.1 Pre-Processor

The pre-processor is a user-friendly interface that provides problem inputs in a form suitable for the flow solver.

A CFD pre-processor provides a

- Definition of the geometry of the computational domain,
- Grid generation or meshing (i.e. subdivision of computation domain into a finite number of non-overlapping sub-domains),
- Choice of time step sizes for unsteady problems,
- Choice of mathematical models for different physical complexities,
- Definition of fluid properties and
- Specification of boundary and initial conditions.

In commercial CFD packages, developing geometry and grid generation for simpler geometry is possible. However, for complex geometries, the geometry can be developed in dedicated CAD software like CATIA and PRO-E, and the geometry can be imported in commercial CFD packages or other dedicated software for generating quality grids.

1.4.2 Solver

Numerical solution of the governing equations consists of the

- Discretization and
- Solution of algebraic equations.

In commercial CFD codes, the solver is a "black box" (i.e. the user has no idea what is going on inside the solver). However, unless the user is familiar with the discretization process, it will be very difficult for him/her to supply the proper input data for solving the cases and interpreting the output data. Hence, proper understanding of the discretization is required.

1.4.3 Post-Processor

The first objective in post-processing is to analyse the quality of the solution. Is the solution independent of the grid size, the convergence criterion and the numerical schemes? Have the proper turbulence model and boundary conditions been chosen, and is the solution strongly dependent on those choices?

Normally, output from the solver is a set of flow parameters (u, v, w, p, ...) corresponding to each point of the mesh. It will be very difficult for the user to interpret the results from these raw data; they need to be manipulated in a visual form so that interpretation becomes easier. For example, we can draw the velocity distribution or velocity vector to understand the flow behavior. Similarly, pressure contours become helpful in understanding the flow phenomena. Commercial CFD packages provide facilities to monitor the residuals during each iteration to check the convergence of the solution so that the user can intervene if required.

Analysis of the final simulation results will then give local information about flow, concentrations, temperatures, reaction rates, etc. The results that can be typically presented are

- Display of domain geometry and grid details,
- Convergence history,
- Velocity diagrams
- Velocity, pressure and concentration contour plots,
- Tracking of a particle and
- Animation and a dynamic display of results.

1.5 GOVERNING EQUATIONS

Fluid dynamics is governed by the conservation of mass, momentum, energy and any additional constituents like transport of particles or concentration of dissolved solids, etc. Governing equations can be written in integral or differential form. The differential form of a governing equation can be written in a conservative or non-conservative form. The details regarding the governing equations will be discussed in Chapter 2.

1.6 DISCRETIZATION

There are three discretization techniques:

1) Finite difference
2) Finite volume and
3) Finite element

1.6.1 Finite Difference Method

Finite difference method (FDM) uses a differential form of governing equations. It uses the Taylor series expansion to obtain the discretized equation at each node. The finite difference method will be discussed in detail in Chapter 3 and the application of FDM to various equations in Chapter 4.

1.6.2 Finite Volume Method

The finite volume method (FVM) uses an integral form of governing equations. It balances the fluxes crossing the faces of a cell, thus enforcing the conservation rigorously and directly related to physical quantities (e.g. mass flux). Unlike FDM, it can be easily applied to arbitrary geometry. The method will be discussed in detail in Chapter 5.

1.6.3 Finite Element Method

The finite element method (FEM) uses an integral form, normally a weak or variation formulation. The domain is discretized by elements. A simple functional form is assumed to approximate the solution. Discretized equations are more complex and difficult for physical comprehension. The development of multi-directional interpolation schemes and bookkeeping for both dependent and independent variables are common difficulties. Furthermore, the conservation of the transport property of advection terms in flow equation is not trivial.

The FEM is popular in solid mechanics because it can be applied to any complex geometry due to its meshing flexibility; hence, general-purpose codes can be used for a variety of problems. However, this method is not as popular for the fluid flow problem. All popular CFD codes use FVM instead because

- It has similar geometrical flexibility to FEM and
- It ensures local and global flux balance and can be directly related to physical quantities (e.g. mass flux).

1.7 DISCRETIZATION PROPERTIES

Once the governing equations are discretized, we need to check that they are

(i) ***Consistent***

As the mesh size approaches zero, if the difference equations arising due to discretization become equivalent to the original partial differential equations, then it is consistent.

$$(\emptyset_E - \emptyset_P)/_{\Delta x} \text{ is a consistent approximation of } {}^{\partial\emptyset}/_{\partial x}$$

(ii) ***Conservative***

When the net flux crossing the control volume is zero, the equation is conservative (i.e. the outgoing flux from one cell becomes the incoming flux for the adjacent cell). We must remember that fluxes are associated with *faces*, not *nodes*. The FVM automatically ensures this property.

(iii) ***Transportive***

In a fluid flow problem, the fluid carries information from the upstream to the downstream side (i.e. the advection scheme involves directional effects). This is known as the transportive property and is achieved when appropriate weightage is provided in the advection scheme.

(iv) ***Bounded***

In an advection-diffusion problem without sources, the solution of flow variables of a node must be between the values of the flow variables of the surrounding nodes.

(v) ***Stable***

Stability indicates whether we can obtain a solution. As we shall see, this is only possible if small errors do not grow into large errors during the solution procedure.

QUESTIONS:

1) What can be simulated in a CFD program?
2) What is discretization? Is it always a necessary step in solving a CFD problem?
3) Distinguish between FDM and FVM.
4) State the desirable properties of a discretization scheme.
5) Discuss the various steps involved in CFD and draw a sketch.
6) What steps are involved in solving a typical CFD problem?

2 Governing Equations and Classification of PDE

GOVERNING EQUATIONS

2.1 INTRODUCTION

As we already know, fluid dynamics is governed by the conservation of mass, momentum, energy and any additional equations describing the scalar transport of species (e.g. transport of particles or the concentration of dissolved solids). In a fluid flow problem, variables can be written in terms of a single generic equation: the *scalar-transport or advection-diffusion equation.* The *advection* is associated with the transport with flow (i.e. a *"bulk"* phenomenon), whereas the *diffusion* is a molecular phenomenon and arises due to the concentration gradient. This *single scalar-transport equation* represents all the physical quantities such as mass, momentum, energy, and we can deal with the discretization of a single equation instead of dealing with all the equations one by one. All these physical principles are expressed mathematically either in differential or integral form.

2.1.1 Integral Form

In the *integral* form of a conservation equation, we consider the net rate of the transport of a property, called flux (e.g. mass, momentum, energy) crossing the surface of an arbitrary control volume with an associated change of that property within the *control volume.* The *finite volume method* uses the numerical approximation of the integral form of the governing equations.

2.1.2 Differential Form

The conservation equation can also be written in differential forms. The *finite difference method* uses the numerical approximation of the differential form of the governing equations.

2.2 CONSERVATIVE FORM OF THE FLOW EQUATIONS

2.2.1 Mass (Continuity)

Let us consider a control volume (or surfaces) as in Figure 2.1.

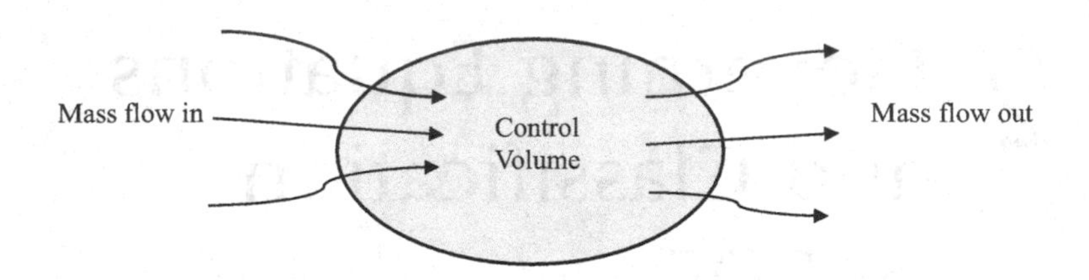

FIGURE 2.1 A Typical Control Volume.

The principle of *conservation of mass* says:

$$\begin{aligned}&\textit{Mass entering per unit time}\\&\quad = \textit{Mass leaving per unit time}\\&\quad + \textit{Increase of mass in the control volume per unit time}\end{aligned}$$

$$\frac{d}{dt}(mass) + net\ outward\ mass\ flux = 0$$

Integral form:

$$\frac{\partial}{\partial t}\int_V \rho dV + \int_A \rho \mathbf{u} \bullet d\mathbf{A} = \mathbf{0}$$

Differential form:

$$\frac{\partial \rho}{\partial t} + \frac{\partial(\rho u)}{\partial x} + \frac{\partial(\rho v)}{\partial y} + \frac{\partial(\rho w)}{\partial z} = 0$$

Vector form:

$$\frac{\partial \rho}{\partial t} + \nabla.(\rho \mathbf{u}) = 0$$

Where $\mathbf{u}$ is a velocity vector given by $u\mathbf{i} + v\mathbf{j} + w\mathbf{k}$; and $\mathbf{i}$, $\mathbf{j}$, $\mathbf{k}$ are unit vectors.

2.2.1.1 Derivation of Continuity Equation

Let us consider a control volume with sides dx, dy, dz in X, Y and Z directions respectively as shown in Figure 2.2.

$$Inlet\ mass\ flow = \rho u\ dydz + \rho v\ dxdz + \rho w\ dydz$$

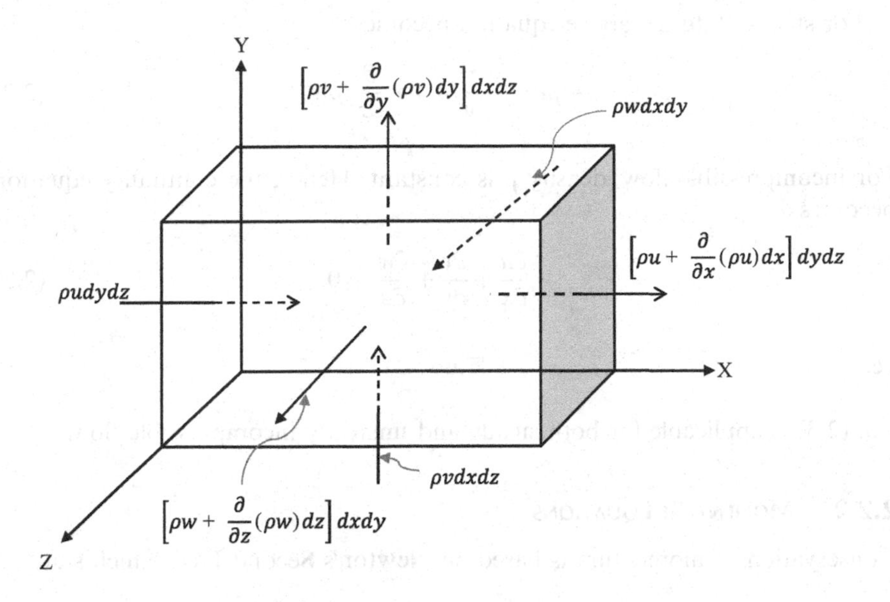

FIGURE 2.2 Control Volume for Mass Conservation.

Outlet mass flow

$$= \left[\rho u+\frac{\partial}{\partial x}(\rho u)dx\right]dydz+\left[\rho v+\frac{\partial}{\partial y}(\rho v)dy\right]dxdz + \left[\rho w+\frac{\partial}{\partial z}(\rho w)dz\right]dxdy$$

$$Accumulation\ of\ mass = \frac{\partial \rho}{\partial t}dxdydz$$

According to the conservation of mass

Mass flow in CV = Mass flow out of CV + Accumulation of mass in CV, if any.

or

$$\frac{\partial \rho}{\partial t}dxdydz+\frac{\partial}{\partial x}(\rho u)dxdydz+\frac{\partial}{\partial y}(\rho v)dxdydz+\frac{\partial}{\partial z}(\rho w)dxdydz=0$$

Dividing by *dxdydz*, we get

$$\frac{\partial \rho}{\partial t}+\frac{\partial}{\partial x}(\rho u)+\frac{\partial}{\partial y}(\rho v)+\frac{\partial}{\partial z}(\rho w)=0 \tag{2.1}$$

Eq. (2.1) gives the generalized equation for conservation of mass.

For steady state the above equation becomes

$$\frac{\partial}{\partial x}(\rho u) + \frac{\partial}{\partial y}(\rho v) + \frac{\partial}{\partial z}(\rho w) = 0 \tag{2.2}$$

For incompressible flow, density ρ is constant. Hence, the continuity equation becomes

$$\frac{\partial u}{\partial x} + \frac{\partial v}{\partial y} + \frac{\partial w}{\partial z} = 0 \tag{2.3}$$

i.e.

$$\nabla.\mathbf{u} = 0$$

Eq. (2.3) is applicable for both steady and unsteady incompressible flow.

2.2.2 Momentum Equations

Conservation of momentum is based on Newton's Second Law, which states

rate of change of momentum = force.

The total rate of change of momentum for fluid passing through any control volume = time rate of change of total momentum inside the control volume + net momentum flux (difference between rate at which momentum leaves and enters).

Hence

$$\frac{d}{dt}(momentum) + net\ outward\ momentum\ flux = force$$

$$\frac{d}{dt}(mass \times \mathbf{u}) + \sum_{faces} mass\ flux \times \mathbf{u} = \mathbf{F}$$

Since, momentum, velocity and force are vectors, we get three component equations in X, Y and Z directions respectively.

The fluid forces consist of *surface forces*, which are proportional to area and act on control-volume faces and *body forces*, proportional to volume. The main surface forces are due to i) pressure p, which always acts normal to a surface and ii) viscous stresses, the frictional forces arising from the relative motion of fluid particles. The main body forces are

1) *Buoyancy forces* arising due to temperature difference and is proportional to ρg.
2) *Centrifugal and Coriolis forces* which are apparent forces arising due to *rotating reference frame.*

Before the derivation of the momentum equation, let us see the integral and differential form of the momentum equation.

Integral form:

$$\frac{\partial}{\partial t}\int_V \rho u_i \, dV + \int_A \rho u_i u_j \, dA = \int_A \sigma_{ij} \, dA + \int_V f_i \, dV$$

The first term of the left-hand side of equation represents the momentum increase in cell, second term the net momentum flux whereas the first term of the right-hand side of equation represents surface forces acting on the control volume and second term the body forces.

Differential form:

$$\frac{\partial(\rho u_i)}{\partial t} + \frac{\partial(\rho u_i u_j)}{\partial x_j} = \frac{\partial \sigma_{ij}}{\partial x_j} + f_i$$

where, $\sigma_{ij} = -p + \tau_{ij}$;

$$\tau_{ij} = \mu\left(\frac{\partial u_i}{\partial x_j} + \frac{\partial u_j}{\partial x_i} - \frac{2}{3}\delta_{ij}\frac{\partial u_k}{\partial x_k}\right)$$

and δ_{ij} *is known as Kronecker delta*; $\delta_{ij} = 1$ for $i = j$ and $= 0$ for $i \neq j$

Vector form:

$$\frac{\partial}{\partial t}(\rho \mathbf{u}) + \nabla.\{\rho \mathbf{u}\mathbf{u}\} = \nabla.\sigma + \mathbf{f_b}$$

2.2.2.1 Derivation of X-Momentum Equation

Let us consider the momentum flux balance in X direction for a control volume as shown in Figure 2.3.

Inlet Momentum: $(\rho uu)dydz + (\rho vu)dxdz + (\rho wu)dxdy$

Outlet Momentum:

$$\left[(\rho uu) + \frac{\partial}{\partial x}(\rho uu)dx\right]dydz + \left[(\rho vu) + \frac{\partial}{\partial y}(\rho vu)dy\right]dxdz + \left[(\rho wu) + \frac{\partial}{\partial z}(\rho wu)dz\right]dxdy$$

Accumulation of Momentum in CV:

$$\frac{\partial}{\partial t}(\rho u)dx\,dy\,dz$$

Hence, Rate of Change of Momentum:

$$\frac{\partial}{\partial t}(\rho u)dx\,dy\,dz + \frac{\partial}{\partial x}(\rho uu)dxdydz + \frac{\partial}{\partial y}(\rho vu)dxdydz + \frac{\partial}{\partial z}(\rho wu)dxdydz \quad (2.4)$$

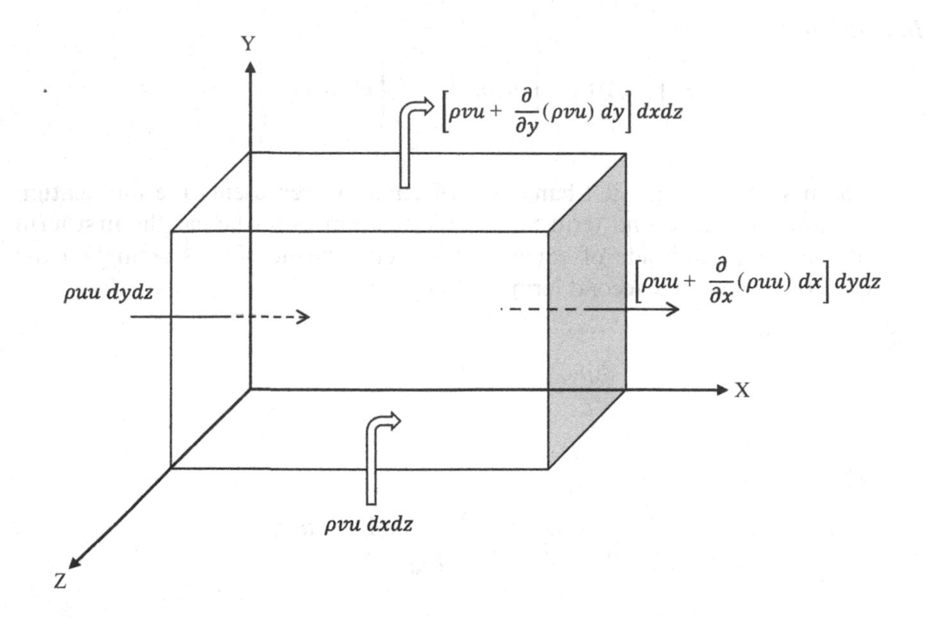

FIGURE 2.3 Momentum Flux Balance in a Control Volume.

Forces: $$\sum F = \sum F_{surface} + Body\ forces$$

The surface forces acting on the control volume are shown in Figure 2.4.

Surface force acting in X- direction:

$$\sum F_x = \left[\frac{\partial}{\partial x}(\sigma_{xx}) + \frac{\partial}{\partial x}(\sigma_{yx}) + \frac{\partial}{\partial x}(\sigma_{zx})\right] dx\, dy\, dz + f_x$$

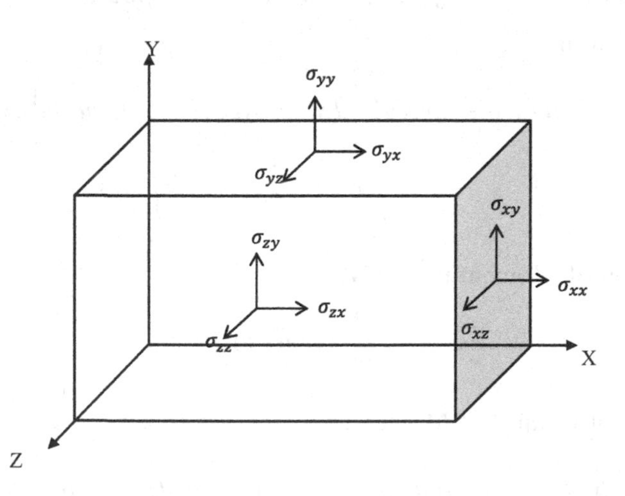

FIGURE 2.4 Definition of Surface Forces.

Now $\sigma_{xx} = -p + \tau_{xx}; \quad \sigma_{yx} = \tau_{yx} = \tau_{xy} \quad \text{and} \quad \sigma_{zx} = \tau_{zx} = \tau_{xz}$

$$\therefore \sum F_x = \left[-\frac{\partial p}{\partial x} + \frac{\partial}{\partial x}(\tau_{xx}) + \frac{\partial}{\partial x}(\tau_{yx}) + \frac{\partial}{\partial x}(\tau_{zx}) \right] dx\, dy\, dz + f_x \tag{2.5}$$

Combining, X-Momentum becomes

$$\begin{aligned} \frac{\partial}{\partial t}(\rho u) + \frac{\partial}{\partial x}(\rho uu) + \frac{\partial}{\partial y}(\rho uv) + \frac{\partial}{\partial z}(\rho uw) &= -\frac{\partial p}{\partial x} \\ &+ \frac{\partial}{\partial x}(\tau_{xx}) + \frac{\partial}{\partial y}(\tau_{yx}) + \frac{\partial}{\partial z}(\tau_{zx}) + f_x \end{aligned} \tag{2.6}$$

Similarly, Y-Momentum becomes:

$$\begin{aligned} \frac{\partial}{\partial t}(\rho v) + \frac{\partial}{\partial x}(\rho uv) + \frac{\partial}{\partial y}(\rho vv) + \frac{\partial}{\partial z}(\rho vw) &= -\frac{\partial p}{\partial y} \\ &+ \frac{\partial}{\partial x}(\tau_{xy}) + \frac{\partial}{\partial y}(\tau_{yy}) + \frac{\partial}{\partial z}(\tau_{zy}) + f_y \end{aligned} \tag{2.7}$$

and Z-Momentum:

$$\begin{aligned} \frac{\partial}{\partial t}(\rho w) + \frac{\partial}{\partial x}(\rho uw) + \frac{\partial}{\partial y}(\rho vw) + \frac{\partial}{\partial z}(\rho ww) &= -\frac{\partial p}{\partial z} \\ &+ \frac{\partial}{\partial x}(\tau_{xz}) + \frac{\partial}{\partial y}(\tau_{yz}) + \frac{\partial}{\partial z}(\tau_{zz}) + f_z \end{aligned} \tag{2.8}$$

Stokes Hypothesis:

In Eqs. (2.6 – 2.8), the values of surface forces are not known. These values are related to the mean velocities by *Stokes hypothesis* which is given below.

$$\begin{aligned} \tau_{xx} &= -\frac{2}{3}\mu\, div\, \mathbf{u} + 2\mu\frac{\partial u}{\partial x} \\ \tau_{yy} &= -\frac{2}{3}\mu\, div\, \mathbf{u} + 2\mu\frac{\partial v}{\partial y} \\ \tau_{zz} &= -\frac{2}{3}\mu\, div\, \mathbf{u} + 2\mu\frac{\partial w}{\partial z} \\ \tau_{yx} = \tau_{xy} &= \mu\left(\frac{\partial v}{\partial x} + \frac{\partial u}{\partial y}\right) \\ \tau_{yz} = \tau_{zy} &= \mu\left(\frac{\partial w}{\partial y} + \frac{\partial v}{\partial z}\right) \\ \tau_{zx} = \tau_{xz} &= \mu\left(\frac{\partial w}{\partial x} + \frac{\partial u}{\partial z}\right) \end{aligned} \tag{2.9}$$

Putting the values of τ_{xx}, τ_{yy}, τ_{zz} etc. we get the generalized equation for viscous flow which is known as *Navier-Stokes Equation.*

For incompressible Flow

$$div\ \mathbf{u} = 0$$

Hence, we get

$$\tau_{xx} = 2\mu\frac{\partial u}{\partial x};\ \tau_{yy} = 2\mu\frac{\partial v}{\partial y};\ \tau_{zz} = 2\mu\frac{\partial w}{\partial z}$$

The X-momentum equation becomes:

$$\begin{aligned}\frac{\partial}{\partial t}(\rho u) + \frac{\partial}{\partial x}(\rho uu) + \frac{\partial}{\partial y}(\rho uv) + \frac{\partial}{\partial z}(\rho uw) &= -\frac{\partial p}{\partial x} \\ + \frac{\partial}{\partial x}\left(\mu\frac{\partial u}{\partial x}\right) + \frac{\partial}{\partial y}\left(\mu\frac{\partial u}{\partial y}\right) + \frac{\partial}{\partial z}\left(\mu\frac{\partial u}{\partial z}\right) &+ f_x\end{aligned} \tag{2.10}$$

Similarly, Y-momentum equation becomes:

$$\begin{aligned}\frac{\partial}{\partial t}(\rho v) + \frac{\partial}{\partial x}(\rho uv) + \frac{\partial}{\partial y}(\rho vv) + \frac{\partial}{\partial z}(\rho vw) &= -\frac{\partial p}{\partial y} \\ + \frac{\partial}{\partial x}\left(\mu\frac{\partial v}{\partial x}\right) + \frac{\partial}{\partial y}\left(\mu\frac{\partial v}{\partial y}\right) + \frac{\partial}{\partial z}\left(\mu\frac{\partial v}{\partial z}\right) &+ f_y\end{aligned} \tag{2.11}$$

and Z-momentum:

$$\begin{aligned}\frac{\partial}{\partial t}(\rho w) + \frac{\partial}{\partial x}(\rho uw) + \frac{\partial}{\partial y}(\rho vw) + \frac{\partial}{\partial z}(\rho ww) &= -\frac{\partial p}{\partial z} \\ + \frac{\partial}{\partial x}\left(\mu\frac{\partial w}{\partial x}\right) + \frac{\partial}{\partial y}\left(\mu\frac{\partial w}{\partial y}\right) + \frac{\partial}{\partial z}\left(\mu\frac{\partial w}{\partial z}\right) &+ f_z\end{aligned} \tag{2.12}$$

All the above equation in vector form can be written as

$$\frac{\partial}{\partial t}(\rho\mathbf{u}) + \nabla.\{\rho\mathbf{uu}\} = -\nabla p + \nabla.\left\{\mu\left[\nabla\mathbf{u} + (\nabla\mathbf{u})^T\right]\right\} + \mathbf{f_b} \tag{2.13}$$

If viscosity is constant, then the above equation becomes

$$\frac{\partial}{\partial t}(\rho\mathbf{u}) + \nabla.\{\rho\mathbf{uu}\} = -\nabla p + \mu\nabla^2\mathbf{u} + \mathbf{f_b} \tag{2.14}$$

2.2.3 Energy Equation

The energy equation is governed by the First Law of Thermodynamics, which states that matter can neither be created nor destroyed. This means that one form of energy is converted into another form and the sum of all energies remains constant in an isolated system.

The total rate of change of energy inside any control volume = the rate of heat addition into the control volume + the rate of work done by body and surface forces on the control volume.

Hence

$$\frac{d}{dt}(energy) = net\ inward\ heat + net\ work\ done\ by\ body\ and\ surface\ forces$$

or

$$\frac{dE}{dt} = \dot{Q} - \dot{W}$$

where, E is the total energy, $\dot{Q}$ is rate of heat added to the system and $\dot{W}$ is the rate of work done by the system.

The total energy consists of internal energy i and kinetic energy. Hence,

$$E = mi + \frac{1}{2}m(u^2 + v^2 + w^2) = mi + \frac{1}{2}mV^2 \text{ where } V^2 = u^2 + v^2 + w^2$$

Let us define

$$e = \frac{E}{m} = i + \frac{1}{2}V^2$$

Integral form:

$$\frac{\partial}{\partial t}\int_V \rho e dV + \int_A \rho u_j \left(e + \frac{p}{\rho}\right) dA = -\int_A \dot{q}_j\, dA + \int_A (u_i \tau_{ij}) dA + \int_V \rho(f_i u_i + \dot{q})\, dV$$

Differential form:

$$\frac{\partial(\rho e_i)}{\partial t} + \frac{\partial}{\partial x_j}\left[\rho u_j \left(e + \frac{p}{\rho}\right)\right] = -\frac{\partial \dot{q}_j}{\partial x_j} + \frac{\partial}{\partial x_j}(u_i \tau_{ij}) + \rho(uf_x + vf_y + wf_z) + \rho\dot{q}$$

where,

$$\tau_{ij} = \mu\left(\frac{\partial u_i}{\partial x_j} + \frac{\partial u_j}{\partial x_i} - \frac{2}{3}\delta_{ij}\frac{\partial u_k}{\partial x_k}\right) \text{ and } q_{ij} = -k\frac{\partial T}{\partial x_j}$$

Vector form:

$$\frac{\partial}{\partial t}(\rho e) + \nabla.\left\{\rho \mathbf{u}\left(e + \frac{p}{\rho}\right)\right\} = -\nabla.\dot{q}_s + \nabla.(\tau.\mathbf{u}) + \rho(\mathbf{f}.\mathbf{u} + \dot{q})$$

2.2.3.1 Derivation of Energy Equation

Let us consider a control volume of size *dx, dy* and *dz* in X, Y and Z directions respectively, as shown in Figure 2.5.

Energy entered in the control volume due to convection $= \rho uedydz + \rho vedxdz + \rho wedxdy$

Energy leaving out of control volume due to convection is given by:

$$\left[\rho ue + \frac{\partial}{\partial x}(\rho ue)\ dx\right] dydz + \left[\rho ve + \frac{\partial}{\partial y}(\rho ve)\ dy\right] dxdz + \left[\rho we + \frac{\partial}{\partial y}(\rho we)\ dy\right] dxdy$$

Rate of accumulation of energy in the control volume is

$$\frac{\partial}{\partial t}(\rho e)dxdydz$$

Hence, rate of change of energy in control volume is

$$\frac{\partial}{\partial t}(\rho e)dxdydz + \frac{\partial}{\partial x}(\rho ue)\ dxdydz + \frac{\partial}{\partial x}(\rho ve)\ dxdydz + \frac{\partial}{\partial x}(\rho we)\ dxdydz \tag{2.15}$$

The net heat flux into the control volume consists of 1) volumetric heat generation and 2) heat transfer through the surface of the control volume due to the temperature difference (i.e. by conduction).

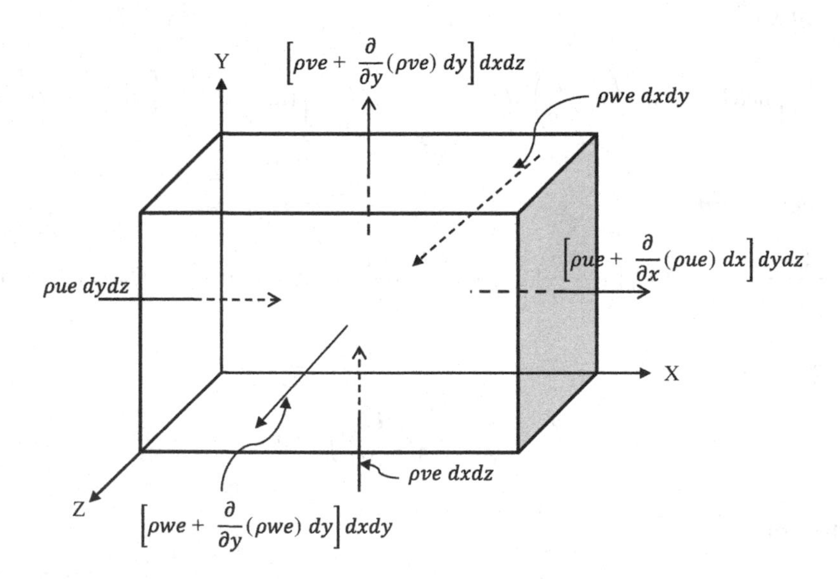

FIGURE 2.5 Energy Flux Balance in a Control Volume.

Volumetric heating of fluid element is

$$\rho \dot{q} dxdydz$$

where $\dot{q}$ is the rate of volumetric heat generation per unit mass.

Net heat transfer by heat conduction in X-direction

$$\left[\dot{q}_x - \left(\dot{q}_x + \frac{\partial \dot{q}_x}{\partial x} dx\right)\right] dydz = -\frac{\partial \dot{q}_x}{\partial x} dxdydz$$

where $\dot{q}_x$ is heat transferred by conduction in X-direction per unit time per unit area.

Similarly, we can find out the heat transferred by conduction in Y and Z directions.

Hence, heating of fluid element by thermal conduction is

$$-\left[\frac{\partial \dot{q}_x}{\partial x} + \frac{\partial \dot{q}_y}{\partial y} + \frac{\partial \dot{q}_z}{\partial z}\right] dxdydz$$

Now we know that

$$\dot{q}_x = -k\frac{\partial T}{\partial x};\ \dot{q}_y = -k\frac{\partial T}{\partial y};\ \dot{q}_z = -k\frac{\partial T}{\partial z}$$

Hence, the rate of heat transfer into control volume by heat is

$$\left[\rho \dot{q} + \frac{\partial}{\partial x}\left(k\frac{\partial T}{\partial x}\right) + \frac{\partial}{\partial y}\left(k\frac{\partial T}{\partial y}\right) + \frac{\partial}{\partial z}\left(k\frac{\partial T}{\partial z}\right)\right] dxdydz \tag{2.16}$$

The net rate of work done consists of 1) rate of work done by surface forces on the fluid element and 2) rate of work done by body forces. Here, we neglected the shaft work.

Rate of work done by body forces $= \rho(uf_x + vf_y + wf_z)dxdydz$.

Now let us consider the rate of work done by surface forces, which is shown in Figure 2.6. For clarity, only forces acting in the X direction are shown.

Net work done by pressure forces in X direction:

$$\left[up - \left(up + \frac{\partial}{\partial x}(up)dx\right)\right] dydz = -\frac{\partial}{\partial x}(up)dxdydz$$

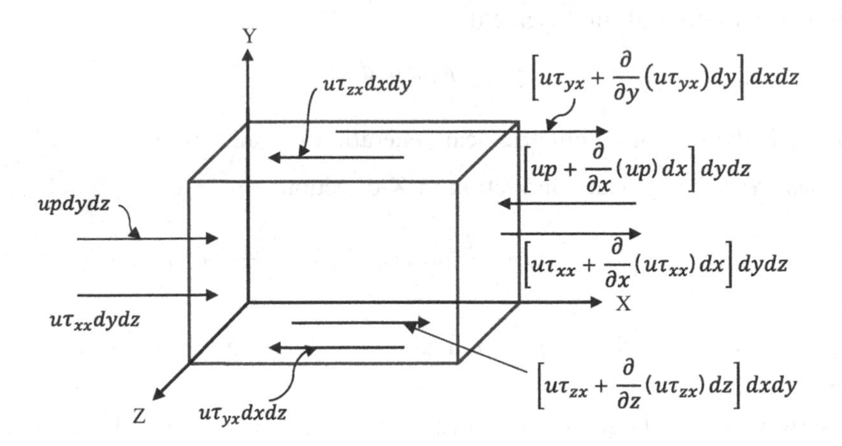

FIGURE 2.6 Work Done by Surface Forces in X Direction.

Net rate of work done by the shear forces in X direction:

$$\left[\left(u\tau_{xx}+\frac{\partial}{\partial x}(u\tau_{xx})dx\right)-u\tau_{xx}\right]dydz+\left[\left(u\tau_{yx}+\frac{\partial}{\partial y}(u\tau_{yx})dy\right)-u\tau_{yx}\right]dxdz$$
$$+\left[\left(u\tau_{zx}+\frac{\partial}{\partial z}(u\tau_{zx})dz\right)-u\tau_{zx}\right]dxdy=\left[\frac{\partial}{\partial x}(u\tau_{xx})+\frac{\partial}{\partial y}(u\tau_{yx})+\frac{\partial}{\partial z}(u\tau_{zx})\right]dxdydz$$

If we consider the other directions, then the net rate of work done by shear forces is

$$\left[\frac{\partial}{\partial x}(u\tau_{xx})+\frac{\partial}{\partial y}(u\tau_{yx})+\frac{\partial}{\partial z}(u\tau_{zx})+\frac{\partial}{\partial x}(v\tau_{xy})+\frac{\partial}{\partial y}(v\tau_{yy})+\frac{\partial}{\partial z}(v\tau_{zy})\right.$$
$$\left.+\frac{\partial}{\partial x}(w\tau_{xz})+\frac{\partial}{\partial y}(w\tau_{yz})+\frac{\partial}{\partial z}(w\tau_{zz})\right]dxdydz$$

Hence, the net rate of work done by surface and body forces is given by

$$\left[-\left(\frac{\partial}{\partial x}(up)+\frac{\partial}{\partial y}(vp)+\frac{\partial}{\partial z}(wp)\right)\right.$$
$$+\left(\frac{\partial}{\partial x}(u\tau_{xx})+\frac{\partial}{\partial y}(u\tau_{yx})+\frac{\partial}{\partial z}(u\tau_{zx})+\frac{\partial}{\partial x}(v\tau_{xy})+\frac{\partial}{\partial y}(v\tau_{yy})\right.$$
$$\left.+\frac{\partial}{\partial z}(v\tau_{zy})+\frac{\partial}{\partial x}(w\tau_{xz})+\frac{\partial}{\partial y}(w\tau_{yz})+\frac{\partial}{\partial z}(w\tau_{zz})\right) \tag{2.17}$$
$$\left.+\rho\left(uf_x+vf_y+wf_z\right)\right]dxdydz$$

Combining Eqs. (2.15), (2.16) and (2.17), we obtain the energy equation, given by

$$
\begin{aligned}
&\frac{\partial}{\partial t}(\rho e)+\frac{\partial}{\partial x}(\rho u e)+\frac{\partial}{\partial x}(\rho v e)+\frac{\partial}{\partial x}(\rho w e) \\
&\quad=\frac{\partial}{\partial x}\left(k\frac{\partial T}{\partial x}\right)+\frac{\partial}{\partial y}\left(k\frac{\partial T}{\partial y}\right)+\frac{\partial}{\partial z}\left(k\frac{\partial T}{\partial z}\right)-\left(\frac{\partial}{\partial x}(up)+\frac{\partial}{\partial y}(vp)+\frac{\partial}{\partial z}(wp)\right) \\
&\qquad+\left(\frac{\partial}{\partial x}(u\tau_{xx})+\frac{\partial}{\partial y}(u\tau_{yx})+\frac{\partial}{\partial z}(u\tau_{zx})+\frac{\partial}{\partial x}(v\tau_{xy})+\frac{\partial}{\partial y}(v\tau_{yy})+\frac{\partial}{\partial z}(v\tau_{zy})\right. \\
&\qquad\left.+\frac{\partial}{\partial x}(w\tau_{xz})+\frac{\partial}{\partial y}(w\tau_{yz})+\frac{\partial}{\partial z}(w\tau_{zz})\right)+\rho\left(uf_x+vf_y+wf_z\right)+\rho\dot{q}
\end{aligned}
\tag{2.18}
$$

The Eq. (2.18) is in terms of total energy: $e=i+\frac{1}{2}V^2 = i+\frac{1}{2}\left(u^2+v^2+w^2\right)$.

To obtain the energy equation in terms of internal energy, i, we need to subtract the kinetic energy equation from Eq. (2.18). Kinetic equation can be obtained by multiplying Eqs. (2.6), (2.7) and (2.8) with u, v, w respectively and adding the results. Subtracting the kinetic energy equation thus obtained, the energy equation in terms of internal energy becomes

$$
\begin{aligned}
&\frac{\partial}{\partial t}(\rho i)+\frac{\partial}{\partial x}(\rho u i)+\frac{\partial}{\partial x}(\rho v i)+\frac{\partial}{\partial x}(\rho w i) \\
&\quad=\frac{\partial}{\partial x}\left(k\frac{\partial T}{\partial x}\right)+\frac{\partial}{\partial y}\left(k\frac{\partial T}{\partial y}\right)+\frac{\partial}{\partial z}\left(k\frac{\partial T}{\partial z}\right)-p\left(\frac{\partial u}{\partial x}+\frac{\partial v}{\partial y}+\frac{\partial w}{\partial z}\right)+\tau_{xx}\frac{\partial u}{\partial x} \\
&\qquad+\tau_{yx}\frac{\partial u}{\partial y}+\tau_{zx}\frac{\partial u}{\partial z}+\tau_{xy}\frac{\partial v}{\partial x}+\tau_{yy}\frac{\partial v}{\partial y}+\tau_{zy}\frac{\partial v}{\partial z}+\tau_{xz}\frac{\partial w}{\partial x}+\tau_{yz}\frac{\partial w}{\partial y}+\tau_{zz}\frac{\partial w}{\partial z}+\rho\dot{q}
\end{aligned}
$$

We know that

$$\tau_{yx}=\tau_{xy};\ \tau_{zx}=\tau_{xz}\ and\ \tau_{zy}=\tau_{yz}.$$

Hence, substituting these values, the equation becomes

$$
\begin{aligned}
&\frac{\partial}{\partial t}(\rho i)+\frac{\partial}{\partial x}(\rho u i)+\frac{\partial}{\partial x}(\rho v i)+\frac{\partial}{\partial x}(\rho w i) \\
&\quad=\frac{\partial}{\partial x}\left(k\frac{\partial T}{\partial x}\right)+\frac{\partial}{\partial y}\left(k\frac{\partial T}{\partial y}\right)+\frac{\partial}{\partial z}\left(k\frac{\partial T}{\partial z}\right)-p\left(\frac{\partial u}{\partial x}+\frac{\partial v}{\partial y}+\frac{\partial w}{\partial z}\right)+\tau_{xx}\frac{\partial u}{\partial x} \\
&\qquad+\tau_{yy}\frac{\partial v}{\partial y}+\tau_{zz}\frac{\partial w}{\partial z}+\tau_{yx}\left(\frac{\partial u}{\partial y}+\frac{\partial v}{\partial x}\right)+\tau_{zx}\left(\frac{\partial u}{\partial z}+\frac{\partial w}{\partial x}\right)+\tau_{zy}\left(\frac{\partial v}{\partial z}+\frac{\partial w}{\partial y}\right)+\rho\dot{q}
\end{aligned}
$$

Substituting the values of τ from Eq. (2.9), we get

$$\frac{\partial}{\partial t}(\rho i)+\frac{\partial}{\partial x}(\rho ui)+\frac{\partial}{\partial x}(\rho vi)+\frac{\partial}{\partial x}(\rho wi)$$
$$=\frac{\partial}{\partial x}\left(k\frac{\partial T}{\partial x}\right)+\frac{\partial}{\partial y}\left(k\frac{\partial T}{\partial y}\right)+\frac{\partial}{\partial z}\left(k\frac{\partial T}{\partial z}\right)-p\left(\frac{\partial u}{\partial x}+\frac{\partial v}{\partial y}+\frac{\partial w}{\partial z}\right)$$
$$+\mu\left[2\left(\frac{\partial u}{\partial x}\right)^2+2\left(\frac{\partial v}{\partial y}\right)^2+2\left(\frac{\partial w}{\partial z}\right)^2+\left(\frac{\partial u}{\partial y}+\frac{\partial v}{\partial x}\right)^2+\left(\frac{\partial u}{\partial z}+\frac{\partial w}{\partial x}\right)^2+\left(\frac{\partial v}{\partial z}+\frac{\partial w}{\partial y}\right)^2\right]$$
$$-\frac{2}{3}\mu\left(\frac{\partial u}{\partial x}+\frac{\partial v}{\partial y}+\frac{\partial w}{\partial z}\right)^2+\rho\dot{q}$$

or

$$\frac{\partial}{\partial t}(\rho i)+\frac{\partial}{\partial x}(\rho ui)+\frac{\partial}{\partial y}(\rho vi)+\frac{\partial}{\partial z}(\rho wi)=\frac{\partial}{\partial x}\left(k\frac{\partial T}{\partial x}\right)+\frac{\partial}{\partial y}\left(k\frac{\partial T}{\partial y}\right)+\frac{\partial}{\partial z}\left(k\frac{\partial T}{\partial z}\right)$$
$$-p\left(\frac{\partial u}{\partial x}+\frac{\partial v}{\partial y}+\frac{\partial w}{\partial z}\right)+\Phi+\rho\dot{q} \tag{2.19}$$

where Φ is the viscous dissipation term.

$$\Phi=\mu\left[2\left(\frac{\partial u}{\partial x}\right)^2+2\left(\frac{\partial v}{\partial y}\right)^2+2\left(\frac{\partial w}{\partial z}\right)^2+\left(\frac{\partial u}{\partial y}+\frac{\partial v}{\partial x}\right)^2+\left(\frac{\partial u}{\partial z}+\frac{\partial w}{\partial x}\right)^2\right.$$
$$\left.+\left(\frac{\partial v}{\partial z}+\frac{\partial w}{\partial y}\right)^2-\frac{2}{3}\mu\left(\frac{\partial u}{\partial x}+\frac{\partial v}{\partial y}+\frac{\partial w}{\partial z}\right)^2\right]$$

The viscous dissipation term indicates the conversion of mechanical energy into heat due to viscosity of the fluid. This term is negligible except at high-speed flows where the velocity gradient is large.

If we want to express this equation in terms of specific enthalpy h, we get

$$\frac{\partial}{\partial t}(\rho h)+\frac{\partial}{\partial x}(\rho uh)+\frac{\partial}{\partial y}(\rho vh)+\frac{\partial}{\partial z}(\rho wh)$$
$$=\frac{\partial}{\partial x}\left(k\frac{\partial T}{\partial x}\right)+\frac{\partial}{\partial y}\left(k\frac{\partial T}{\partial y}\right)+\frac{\partial}{\partial z}\left(k\frac{\partial T}{\partial z}\right)+\frac{Dp}{Dt}+\Phi+\rho\dot{q} \tag{2.20}$$

where

$$h=i+\frac{p}{\rho}\ \textit{and}\ \frac{Dp}{Dt}=\frac{\partial p}{\partial t}+u\frac{\partial p}{\partial x}+v\frac{\partial p}{\partial y}+w\frac{\partial p}{\partial z}.$$

For incompressible flow with constant property, Eq. (2.20) becomes

$$C_P\left[\frac{\partial}{\partial t}(\rho T)+\frac{\partial}{\partial x}(\rho uT)+\frac{\partial}{\partial y}(\rho vT)+\frac{\partial}{\partial z}(\rho wT)\right]=k\left(\frac{\partial^2 T}{\partial x^2}+\frac{\partial^2 T}{\partial y^2}+\frac{\partial^2 T}{\partial z^2}\right)+\rho\dot{q} \tag{2.21}$$

2.2.4 General Scalar

As already discussed, we can write a single transport equation for any physical quantity ($\emptyset$) for an advection-diffusion problem. Examples are the transport of salt concentration, sediment and a chemical constituent. Diffusion takes place due to variations of concentration with position and involves transport from regions of high concentration to regions of low concentration.

By balancing the rate of change, the net flux through the boundary and rate of production, we get

$$\frac{d}{dt}(mass\times\emptyset)+\sum_{faces}\left(mass\,flux\times\emptyset-\Gamma\frac{\partial\emptyset}{\partial n}A\right)=S \tag{2.22}$$

The Eq. (2.22) for incompressible flow can be expressed as

$$\frac{\partial}{\partial t}(\rho\emptyset)+\frac{\partial}{\partial x}(\rho u\emptyset)+\frac{\partial}{\partial y}(\rho v\emptyset)+\frac{\partial}{\partial z}(\rho w\emptyset)=\Gamma\left(\frac{\partial^2\emptyset}{\partial x^2}+\frac{\partial^2\emptyset}{\partial y^2}+\frac{\partial^2\emptyset}{\partial z^2}\right)+S \tag{2.23}$$

Comparing continuity, X-momentum, Y-momentum, Z-momentum and energy equations with Eq. (2.23), the values of $\emptyset$ for various equations thus obtained are shown in Table 2.1.

where, μ = viscosity; $\dot{Q}=\rho\dot{q}$ = total internal heat generation; k = thermal conductivity; C_P = specific heat.

TABLE 2.1
Values of $\emptyset$ in various equations

Equation	$\emptyset$	Γ	S
Continuity	**1**	**0**	**0**
X Momentum	u	μ	$-\frac{\partial p}{\partial x}+f_x$
Y Momentum	v	μ	$-\frac{\partial p}{\partial y}+f_y$
Z Momentum	w	μ	$-\frac{\partial p}{\partial z}+f_z$
Energy	T	k/C_P	$\dot{Q}/C_P$

2.3 SOME COMMENTS

2.3.1 Conservative and Non-Conservative Forms of Equations

The conservative form of a continuity equation in one dimension is given by

$$\frac{\partial \rho}{\partial t}+\frac{\partial}{\partial x}(\rho u)=0$$

That is in the conservative form of equation, the derivative terms contain all the variables and constants in themselves and they do not have any additional preceding term. The above equation in the non-conservative form can be written as

$$\frac{\partial \rho}{\partial t}+\rho\frac{\partial u}{\partial x}+u\frac{\partial \rho}{\partial x}=0$$

The implication of using conservative vs non-conservative forms will be explored in detail in the discussion of the discretization procedure.

2.3.2 Compressible and Incompressible Flow

Compressibility becomes important when significant changes in density occur due to variations in pressure or temperature. Normally this happens in high-speed flow or in cases of significant heat input.

Normally, liquids are treated as incompressible; gases can also be treated as incompressible if speed is much less than that of sound (Ma $\ll$ 1), where Ma is Mach number. There may be small variations of density due to concentration gradient or temperature variations, but these cases do not lead to compressibility effects. Based on whether flow is compressible or incompressible, the solution method changes.

For compressible flow:

- Density, ρ is calculated from Mass conservation equation.
- Velocities, u, v and w, are calculated from Momentum conservation.
- The temperature is calculated from the relations

$$e=c_v T \quad or \quad h=c_P T$$

where c_v and c_P are specific heat capacities at constant volume and constant pressure respectively.

Now the internal energy e or enthalpy $h=e+{}^{P}\!/_{\rho}$ is obtained by solving the energy equation.

- Pressure is derived from *equation of state*, e.g. the ideal-gas law

$$p = \rho RT$$

- *Density-based solution methods* are used for the numerical solution.

For incompressible flows:

- It is not essential to solve separate energy equation for obtaining the velocities.
- Incompressibility means that density remains constant and does not change with pressure. However, there will be small variations in density due to temperature variations in the flow field.
- Velocities, u, v and w, are calculated from Momentum conservation.
- There is no equation for pressure. The pressure is derived from the requirement that the velocities calculated from the momentum equations are to be mass consistent.
- Normally *pressure-based methods* are used for numerical solution.

PHYSICAL AND MATHEMATICAL CLASSIFICATION OF PARTIAL DIFFERENTIAL EQUATIONS

Many important processes in nature are governed by Partial Differential Equations (PDEs). It is necessary to understand the physical behavior of the model presented by PDE. Knowledge of mathematical character and properties strongly depends on them.

2.4 EQUILIBRIUM PROBLEMS

In equilibrium problems, the solution of a given PDE is desired in a closed domain subject to prescribed BC. Equilibrium problems are boundary value problems. Steady state heat conduction and incompressible inviscid flows are two examples of equilibrium problems. The solution of PDE at every point in the domain depends upon the BC of every point on B. Mathematically, equilibrium problems are governed by elliptic PDEs. The schematic diagram of an equilibrium problem is shown in Figure 2.7.

2.5 MARCHING PROBLEMS

Marching or propagation problems are transient or transient-like problems. In a marching problem, the PDE is solved in an open domain satisfying a set of initial conditions and a set of BCs. The solution is computed from the given set of initial data by marching outward while satisfying the BCs. Mathematically, these problems are governed by hyperbolic or parabolic PDEs. Figure 2.8 shows the schematic diagram of marching problem.

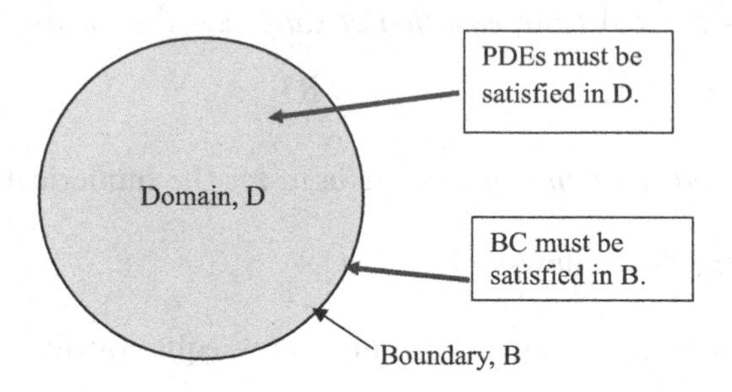

FIGURE 2.7 Schematic Diagram of an Equilibrium Problem.

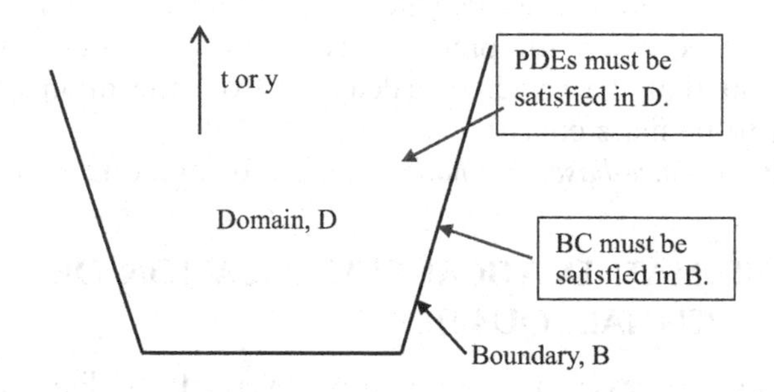

FIGURE 2.8 Schematic Diagram of a Marching Problem.

2.6 MATHEMATICAL CLASSIFICATION

The mathematical classification of PDEs is based on the concepts of lines or surfaces along which certain properties remain same or the derivatives become discontinuous. Examples are the Mach cone and the shock wave. These lines provide us the direction of the information, which is transmitted in physical systems governed by PDEs.

Let us consider a second-order linear PDE in two co-ordinates x and y

$$a\frac{\partial^2 \emptyset}{\partial x^2} + b\frac{\partial^2 \emptyset}{\partial x \partial y} + c\frac{\partial^2 \emptyset}{\partial y^2} + d\frac{\partial \emptyset}{\partial x} + e\frac{\partial \emptyset}{\partial y} + f\emptyset + g = 0 \tag{2.24}$$

where a, b, c, d, e and f are functions of (x, y) (i.e. a linear equation).

Let us consider the following one-dimensional equations in which $\emptyset = \emptyset(x, t)$

$$\frac{\partial \emptyset}{\partial t} + c\frac{\partial \emptyset}{\partial x} = 0 \tag{2.25}$$

where c is a constant.

and

$$\frac{\partial \emptyset}{\partial t} + \emptyset \frac{\partial \emptyset}{\partial x} = 0 \tag{2.26}$$

Eq. (2.25) is linear as the coefficient of the derivative of ∅ is constant (i.e. independent of dependent variable x and t). However, in Eq. (2.26), the coefficient ∅ of the derivative (∂∅)/∂x is a function of x and t and hence non-linear. In a linear PDE, coefficients can depend on independent variables.

The distinction between linear and non-linear PDEs is extremely important. Many linear PDE problems can be solved analytically using, for example, separation of variables or principles of superposition. Exact solutions are valuable because they can be used for validation. However, non-linear PDEs generally are difficult to solve analytically and in most cases require numerical techniques. Since most practical problems involve non-linear PDEs, a great deal of effort is directed toward obtaining numerical solutions.

The classification of a linear second-order equation can be determined by
if $b^2-4ac > 0$ the equation is hyperbolic,
$= 0$ the equation is parabolic,
< 0 the equation is elliptic.

It may be noted that in the solution domain, we may have different types of equations in various regions. For example, consider the equation

$$y\frac{\partial^2 \emptyset}{\partial x^2} + \frac{\partial^2 \emptyset}{\partial y^2} = 0$$

Let us look at the behavior of the equation within the region $-1 < y < 1$.

Comparing it with Eq. (2.24), we get the coefficients $a = y$, $b = 0$ and $c = 1$ so that $b^2-4ac = -4y$.

Now, when
$y < 0$, b^2-4ac becomes greater than 0, so the equation is hyperbolic,
when $y = 0$, b^2-4ac is 0, so the equation is parabolic,
when $y > 0$, b^2-4ac becomes less than 0, so the equation is elliptic.

Hence, we can see that the equation is locally hyperbolic, parabolic or elliptic depending on the value of y.

2.7 IMPORTANT EQUATIONS

Normally, when somebody is trying to develop a discretization procedure, it is better to adopt the simplified equation instead of the full governing equation (e.g. the Navier-Stokes equation) to test the discretization properties. The following simplified equations govern important physical phenomena.

Linear Wave Equation

$$\frac{\partial u}{\partial t}+c\frac{\partial u}{\partial x}=0 \tag{2.27}$$

This equation governs the wave propagation with a constant speed.

Inviscid Burgers' Equation

$$\frac{\partial u}{\partial t}+u\frac{\partial u}{\partial x}=0 \tag{2.28}$$

This is a non-linear wave equation.

Burgers' Equation

$$\frac{\partial u}{\partial t}+u\frac{\partial u}{\partial x}=\nu\frac{\partial^2 u}{\partial x^2} \tag{2.29}$$

This is a non-linear transport equation. This equation is a simplified form of equations governing fluid flow problems.

Poisson Equation

$$\frac{\partial^2 u}{\partial x^2}+\frac{\partial^2 u}{\partial y^2}=f(x,y) \tag{2.30}$$

Example: Heat conduction equation with heat generation.

Helmholtz Equation

$$\frac{\partial^2 u}{\partial x^2}+\frac{\partial^2 u}{\partial y^2}+k^2u=0 \tag{2.31}$$

This equation governs the motion of time-dependent harmonic waves where k is frequency parameter.

Note that Eqs. (2.27), (2.30) and (2.31) are linear PDEs while Eqs. (2.28) and (2.29) are non-linear.

2.8 BOUNDARY CONDITIONS (BCS)

Appropriate BCs and initial conditions are needed for solving the Navier-Stokes equation. The proper specification of BCs is very important for obtaining correct solutions. The common boundary types normally encountered in CFD problems are discussed here.

2.8.1 Inlet Boundary

2.8.1.1 Inflow

At the inlet, velocity or mass flow rate can be defined. One can specify average inlet velocity perpendicular to the inlet surface. Alternately, velocity-component distribution can be specified over the inlet surface based on the one-dimensional fully developed-flow calculation. In some cases, where the inlet pressure is known, instead of specifying inlet velocities, the pressure inlet BC can be used.

2.8.1.2 Stagnation (Reservoir)

In cases in which the inlet is connected to a large reservoir, stagnation conditions like total pressure and total temperature can be specified as the inlet BC.

2.8.2 Outlet Boundary

2.8.2.1 Outflow

Normally, at the outlet, zero normal gradient *for all variables* except pressure is applied. This means that the conditions of the outlet plane are obtained from the upstream flow conditions and do not influence upstream flow.

2.8.2.2 Pressure

At the outlet, static pressure is usually prescribed as the pressure outlet BC. When backflow occurs during iteration, prescribing a pressure outlet BC instead of an outflow condition is very helpful in achieving a better convergence rate. If there are several outflows, the pressure BC varies. For subsonic compressible flow at the outlet, this is a usual outlet condition.

2.8.2.3 Radiation (Convection)

For supersonic flow, a simplified first-order convective equation with outward-directed velocity is specified to prevent reflection of wave-like motions from the outflow boundaries. This also prevents shock waves being reflected from the outlet boundary.

2.8.3 Wall Boundaries

2.8.3.1 No-Slip Wall

For Newtonian fluid (i.e. viscous fluid), velocity relative to solid wall is set as zero. For turbulent flow, velocity is normally specified through wall-function expressions. For heat transfer, insulated, heat flux, fixed temperature, convective heat transfer, radiation heat transfer or a combination of these BCs may be specified.

2.8.3.2 Slip Wall

In the case of inviscid flow, only the velocity component normal to the wall is set to zero.

2.8.4 Other Boundary Conditions

2.8.4.1 Symmetry Plane and Axis Boundary

In the geometric plane of symmetry, normal velocity component to the symmetry plane is set to zero and normal gradient of all other variables is set to zero (i.e. $u_n = 0$ *and* $\partial\emptyset/\partial n = 0$). This is used to reduce computational effort. This is also used as a far-field BC, because it ensures that there is no flow crossing the boundary.

2.8.4.2 Periodic

For rotating machinery and regular arrays like heat exchanger tubes, physical geometry and flow patterns are periodic in nature. This BC is used in all cases to reduce the computational effort. In periodic boundaries, the inlet is set to the outlet.

So far, we have discussed the governing equations and BCs associated with CFD problems. In consequent chapters, we shall discuss finite difference and finite volume discretization methods for incompressible flow only.

2.9 SUMMARY

- Fluid dynamics is governed by conservation of mass, momentum, energy (and, for a non-homogeneous fluid, the balancing of individual constituents).
- The governing equations can be expressed in equivalent integral (control-volume) or differential forms.
- While the finite difference method uses discretization of the differential form of governing equations, the finite volume method uses the integral form of equation.
- Differential forms of the flow equations may be *conservative or non-conservative.*

QUESTIONS

1) Discuss the circumstances under which a fluid flow can be approximated as a) Incompressible or b) inviscid.
2) Derive the three-dimensional mass conservation equation in generalized form. What shall be the form of the equation for unsteady incompressible flow?
3) Distinguish between the conservative and non-conservative forms of the Navier-Stokes Equation.
4) Write down conservative and non-conservative forms of the three-dimensional equations for mass and x momentum.
5) Write down two-dimensional momentum equations for unsteady incompressible flow.
6) Define what is meant by the statement that a flow is incompressible. To what does the continuity equation reduce in incompressible flow?

7) Derive the x-momentum equation in two dimensions.
8) Propose a general expression for the transport of the entity $\emptyset$.
9) Explain why the three equations for the components of momentum cannot be treated as independent scalar equations.
10) If the continuity (mass-conservation) equation were to be regarded as a special case of the general scalar-transport equation, what would be the expressions for $\emptyset$, Γ and S?
11) How do you classify a PDE to be *linear* or *nonlinear*?
12) Consider the equation $u\,{\partial u}/{\partial x} = \mu\,{\partial^2 u}/{\partial y^2}$, where μ is a constant and u the x component of velocity. The normal direction is y.

    a) Is this equation in conservative form? If not, suggest a conservative form of the equation. Give explanations for both cases.
    b) Is the equation linear or non-linear? Explain.

13) In two-dimensional flow, continuity and x-momentum equations can be written in conservative form (and with compressibility neglected in the viscous forces) as

$$\frac{\partial \rho}{\partial t} + \frac{\partial}{\partial x}(\rho u) + \frac{\partial}{\partial y}(\rho v) = 0$$

$$\frac{\partial}{\partial t}(\rho u) + \frac{\partial}{\partial x}(\rho u u) + \frac{\partial}{\partial y}(\rho u v) = -\frac{\partial p}{\partial x} + \mu \nabla^2 u$$

respectively.

    a) By expanding derivatives of products, show that these can be written in equivalent non-conservative forms:

$$\frac{D\rho}{Dt} + \rho\left(\frac{\partial u}{\partial x} + \frac{\partial v}{\partial y}\right) = 0$$

$$\rho \frac{Du}{Dt} = -\frac{\partial p}{\partial x} + \mu \nabla^2 u$$

where,

$$\frac{D}{Dt} = \frac{\partial}{\partial t} + u\frac{\partial}{\partial x} + v\frac{\partial}{\partial y}$$

14) How do you classify a partial differential equation?
15) How do you classify the partial differential equation a) physically and b) mathematically? Give examples. Why is the classification needed?
16) Classify the following partial differential equations

$$\frac{\partial^2 u}{\partial x^2} + 3\frac{\partial^2 u}{\partial x \partial y} + 2\frac{\partial^2 u}{\partial y^2} = 0$$

$$\frac{\partial^2 u}{\partial x^2} - 2\frac{\partial^2 u}{\partial x \partial y} + \frac{\partial^2 u}{\partial y^2} = 0$$

$$2\frac{\partial^2 u}{\partial x^2} + 4\frac{\partial^2 u}{\partial x \partial y} + 3\frac{\partial^2 u}{\partial y^2} = 0$$

17) What are initial and boundary value problems? Give examples.
18) Discuss the various BCs.
19) What are standard outlet conditions?
20) What is a no-slip condition at the wall, and when can it be used?
21) What is a periodic BC?
22) What are symmetry and axisymmetric BCs?

3 Finite Difference Method

Fundamentals

3.1 INTRODUCTION

Fluid dynamics is governed by conservation of mass, momentum, energy and any other constituents that are continuous functions (Partial Differential Equations, PDE). *Discretization* is the process of approximating a continuously varying function in terms of values at a finite number of points. In this chapter, the basic concepts of discretization by the *finite difference method (FDM)* are discussed.

As discussed in Chapter 1, the differentials of the dependent variables appearing in PDE must be expressed as approximate expressions (difference equations), so that a digital computer can be employed to obtain a solution. The continuous problem domain is approximated so that dependable variables are considered to exist at discrete points. Derivatives are approximated (i.e. discretized) in an algebraic representation of PDE. The nature of the resulting algebraic system depends on the character of the original PDE.

The first step in establishing a finite difference procedure for solving PDE is to replace a continuous problem domain with a finite-difference mesh or grid as shown in the Figure 3.1.

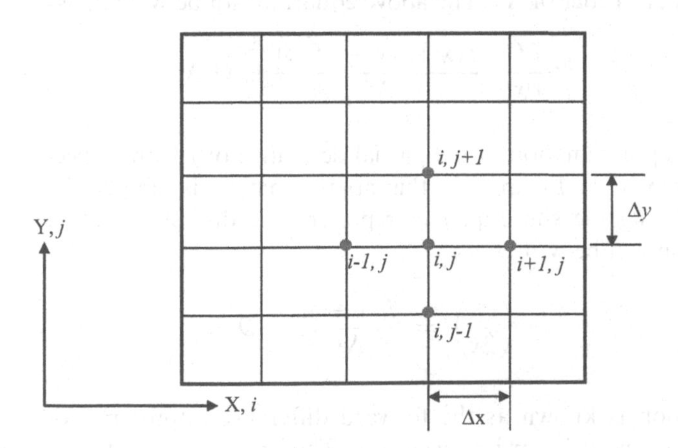

FIGURE 3.1 Finite Difference Representation.

Here the grid points are identified by i *and* j; Δx and Δy represent grid spacing in the X and Y directions, respectively. This is for two dimensions. For three dimensions, the grid points are identified by i, j and k and grid spaces by Δx, Δy and Δz in X, Y and Z directions, respectively.

Two methods are employed to obtain the difference equations: the Taylor series expansion of the function f and polynomials of degree n. In this chapter, the Taylor series expansion for obtaining the difference equations will be discussed.

3.2 TAYLOR SERIES EXPANSION

Given a continuous function $f(x), f(x+\Delta x)$ can be expanded in Taylor series as

$$\begin{aligned} f(x+\Delta x) &= f(x) + (\Delta x)\frac{\partial f}{\partial x} + \frac{(\Delta x)^2}{2!}\frac{\partial^2 f}{\partial x^2} + \frac{(\Delta x)^3}{3!}\frac{\partial^3 f}{\partial x^3} + \dots\dots\dots \\ &= f(x) + \sum_{n=1}^{\infty}\frac{(\Delta x)^n}{n!}\frac{\partial^n f}{\partial x^n} \end{aligned} \tag{3.1}$$

Solving for $\partial f/\partial x$, we get

$$\frac{\partial f}{\partial x} = \frac{f(x+\Delta x) - f(x)}{\Delta x} - \frac{\Delta x}{2!}\frac{\partial^2 f}{\partial x^2} - \frac{(\Delta x)^2}{3!}\frac{\partial^3 f}{\partial x^3} + \dots\dots\dots \tag{3.2}$$

i.e. $$\text{PDE} = \text{FDE} + \text{HOT(TE)}$$

So, the PDE is expressed as Finite difference equation (FDE) plus Higher Order Terms (HOT), which is also known as truncation error (TE). The first term of the HOT indicates the accuracy of the difference scheme. Since the first term of HOT in Eq. (3.2) contains Δx, it is firstorder accurate and represents as $O(\Delta x)$, read as order of Δx. The above equation can be written as

$$\frac{\partial f}{\partial x} = \frac{f(x+\Delta x) - f(x)}{\Delta x} + O(\Delta x) \tag{3.3}$$

This is the approximation of the partial derivative of f with respect to x. Graphically, as shown in Figure 3.2, this approximation is simply the slope of the function. Using the subscript i to represent the discrete points in x direction, the equation can be written as

$$\left(\frac{\partial f}{\partial x}\right)_i = \frac{f_{i+1} - f_i}{\Delta x} + O(\Delta x) \tag{3.4}$$

This equation is known as the forward difference approximation of $\partial f/\partial x$ at point i and is first order accurate. The accuracy of the approximation increases as the step size Δx decreases.

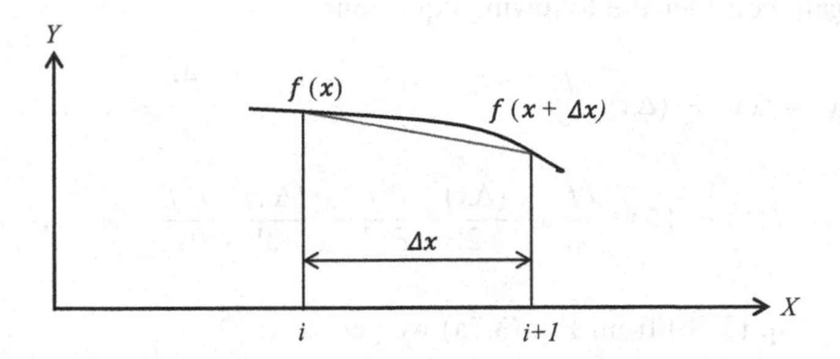

FIGURE 3.2 Forward Difference Approximation of the First Derivative.

Now, considering Taylor series expansion of function $f(x - \Delta x)$ about x, we get

$$f\,(x - \Delta x) = f(x) - (\Delta x)\,\frac{\partial f}{\partial x} + \frac{(\Delta x)^2}{2!}\,\frac{\partial^2 f}{\partial x^2} - \frac{(\Delta x)^3}{3!}\,\frac{\partial^3 f}{\partial x^3} + \ldots\ldots\ldots$$

$$= f(x) + \sum_{n=1}^{\infty}(-1)^n\frac{(\Delta x)^n}{n!}\,\frac{\partial^n f}{\partial x^n} \tag{3.5}$$

where + sign is for even n and – sign for odd n.

Solving for $\partial f/\partial x$, we get

$$\frac{\partial f}{\partial x} = \frac{f(x) - f(x - \Delta x)}{\Delta x} + O(\Delta x)$$

or

$$\left(\frac{\partial f}{\partial x}\right)_i = \frac{f_i - f_{i-1}}{\Delta x} + O(\Delta x) \tag{3.6}$$

The above equation is the Backward Difference approximation of $\partial f/\partial x$ of order Δx as shown in Figure 3.3 and is first-order accurate.

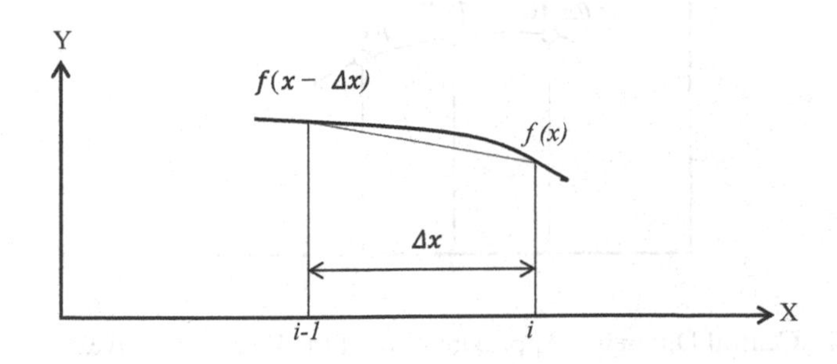

FIGURE 3.3 Backward Difference Approximation of the First Derivative.

Now again consider the following equations

$$f\ (x+\ \Delta x)=f(x)+\ (\Delta x)\ \frac{\partial f}{\partial x}+\ \frac{(\Delta x)^2}{2!}\frac{\partial^2 f}{\partial x^2}+\ \frac{(\Delta x)^3}{3!}\frac{\partial^3 f}{\partial x^3}+\ldots\ldots\ldots \tag{3.7a}$$

$$f\ (x-\ \Delta x)=f(x)-\ (\Delta x)\ \frac{\partial f}{\partial x}+\ \frac{(\Delta x)^2}{2!}\frac{\partial^2 f}{\partial x^2}-\ \frac{(\Delta x)^3}{3!}\frac{\partial^3 f}{\partial x^3}+\ldots\ldots\ldots \tag{3.7b}$$

Subtracting Eq. (3.7b) from Eq. (3.7a) we get

$$\frac{\partial f}{\partial x}=\frac{f(x+\Delta x)-f(x-\ \Delta x)}{2\Delta x}+\frac{(\Delta x)^2}{3!}\frac{\partial^3 f}{\partial x^3}+\ldots\ldots\ldots$$

or

$$\frac{\partial f}{\partial x}=\frac{f(x+\ \Delta x)-f(x-\Delta x)}{2\Delta x}+\ O(\Delta x)^2$$

or

$$\left(\frac{\partial f}{\partial x}\right)_i=\frac{f_{i+1}-f_{i-1}}{2\Delta x}+\ O(\Delta x)^2 \tag{3.8}$$

This is the central difference approximation of $\partial f/\partial x$ of order $(\Delta x)^2$ as shown in Figure 3.4.

Hence, the central difference is second order accurate scheme as the first term of HOT is $(\Delta x)^2$and is represented by $O(\Delta x)^2$ whereas both forward and backward differences are first order accurate schemes and represented by $O(\Delta x)$.

So far, we have considered the approximation of the first derivatives. Now we shall consider the methods of finding the approximations of the higher derivatives. Again, consider the Taylor series expansion

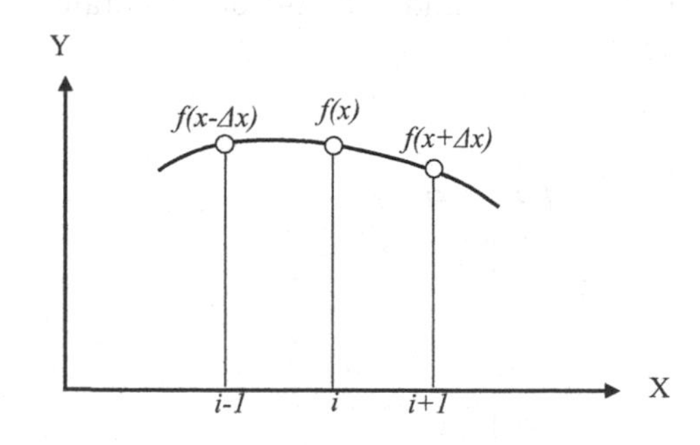

FIGURE 3.4 Central Difference Approximation of the First Derivative.

$$f\ (x+\ \Delta x) = f(x) +\ (\Delta x)\ \frac{\partial f}{\partial x} + \frac{(\Delta x)^2}{2!}\frac{\partial^2 f}{\partial x^2} + \frac{(\Delta x)^3}{3!}\frac{\partial^3 f}{\partial x^3} + \dots\dots\dots$$

and

$$f\ (x+2\ \Delta x) = f(x) +\ (2\Delta x)\ \frac{\partial f}{\partial x} + \frac{(2\Delta x)^2}{2!}\frac{\partial^2 f}{\partial x^2} + \frac{(2\Delta x)^3}{3!}\frac{\partial^3 f}{\partial x^3} + \dots\dots\dots$$

Multiplying the first equation by 2 and then subtracting it from the second equation we get

$$-2f(x+\Delta x) +\ f(x+2\Delta x) = -f(x) + (\Delta x)^2\ \frac{\partial^2 f}{\partial x^2} + (\Delta x)^3\frac{\partial^3 f}{\partial x^3} + \dots$$

Solving for $\partial^2 f / \partial x^2$, we get

$$\frac{\partial^2 f}{\partial x^2} = \frac{f(x+2\Delta x) - 2f(x+\Delta x) + f(x)}{(\Delta x)^2} + O(\Delta x) \quad (3.9a)$$

$$\left(\frac{\partial^2 f}{\partial x^2}\right)_i = \frac{f_{i+2} - 2f_{i+1} + f_i}{(\Delta x)^2} + O(\Delta x) \quad (3.9b)$$

This equation is the forward difference approximation of $\partial^2 f / \partial x^2$ and is the order of $O(\Delta x)$ i.e. first order accurate.

A similar approximation for the backward difference of the second derivative can be obtained by Taylor series expansion of $f(x-\Delta x)$ *and* $f(x-2\Delta x)$. The result is

$$\left(\frac{\partial^2 f}{\partial x^2}\right)_i = \frac{f_i - 2f_{i-1} + f_{i-2}}{(\Delta x)^2} + O(\Delta x) \quad (3.10)$$

To obtain a central difference approximation of the second derivative, simply add Taylor series expansion of $f(x+\Delta x)$ *and* $f(x-\Delta x)$ (i.e. Eqs. (3.7a) and (3.7b)). Thus

$$\frac{\partial^2 f}{\partial x^2} = \frac{f(x+\Delta x) - 2f(x) + f(x-\Delta x)}{(\Delta x)^2} + O(\Delta x)^2$$

$$\left(\frac{\partial^2 f}{\partial x^2}\right)_i = \frac{f_{i+1} - 2f_i + f_{i-1}}{(\Delta x)^2} + O(\Delta x)^2 \quad (3.11)$$

Approximations of the higher-order derivatives of function f with respect to x can be obtained by a similar procedure.

So far, the first and higher order derivatives have been expressed using forward and backward differencing of order (Δx) and central difference of order $(\Delta x)^2$. By considering additional terms in the Taylor series expansions, a more accurate approximation of the derivatives can be obtained. Consider the Taylor series expansion

$$f\ (x+\ \Delta x)=f(x)+\ (\Delta x)\ \frac{\partial f}{\partial x}+\frac{(\Delta x)^2}{2!}\ \frac{\partial^2 f}{\partial x^2}+\frac{(\Delta x)^3}{3!}\ \frac{\partial^3 f}{\partial x^3}+\ldots\ldots\ldots$$

Solving for $\partial f/\partial x$, we get

$$\frac{\partial f}{\partial x}=\frac{f(x+\Delta x)-f(x)}{\Delta x}-\frac{\Delta x}{2!}\ \frac{\partial^2 f}{\partial x^2}-\frac{(\Delta x)^2}{3!}\ \frac{\partial^3 f}{\partial x^3}+\ldots\ldots\ldots$$

Now, substitute a forward difference expression for $\partial^2 f/\partial x^2$ from Eq. (3.9a), i.e.

$$\frac{\partial^2 f}{\partial x^2}=\frac{f(x+2\Delta x)-2f(x+\Delta x)+f(x)}{(\Delta x)^2}+O(\Delta x)$$

we obtain,

$$\frac{\partial f}{\partial x}=\frac{f(x+\Delta x)-f(x)}{\Delta x}-\frac{\Delta x}{2!}\left[\frac{f(x+2\Delta x)-2f(x+\Delta x)+f(x)}{(\Delta x)^2}+O(\Delta x)\right]$$
$$-\frac{(\Delta x)^2}{3!}\ \frac{\partial^3 f}{\partial x^3}+\ldots\ldots\ldots$$

or

$$\frac{\partial f}{\partial x}=\frac{-f(x+2\Delta x)+4f(x+\Delta x)-3f(x)}{2\Delta x}+O(\Delta x)^2 \tag{3.12}$$

Thus, a second order approximation of $\partial f/\partial x$ has been obtained. Similarly, a second order approximation can be obtained for the backward difference by replacing the second order derivative in the backward difference of first order.

3.3 UNEQUAL GRID SPACING

Refer to Figure 3.5.

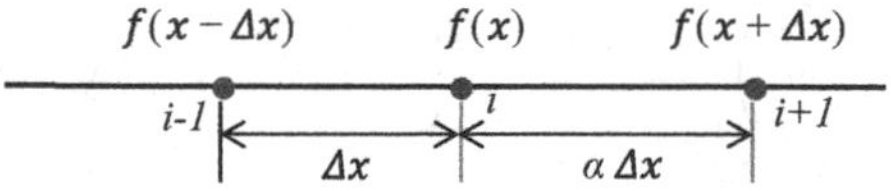

$f(x-\Delta x)$ $f(x)$ $f(x+\Delta x)$

i-1 i i+1

Δx $\alpha\,\Delta x$

FIGURE 3.5 Unequal Grid Spacing.

We know that

$$f\ (x+\ \Delta x)=f(x)+\ (\Delta x)\ \frac{\partial f}{\partial x}+\ \frac{(\Delta x)^2}{2!}\ \frac{\partial^2 f}{\partial x^2}+\ \frac{(\Delta x)^3}{3!}\ \frac{\partial^3 f}{\partial x^3}+\ldots\ldots\ldots \tag{3.13a}$$

$$f\ [x+(1+\alpha)\ \Delta x]=f(x)+\ (1+\alpha)(\Delta x)$$

$$\frac{\partial f}{\partial x}+\ \frac{(1+\alpha)^2(\Delta x)^2}{2!}\ \frac{\partial^2 f}{\partial x^2}+\ \frac{(1+\alpha)^3(\Delta x)^3}{3!}\ \frac{\partial^3 f}{\partial x^3}+\ldots\ldots\ldots \tag{3.13b}$$

Multiplying Eq. (3.13a) by – $(1+\alpha)$ and adding with Eq. (3.13b) we get

$$\frac{\partial^2 f}{\partial x^2}=\frac{f_{i+2}-(1+\alpha)f_{i+1}+\alpha f_i}{\frac{1}{2}\alpha(1+\alpha)(\Delta x)^2}+O(\Delta x)$$

Substituting in Eq. (3.13a) we get

$$f_{i+1}=f_i+(\Delta x)\frac{\partial f}{\partial x}+\frac{(\Delta x)^2}{2!}\frac{f_{i+2}-(1+\alpha)f_{i+1}+\alpha f_i}{\frac{1}{2}\alpha(1+\alpha)(\Delta x)^2}+O(\Delta x)^3$$

or

$$\frac{\partial f}{\partial x}=\frac{-f_{i+2}+(1+\alpha)^2 f_{i+1}-\alpha(\alpha+2)f_i}{\alpha(1+\alpha)\Delta x}+O(\Delta x)^2 \tag{3.14}$$

3.4 DIFFERENCE REPRESENTATION OF PDE

3.4.1 Errors

Essentially, two types of errors may appear in difference representation.

3.4.1.1 Truncation Error (TE)

Truncation error is introduced because of approximation of derivatives by differences. These are essentially created when we drop the HOT terms in the expression of FDE. They are called *discretization errors* or *truncation errors (TE)*.

Let us consider the Heat Conduction Equation

$$\frac{\partial T}{\partial t}=\alpha\frac{\partial^2 T}{\partial x^2} \tag{3.15}$$

where $\alpha =$ *thermal diffusivity* $= {}^{k}/_{(\rho C_P)}$.

Using a forward-difference representation for the time derivative $(t = n\Delta t)$ and central difference for the second derivative we get

$$\frac{\partial T}{\partial t} = \frac{T_i^{n+1} - T_i^n}{\Delta t} + \frac{\partial^2 T}{\partial t^2}\frac{\Delta t}{2} + \ldots$$

and

$$\alpha \frac{\partial^2 T}{\partial x^2} = \frac{\alpha}{\Delta x^2}\left(T_{i+1}^n - 2T_i^n + T_{i-1}^n\right) + \alpha \frac{\partial^4 T}{\partial x^4}\frac{(\Delta x)^2}{12} + \ldots$$

Hence Eq. (3.15) becomes

$$\begin{aligned}\frac{\partial T}{\partial t} - \alpha \frac{\partial^2 T}{\partial x^2} = {} & \frac{T_i^{n+1} - T_i^n}{\Delta t} - \frac{\alpha}{\Delta x^2}\left(T_{i+1}^n - 2T_i^n + T_{i-1}^n\right) \\ & + \left[\frac{\partial^2 T}{\partial t^2}\frac{\Delta t}{2} - \alpha \frac{\partial^4 T}{\partial x^4}\frac{(\Delta x)^2}{12} + \ldots\right]\end{aligned} \tag{3.16}$$

i.e. $PDE = FDE + TE$

Here TE is $O\left[\Delta t, (\Delta x)^2\right]$.

We solve only FDE and hope that TE is small. Question is how do we know that the solution of FDE is acceptable? For solution to be acceptable, FDE should meet the conditions of Consistency and Stability which is discussed later.

3.4.1.2 Round-off Error

Normally, calculations are carried out to a finite number of decimal places or significant digits. Errors that arise because of this finite number of decimal places are called *round-off errors.*

For example:

Let A = the exact solution of a PDE (analytical)

D = the exact solution of a FDE

Discretization error or TE = PDE – FDE

The exact solution D is obtained by a machine with infinite accuracy. However, the numerical solution is obtained by a machine with finite accuracy. Let this solution be N.

Hence, Round-off error, ε = N – D.

3.4.2 Consistency

Consistency deals with the extent to which FDE approximates PDE.

As we know, PDE = FDE + TE.

A FDE is said to be consistent if we can show that the difference between the PDE and the FDE vanishes as the mesh is refined:

$$\lim_{\Delta x \to 0}(PDE - FDE) = \lim_{\Delta x \to 0}(TE) = 0$$

It is possible that the truncation error tends toward zero as the mesh size approaches zero, but the round-off error does not.

3.4.3 Stability

A FDE may be consistent, but the solution need not necessarily converge to the solution. The term, *stability*, is used to describe the decay or amplification of numerical errors that somehow entered the calculations. The most common such error is the round-off error. The question of the stability of a numerical scheme examines the growth of error while computations are performed. A numerical scheme is said to be stable if an error introduced in the FDE does not grow with the solution of the FDE.

If a specific scheme of numerical analysis is chosen, then the TE of each term in the equation is uniquely defined. There are two well-defined solutions: the solution to the differential equation, A, and the solution to the FDE, D. The last may depend on the choice of Δx and Δt. Moreover, the final expression for the discretization error is generally given in terms of unknown derivatives for which no upper and lower bounds can be estimated.

O' Brien et al. [1950] poses the question of stability in the following manner:

1. If, due to round off, the overall error grows during iteration, then we have *strong instability.* If the error does not grow, then the scheme is *strongly stable.*
2. If, due to round off, a single general error grows during iteration, then the scheme is *weakly instable.* If the error decays, then the scheme is *weakly stable.*

The second point is the one most frequently answered because it can be treated much more easily.

Consider the FDE for the one-dimensional unsteady diffusion equation

$$T_i^{n+1} = T_i^n + \frac{\alpha\,\Delta t}{\Delta x^2}\left(T_{i+1}^n - 2T_i^n + T_{i-1}^n\right)$$

Let us assume that a certain number of errors were committed at some time $\bar{n}$ and we need to know their propagation for larger n. Let ε represents the error in the numerical solution arising due to round-off error. The numerical solution computed may be written as N = D + ε. This computed numerical solution must satisfy the difference equation.

Substituting equation for N in the FDE we get:

$$\frac{D_i^{n+1} + \varepsilon_i^{n+1} - D_i^n - \varepsilon_i^n}{\Delta t} = \frac{\alpha}{\Delta x^2}\left(D_{i+1}^n + \varepsilon_{i+1}^n - 2D_i^n - 2\varepsilon_i^n + D_{i-1}^n + \varepsilon_{i-1}^n\right)$$

Since the exact solution, D must satisfy the difference equation, same is true for the error, i.e.

$$\frac{\varepsilon_i^{n+1} - \varepsilon_i^n}{\Delta t} = \frac{\alpha}{\Delta x^2}\left(\varepsilon_{i+1}^n - 2\varepsilon_i^n + \varepsilon_{i-1}^n\right)$$

This means, the numerical error and exact solution exhibits the similar behavior. Hence, for a stable system any perturbation of input values at nth time will remain within bounds whereas for an unstable system it will grow.

Various methods are used for stability analysis:

- Matrix method (Smith)
- Discrete perturbation analysis method (Thom & Apett)
- Hirt's Method
- Von Neumann's Method

We shall discuss Von Neumann's method.

3.4.3.1 Von Neumann's Method

An initial line of errors can be expressed in terms of finite Fourier series. The growth of a function that for $t = 0$ reduces to Fourier series is then studied by the method of separation of variables. Von Neumann's analysis is therefore applicable for linear problems only.

Consider a set of errors $E_{k,n}$ that crept into our calculation at time step n. This arbitrary distribution of values by Fourier analysis can be finally expressed in terms of a sine and a cosine series of the form

$$E_{k,n} = \sum a_m \cos\frac{m\pi x}{L} + \sum b_m \sin\frac{m\pi x}{L}$$

where L is domain of integration in x-direction. In complex notation, the above equation can be written as

$$E_{k,n} = \sum_{m=0}^{N} (A_m)_n e^{\frac{im\pi x}{L}}$$

where i is imaginary unit and A_m's are arbitrary complex coefficients.

We shall consider only one term since the system is linear. Drop index m from $(A_m)_n$, and take $k_x = \frac{m\pi}{L}$, where k_x is the wave propagation in the X direction

$$E_{k,n} = A_n\ e^{ik_x x} = A_n\ e^{ik\theta}$$

where, k = wave number and θ is the phase angle.

We need to investigate how coefficient A_n behaves for higher values of n. In other words, the Fourier analysis is used to determine all A_n's at the nth time

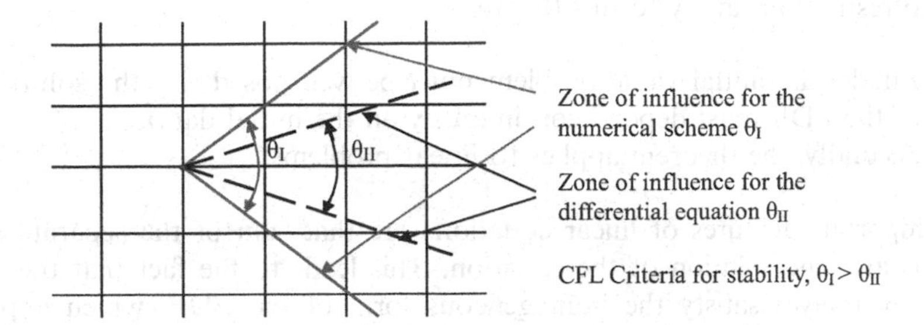

FIGURE 3.6 CFL Criteria.

level. We need to determine whether the amplitude of error grows or decays as we proceed to time level $n+1$.

If we express A_n in terms of exponential $A_n = e^{at} = e^{an\Delta t}$ then, the equation for $E_{k,n}$ takes more familiar form of wave propagation.

The amplification factor is defined as $\xi = \frac{A_{n+1}}{A_n}$ and the error will decrease if

$$|\xi| = \left|\frac{A_{n+1}}{A_n}\right| \leq 1.$$

This is Von Neumann's criterion for stability.

Now
$$E_{k,n+1} = A_{n+1}e^{ik\theta}$$
$$E_{k\pm1,n} = A_{n+1}e^{i(k\pm1)\theta}$$

The stability of a numerical scheme is also restricted by the Courant condition. This is also known as the Courant-Friedrichs-Lewy (CFL) condition. It states: *"Finite-difference domain of influence should include continuum domain of influence"* (i.e. the local numerical net of influence of each grid point should contain the domain of influence of the differential equation defined at that point). Figure 3.6 explains the CFL criteria.

3.4.4 Convergence

Let $u(x,t)$ is the exact solution of a differential equation, and $u_{k,n}$ the exact solution of the corresponding difference equation.

Finite difference equation is said to be convergent if

$$u_{k,n} \rightarrow u(k\Delta x, n\Delta t) \text{ as } \Delta t \rightarrow 0 \text{ and } \Delta x \rightarrow 0.$$

3.4.5 Lax's Equivalence Theorem

Lax's Equivalence Theorem states that for linear initial value problems, *consistency* and *stability* are necessary conditions for *convergence.*

Two restrictions apply to this theorem.

- Firstly, the initial value problem must be well posed (i.e. the solution of the PDE must depend continuously on the initial data).
- Secondly, the theorem applies to linear problems.

The important features of linear equations are that sum of the separate solutions is also the solution of the equation. This leads to the fact that the error terms themselves satisfy the homogeneous form of the FDE, which approximates the given difference equation.

3.4.6 Courant Number

The *Courant number c is defined by*

$$c = \frac{u\Delta t}{\Delta x}.$$

This is the ratio of distance travelled by a fluid particle with speed *u in one time step* ($u\Delta t$) to the mesh spacing Δx.

3.5 EXAMPLES

Exercise 3.1 Let a function be $f(x) = x^2$. Find the forward difference approximation of $f'(x)$ at x = 3 for I) $\Delta x = 0.1$; II) $\Delta x = 0.05$ *and* III) $\Delta x = 0.01$.

Answer: The exact answer of $\partial f/\partial x$ at x = 3 is *6*. The forward difference is given as

$$\left(\frac{\partial f}{\partial x}\right)_i = \frac{f_{i+1} - f_i}{\Delta x}$$

Now, for the given problem, i = 3 and i+1 = 3.1 for case I, 3.05 for case II and 3.01 for case III.
Substituting the values, we get
Case I

$$\left(\frac{\partial f}{\partial x}\right)_3 = \frac{3.1^2 - 3^2}{0.1} = 6.1$$

Similarly, for Case II

$$\left(\frac{\partial f}{\partial x}\right)_3 = \frac{3.05^2 - 3^2}{0.05} = 6.05$$

and, for Case III

$$\left(\frac{\partial f}{\partial x}\right)_3 = \frac{3.01^2 - 3^2}{0.01} = 6.01$$

Hence, the error for Case I is 6.1 − 6 = 0.1; for Case II = 0.05 and for Case III 0.01. So, we can see that as the gird size reduces, error also reduces. Now find out whether you can reduce the error to zero by reducing the grid size.

Exercise 3.2 Let a function is $f(x) = e^x$. Using $\Delta x = 0.1$, determine $\partial f/\partial x$ at x = 2 with forward difference, central difference and second order three-point formula given in Eq. (3.12). Compare the results with exact value.

Answer: The exact answer of $\partial f/\partial x$ at x = 2 is 7.389.
Case I: Forward difference

$$\left(\frac{\partial f}{\partial x}\right)_i = \frac{f_{i+1} - f_i}{\Delta x}$$

Here $i+1 = 2+0.1 = 2.1$ *and* $i=2$.
So

$$\left(\frac{\partial f}{\partial x}\right)_2 = \frac{e^{2.1} - e^2}{0.1} = \frac{8.166 - 7.389}{0.1} = 7.77$$

Hence, *error* = 7.77 − 7.389 = **0.381**

Case II: Central difference

$$\left(\frac{\partial f}{\partial x}\right)_i = \frac{f_{i+1} - f_{i-1}}{2\Delta x}$$

Here, $i-1 = 1.9$

$$\left(\frac{\partial f}{\partial x}\right)_2 = \frac{e^{2.1} - e^{1.9}}{0.2} = \frac{8.166 - 6.686}{0.1} = 7.4$$

Hence, *error* = 7.4 − 7.389 = **0.011**

Case III: Second order three-point formula

$$\frac{\partial f}{\partial x} = \frac{-f_{i+2} + 4f_{i+1} - 3f_i}{2\Delta x}$$

Here, $i+2 = 2 + 0.2 = 2.02$ *and* $i+1 = 2 + 0.1 = 2.1$

$$\left(\frac{\partial f}{\partial x}\right)_i = \frac{-e^{2.2} + 4e^{2.1} - 3e^2}{0.2} = \frac{-9.025 + 4 \times 8.166 - 3 \times 7.389}{0.2} = 7.36$$

Hence, $error = 7.36 - 7.389 = \mathbf{-0.029}$

3.6 SUMMARY

- Fluid dynamics is governed by conservation equations for mass, momentum, energy etc.
- The governing equations can be expressed in either integral or differential form.
- The finite-difference method uses a differential form of governing equations for discretization.
- *FDM mainly uses Taylor's series expansion to obtain discretized equations.*
- To be acceptable, FDE should meet the conditions of consistency and stability.
- For consistency, the TE should tend to zero as mesh sizes tend to zero.
- Stability is involved with round-off errors.
- Achieving stability requires satisfying the CFL condition.
- For convergence, Lax's Equivalence Theorem applies.

QUESTIONS:

1) State Lax's Equivalence Theorem.
2) Define the CFL condition for stability.
3) Name the important errors that commonly occur in numerical solutions.
4) What is meant by the Courant number?
5) What sources of error are introduced during discretization?
6) Let a function be $f(x) = x^2$. Find the backward and central difference approximations of $\partial f/\partial x$ at x = 4 for i) $\Delta x = 0.1$; ii) $\Delta x = 0.05$ *and* iii) $\Delta x = 0.01$ and compare them with exact values.
7) Find the FDE for $\frac{\partial f}{\partial y}$ involving j, j+1 and j+2 nodes. Also, state the accuracy of the system. Using this equation solve the problem in Exercise 3.2.
8) For problem 6, find the central difference approximation of $f''(x)$ at x = 4 for i) $\Delta x = 0.1$; ii) $\Delta x = 0.05$ *and* iii) $\Delta x = 0.01$ and compare them with exact values.
9) Find the central difference approximation of $f''(x)$ at x = 4 if $f(x) = e^x$ Take i) $\Delta x = 0.1$; ii) $\Delta x = 0.05$ *and* iii) $\Delta x = 0.01$ and compare them with exact values.
10) Find the backward difference approximation of $\frac{\partial f}{\partial x}$ of second-order accuracy using the Taylor series expansion. Assume uniform grid spacing.
11) Find the finite difference approximation of

$$\frac{\partial^2 u}{\partial x^2} + \frac{\partial^2 u}{\partial y^2} = 0$$

Also, find out the expression if $\Delta x = \Delta y = h$.

12) Consider the equation

$$u\frac{\partial u}{\partial x} = \mu\frac{\partial^2 u}{\partial x^2}$$

where μ is a constant

Is the equation in conservative form? If not, suggest a conservative form of this equation.

13) Consider the continuity equation

$$\frac{\partial u}{\partial x} + \frac{\partial v}{\partial y} = 0$$

Write the central difference approximation of this equation. Check whether the FDE is consistent.

4 Finite Difference Method

Application

4.1 INTRODUCTION

In this section, we shall show the application of finite difference methods (FDM) for various linear equations like diffusion equations, wave equation and non-linear inviscid and viscid Burgers' equations. For all cases, we shall check whether the basic discretization properties are satisfied.

4.2 ONE-DIMENSIONAL DIFFUSION EQUATIONS

PDE governing one-dimensional diffusion equations:

$$\frac{\partial u}{\partial t} = D\frac{\partial^2 u}{\partial x^2} \tag{4.1a}$$

where $u\ (x,\ t)$ is dependent variable and D is diffusion co-efficient. Corresponding PDE at grid points:

$$\left(\frac{\partial u}{\partial t}\right)_k^n = \left(D\frac{\partial^2 u}{\partial x^2}\right)_k^n \tag{4.1b}$$

where n denotes the time and k is the grid point. We shall now concentrate on the various discretisation procedures. This equation is parabolic in nature and in its present form, represents one-dimensional heat conduction or diffusion in isotropic medium. This equation can be used in an elementary way to model the parabolic boundary layer equations. The exact solution of this equation for initial condition $u(x,0) = f(x)$, and boundary conditions $u(0,t) = u(1,t) = 0$ is

$$u_n = \sum_{n=1}^{\infty} A_n e^{-Dk^2 t}\ sin(kx)$$

where

$$A_n = 2\int_0^1 f(x)\ sin(kx)dx \qquad and\ \ k = n\pi$$

Let us now find the finite difference formulation of Eq. (4.1b).

4.2.1 Explicit Methods

Explicit methods are those in which the value of u at new time step n+1 can be found from old values of u, as shown in Figure 4.1.

4.2.1.1 The Forward Time, Central Space (FTCS)

$$\frac{u_k^{n+1} - u_k^n}{\Delta t} = D\,\frac{u_{k+1}^n - 2u_k^n + u_{k-1}^n}{(\Delta x)^2} \tag{4.2a}$$

or

$$u_k^{n+1} = \lambda u_{k-1}^n + (1 - 2\lambda)u_k^n + \lambda\, u_{k+1}^n\,; \qquad where\ \lambda = \frac{D\Delta t}{(\Delta x)^2} \tag{4.2b}$$

Consistency:

$$u_k^n + \Delta t\left(\frac{\partial u}{\partial t}\right)_k^n + \frac{(\Delta t)^2}{2}\left(\frac{\partial^2 u}{\partial t^2}\right)_k^n + O\left[(\Delta t)^3\right] + .. =$$

$$\lambda\left[u_k^n - \Delta x\left(\frac{\partial u}{\partial x}\right)_k^n + \frac{(\Delta x)^2}{2}\left(\frac{\partial^2 u}{\partial x^2}\right)_k^n + O\left[(\Delta x)^3\right] + ..\right] + (1 - 2\,\lambda)u_k^n$$

$$+\lambda\left[u_k^n + \Delta x\left(\frac{\partial u}{\partial x}\right)_k^n + \frac{(\Delta x)^2}{2}\left(\frac{\partial^2 u}{\partial x^2}\right)_k^n + O\left[(\Delta x)^3\right] + ..\right]$$

or

$$\left(\frac{\partial u}{\partial t}\right)_k^n = \frac{\lambda(\Delta x)^2}{\Delta t}\left(\frac{\partial^2 u}{\partial x^2}\right)_k^n + HOT = D\left(\frac{\partial^2 u}{\partial x^2}\right)_k^n + HOT$$

where

$$HOT = -\frac{\Delta t}{2}\left(\frac{\partial^2 u}{\partial t^2}\right)_k^n + \frac{D(\Delta x)^2}{12}\left(\frac{\partial^4 u}{\partial x^4}\right)_k^n + \ldots$$

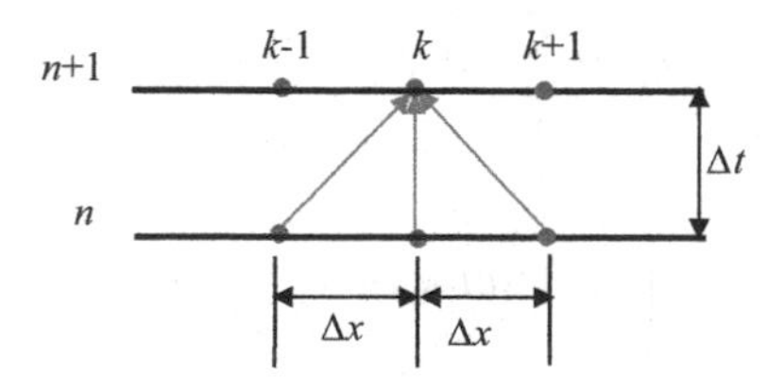

FIGURE 4.1 Schematic Diagram of Explicit Methods.

The term HOT is a measure of the error introduced using finite difference equation to solve the PDE at grid point *(n, k)*. This is known as truncation error. Now, in the limit as $\Delta x \to 0$ and $\Delta t \to 0$, the HOT $\to 0$ and hence, it is consistent.

Stability:

Von Neumann's spectral analysis

Error $\qquad E_{k,n} = A^n\, e^{ik\theta}$

where, $\theta = phase\ \ angle = \ k_x \Delta x$

Hence, $\qquad A^{n+1} e^{ik\theta} = \lambda A^n e^{i(k-1)\theta} + (1-2\lambda) A^n e^{ik\theta} + \lambda A^n e^{i(k+1)\theta}$

or $\quad A^{n+1} = \lambda\, A^n e^{-i\theta} + (1-2\lambda)\, A^n + \lambda\, A^n\, e^{i\theta} = \ A^n\, \left(\lambda\, e^{-i\theta} + (1-2\lambda) + \lambda\, e^{i\theta}\right)$

Amplification factor

$$\xi = \frac{A^{n+1}}{A^n} = \lambda\, e^{-i\theta} + (1-2\lambda) + \lambda\, e^{i\theta}$$

For stability

$$|\xi| = \left|\frac{A^{n+1}}{A^n}\right| \le 1\ for\ all\ \theta$$

i.e. $\qquad \left|\lambda\left(e^{-i\theta} + e^{i\theta}\right) + (1-2\lambda)\right| \le 1$

or $\qquad |1 - 2\lambda + 2\lambda\ \cos\theta| \le 1\ for\ all\ \theta$

This inequality should be satisfied for all λ and θ. Now for $\theta = 0$, $\pi/2$ and π we get

$$|\xi(0)| = 1$$
$$|\xi(\pi/2)| = 1 - 2\lambda\ \therefore\ |1 - 2\lambda\ | \le 1 or \lambda \le 1$$
$$|\xi(\pi)| = 1 - 4\lambda\ \therefore\ |1 - 4\lambda\ | \le 1 or \lambda \le \frac{1}{2}$$

Hence, the FTCS for 1-D diffusion equation is stable for $\lambda \le \frac{1}{2}$

4.2.1.2 The Richardson's Method [1910]

In this scheme, the forward difference approximation for the time derivative is replaced by a central difference form. Hence, for the diffusion equation, we get

$$\frac{u_k^{n+1} - u_k^{n-1}}{2\Delta t} = D\ \frac{u_{k+1}^n - 2u_k^n + u_{k-1}^n}{(\Delta x)^2} \tag{4.3a}$$

or

$$u_k^{n+1} = \lambda u_k^{n-1} + 2\lambda\ \left(u_{k+1}^n - 2u_k^n +\ u_{k-1}^n\right)\ \ where\ \lambda = \frac{D\Delta t}{(\Delta x)^2} \tag{4.3b}$$

Consistency:

$$u_k^n + \Delta t\left(\frac{\partial u}{\partial t}\right)_k^n + \frac{(\Delta t)^2}{2!}\left(\frac{\partial^2 u}{\partial t^2}\right)_k^n + \frac{(\Delta t)^3}{3!}\left(\frac{\partial^2 u}{\partial t^2}\right)_k^n$$

$$+ O\left[(\Delta t)^4\right] + .. = \left[u_k^n - \Delta t\left(\frac{\partial u}{\partial t}\right)_k^n + \frac{(\Delta t)^2}{2!}\left(\frac{\partial^2 u}{\partial x^2}\right)_k^n - \frac{(\Delta t)^3}{3!}\left(\frac{\partial^3 u}{\partial t^3}\right)_k^n + ..\right]$$

$$+ 2\lambda\left[u_k^n + \Delta x\left(\frac{\partial u}{\partial x}\right)_k^n + \frac{(\Delta x)^2}{2!}\left(\frac{\partial^2 u}{\partial x^2}\right)_k^n + O\left[(\Delta x)^3\right] + .. - 2u_k^n + u_k^n\right.$$

$$\left. - \Delta x\left(\frac{\partial u}{\partial x}\right)_k^n + \frac{(\Delta x)^2}{2!}\left(\frac{\partial^2 u}{\partial x^2}\right)_k^n + O\left[(\Delta x)^3\right] + ..\right]$$

or

$$2\Delta t\left(\frac{\partial u}{\partial t}\right)_k^n = -\frac{\Delta t^3}{3}\left(\frac{\partial^3 u}{\partial t^3}\right)_k^n + 2\lambda\left[(\Delta x)^2\left(\frac{\partial^2 u}{\partial x^2}\right)_k^n + (\Delta x)^4\left(\frac{\partial^4 u}{\partial x^4}\right)_k^n + \ldots\right]$$

or

$$\left(\frac{\partial u}{\partial t}\right)_k^n = D\left(\frac{\partial^2 u}{\partial x^2}\right)_k^n - \frac{\Delta t^2}{6}\left(\frac{\partial^3 u}{\partial t^3}\right)_k^n + \frac{D\Delta x^2}{12}\left(\frac{\partial^4 u}{\partial x^4}\right)_k^n + \ldots$$

Here,

$$HOT = -\frac{\Delta t}{2}\left(\frac{\partial^2 u}{\partial t^2}\right)_k^n + \frac{D(\Delta x)^2}{12}\left(\frac{\partial^4 u}{\partial x^4}\right)_k^n + \ldots$$

Now, in the limit as $\Delta x \to 0$ and $\Delta t \to 0$ the HOT $\to 0$ and hence, Richardson's method is consistent.

Stability:

From Von Neumann's spectral analysis, we get

Error $\quad E_{k,n} = A^n\, e^{ik\theta}$

where, $\theta = phase\;\; angle = k_x \Delta x$

$$E_k^{n+1} = E_k^{n-1} + 2\lambda\left(E_{k+1}^n - 2E_k^n + E_{k-1}^n\right)$$

Hence, $\quad A^{n+1}e^{ik\theta} = A^{n-1}e^{ik\theta} + 2\lambda\left(A^n e^{i(k+1)\theta} - 2A^n e^{ik\theta} + A^n e^{i(k-1)\theta}\right)$

or $\quad A^{n+1} = A^{n-1} + 2\lambda\, A^n\left(e^{i\theta} - 2 + e^{-i\theta}\right) = A^{n-1} + 4\lambda\, A^n(\cos\theta - 1)$

Amplification factor

$$\xi = \frac{A^{n+1}}{A^n} \quad \therefore\; A^{n+1} = \xi A^n$$

$$\text{Again } \xi = \frac{A^n}{A^{n-1}} \quad \therefore\; A^{n-1} = \frac{A^n}{\xi}$$

Substituting the values of $A^{n+1}$and A^{n-1} and solving we get

$$\xi - \frac{1}{\xi} = 4\lambda\ (\cos\theta - 1)$$

$$\xi^2 - 4\lambda\xi\ (\cos\theta - 1) - 1 = 0$$

or
$$\xi = 2\lambda\ (\cos\theta - 1) \pm \left[1 + 4\lambda^2(\cos\theta - 1)^2\right]^{\frac{1}{2}}.$$

Now the two roots ξ_1 and ξ_2 should satisfy the following equations

$$\xi_1\xi_2 = -1$$

or
$$\xi_1 + \xi_2 = 4\lambda(\cos\theta - 1) = 8\lambda\ sin^2\theta/_2.$$

From first equation, we get either

$$|\xi_1| > 1\ or\ |\xi_2| > 1$$

in which case the error at a given time level will contain some Fourier component which will grow without bound as n increases. Otherwise, $\xi_1 = \xi_2 = 1$ *(say)*, in which case from the second equation λ should be equal to zero i.e. $\Delta t = 0$ and the method has no practical value. Hence, Richardson's method is unconditionally unstable and has little practical value.

4.2.1.3 The DuFort-Frankel Method (D-F Leap-Frog Method) [1953]

This is modification of Richardson's method in which the central grid-point value u_k^n in the finite difference approximation for the diffusion term is replaced by its average at the (n-1) and (n+1) time levels giving

$$\frac{u_k^{n+1} - u_k^{n-1}}{2\Delta t} = D\ \frac{u_{k+1}^n - \left(u_k^{n+1} + u_k^{n-1}\right) + u_{k-1}^n}{(\Delta x)^2} \tag{4.4a}$$

or

$$or\ u_k^{n+1} = \frac{2\lambda\ \left(u_{k+1}^n + u_{k-1}^n\ \right) + (1 - 2\lambda)\ u_k^{n-1}}{1 + 2\lambda} \tag{4.4b}$$

Consistency:

$$(1 + 2\lambda)\left[u_k^n + \Delta t\left(\frac{\partial u}{\partial t}\right)_k^n + \frac{(\Delta t)^2}{2!}\left(\frac{\partial^2 u}{\partial t^2}\right)_k^n + \frac{(\Delta t)^3}{3!}\left(\frac{\partial^2 u}{\partial t^2}\right)_k^n + \ldots\right]$$

$$= (1 + 2\lambda)\left[u_k^n - \Delta t\left(\frac{\partial u}{\partial t}\right)_k^n + \frac{(\Delta t)^2}{2!}\left(\frac{\partial^2 u}{\partial x^2}\right)_k^n - \frac{(\Delta t)^3}{3!}\left(\frac{\partial^3 u}{\partial t^3}\right)_k^n + ..\right]$$

$$+ 2\lambda\left[u_k^n + \Delta x\left(\frac{\partial u}{\partial x}\right)_k^n + \frac{(\Delta x)^2}{2!}\left(\frac{\partial^2 u}{\partial x^2}\right)_k^n + \frac{(\Delta x)^3}{3!}\left(\frac{\partial^3 u}{\partial x^3}\right)_k^n + .. + u_k^n \right.$$

$$\left. - \Delta x\left(\frac{\partial u}{\partial x}\right)_k^n + \frac{(\Delta x)^2}{2}\left(\frac{\partial^2 u}{\partial x^2}\right)_k^n - \frac{(\Delta x)^3}{3!}\left(\frac{\partial^3 u}{\partial x^3}\right)_k^n + ..\right]$$

or

$$2\Delta t\left(\frac{\partial u}{\partial t}\right)_k^n = 2\,\lambda\,(\Delta x)^2\left(\frac{\partial^2 u}{\partial x^2}\right)_k^n - 2\lambda\,(\Delta t)^2\left(\frac{\partial^2 u}{\partial x^2}\right)_k^n - \frac{(\Delta t)^3}{3}\left(\frac{\partial^3 u}{\partial t^3}\right)_k^n + 2\,\lambda\frac{(\Delta x)^4}{12}\left(\frac{\partial^4 u}{\partial x^4}\right)_k^n$$

or

$$\left(\frac{\partial u}{\partial t}\right)_k^n = D\left(\frac{\partial^2 u}{\partial x^2}\right)_k^n - D\left(\frac{\Delta t}{\Delta x}\right)^2\left(\frac{\partial^2 u}{\partial x^2}\right)_k^n - \frac{(\Delta t)^2}{6}\left(\frac{\partial^3 u}{\partial t^3}\right)_k^n + \ldots$$

So, the DuFort-Frankel Method is consistent with 1-D diffusion equation with truncation error going to zero only if $\Delta t/\Delta x \rightarrow 0$ as both Δt and Δx tends to zero. Hence, this method puts the restriction $\Delta t \ll \Delta x$ for consistency. Note that if Δx *and* $\Delta t \rightarrow 0$ in such way that $\Delta t/\Delta x \rightarrow C$, a constant, then the finite difference equation is consistent with the hyperbolic equation

$$\frac{\partial u}{\partial t} = D\left(\frac{\partial^2 u}{\partial x^2} - C^2\frac{\partial^2 u}{\partial t^2}\right)$$

and not with the 1-D parabolic equation given by Eq. (4.1a).

Stability:

To investigate the stability of DuFort-Frankel method, the nth Fourier component of the error distribution, $E_{k,n} = A^n e^{ik\theta}$ is substituted into the error equation component of FDE

$$A^{n+1}e^{ik\theta} = \frac{\left[2\lambda\left(A^n e^{i(k+1)\theta} + A^n e^{i(k-1)\theta}\right) + (1-2\lambda)\,A^{n-1}e^{ik\theta}\right]}{(1+2\lambda)}$$

or

$$A^{n+1} = \frac{A^n\,4\lambda\,cos\theta + (1-2\lambda)\,A^{n-1}}{(1+2\lambda)}$$

or

$$(1+2\lambda)\xi^2 - 4\lambda\xi\,cos\theta - (1-2\lambda) = 0$$

i.e.

$$\xi = \frac{2\lambda\,cos\theta \pm \left(1 - 4\lambda^2 sin^2\theta\right)^{1/2}}{(1+2\lambda)}$$

If λ and θ are such that $1 - 4\lambda^2 sin^2\theta \geq 0$, then both the terms in the numerator are real. Also, $1 + 2\lambda \geq 0$ so that

$$|\xi| \leq \frac{|2\lambda\ cos\theta| + \left|(1 - 4\lambda^2 sin^2\theta)^{1/2}\right|}{(1 + 2\lambda)}$$

And so $|2\lambda\ cos\theta| \leq 2\lambda$ and $0 \leq (1 - 4\lambda^2 sin^2\theta)^{1/2} \leq 1$

Hence, DuFort-Frankel method is unconditionally stable for $\lambda \geq 0$.

4.2.2 Implicit Methods

Each of the method previously described is Explicit, i.e. the finite difference equation at the new time level contains only one unknown value u_k^{n+1}, which is calculated explicitly from the values of u known at previous time. These methods are easy to programme and require few computations at each time step. Unfortunately, each of the methods is restricted to very small time-steps, Δt, in classical FTCS method because of stability requirements and DuFort-Frankel method because of consistency requirement. For instance, if $\Delta x = 10^{-2}$, then use of FTCS method with $D = 5$ implies that Δt must be chosen to satisfy

$$\lambda = \frac{5\ \Delta t}{\Delta x^2} \leq \frac{1}{2}$$

or

$$\Delta t \leq 10^{-5}$$

To generate a numerical solution at $t = 10$ would therefore require computation of at least one million rows of grid points in the solution region. Hence, implicit methods are developed for reducing the computational time.

4.2.2.1 The Classical Implicit Method (BTCS)

An implicit method is one in which two or more unknown values of $(n+1)$ time step are specified in terms of known values in the n row. This is shown in Figure 4.2.

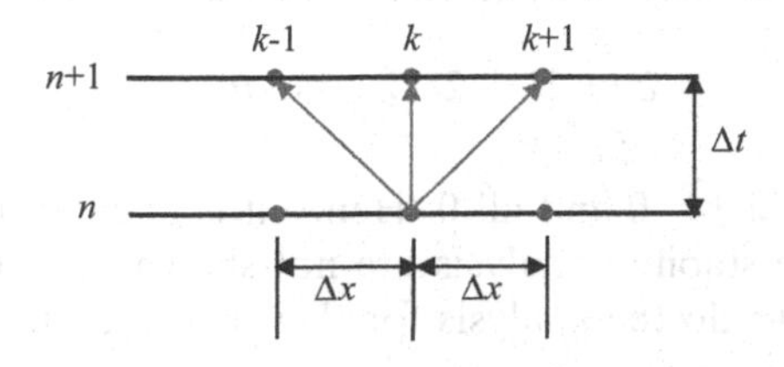

FIGURE 4.2 Schematic Diagram of Implicit Methods.

By substituting the backward difference approximation for the time derivative and central difference approximation for the space derivative in the diffusion equation at the grid point $(k,\ n+1)$ we get

$$\frac{u_k^{n+1} - u_k^n}{\Delta t} = D\left[\frac{u_{k+1}^{n+1} - 2u_k^{n+1} + u_{k-1}^{n+1}}{\Delta x^2}\right] \tag{4.5a}$$

or

$$-\lambda u_{k-1}^{n+1} + (1 + 2\lambda)u_k^{n+1} - \lambda u_{k+1}^{n+1} = u_k^n \tag{4.5b}$$

in which there are three unknown values of u at the $(n+1)^{\text{th}}$ time level. With $n = 0$, substitution of $k = 1, 2, 3 \ldots .$ $(k\text{-}1)$ gives $(k\text{-}1)$ simultaneous equations for $(k\text{-}1)$ unknown values of u at internal grid points along the first-time level, in terms of known initial and boundary values. Similarly, $n = 1$ gives $(k\text{-}1)$ equations for the $(k\text{-}1)$ unknown values of u at the second-time level and so on.

Assuming values are known at the nth level, we may substitute $k = 1, 2, \ldots,$ $(k\text{-}1)$ in Eq. (4.5b) to obtain the following set of simultaneous linear algebraic equations

$$\begin{bmatrix} (1+2\lambda)u_1^{n+1} & -\lambda u_2^{n+1} & \cdots & \cdots & \cdots & \cdots \\ -\lambda u_1^{n+1} & (1+2\lambda)u_2^{n+1} & -\lambda u_3^{n+1} & \cdots & \cdots & \cdots \\ \cdots & -\lambda u_2^{n+1} & (1+2\lambda)u_3^{n+1} & -\lambda u_4^{n+1} & \cdots & \cdots \\ & & \vdots & & & \\ \cdots & & & -\lambda u_{k-2}^{n+1} & (1+2\lambda)u_{k-1}^{n+1} & \end{bmatrix} = \begin{bmatrix} d_1 \\ d_2 \\ \vdots \\ \vdots \\ d_{k-1} \end{bmatrix}$$

$$\text{where} \quad \begin{aligned} d_1 &= u_1^n + u_0^{n+1};\ d_k = u_k^n,\ k = 1,\ 2, \ldots .(k-2); \\ d_{k-1} &= u_{k-1}^n + \lambda u_k^{n+1} \end{aligned} \tag{4.6}$$

This tri-diagonal matrix system can be solved efficiently by the Thomas algorithm. Eq. (4.6) must be solved at every time step to obtain the solution.

Von Neumann's stability analysis yields the amplification factor

$$\xi = [1 + 2\lambda(1 - \cos\theta)]^{-1}.$$

So, that $|\xi| \leq 1$ for all $\lambda > 0$ and all θ. Hence it is unconditionally stable.

The details of the stability analysis are not shown here. However, interested readers are advised to do the analysis for their better understanding.

Consistency:

$$u_{k+1}^{n+1} = u_k^n + \Delta t\left(\frac{\partial u}{\partial t}\right)_k^n + \Delta x\left(\frac{\partial u}{\partial x}\right)_k^n + \frac{(\Delta t)^2}{2!}\left(\frac{\partial^2 u}{\partial t^2}\right)_k^n + \frac{(\Delta x)^2}{2!}\left(\frac{\partial^2 u}{\partial x^2}\right)_k^n$$
$$+ \frac{\Delta t}{1!}\frac{\Delta x}{1!}\left(\frac{\partial^2 u}{\partial t \partial x}\right)_k^n + \frac{(\Delta t)^3}{3!}\left(\frac{\partial^3 u}{\partial t^3}\right)_k^n + \frac{\Delta t^2}{2!}\frac{\Delta x}{1!}\left(\frac{\partial^3 u}{\partial t^2 \partial x}\right)_k^n$$
$$+ \frac{\Delta t}{1!}\frac{\Delta x^2}{2!}\left(\frac{\partial^3 u}{\partial t \partial x^2}\right)_k^n + \frac{(\Delta x)^3}{3!}\left(\frac{\partial^3 u}{\partial x^3}\right)_k^n$$

$$u_{k+1}^{n+1} = u_k^n + \Delta t\left(\frac{\partial u}{\partial t}\right)_k^n - \Delta x\left(\frac{\partial u}{\partial x}\right)_k^n + \frac{(\Delta t)^2}{2!}\left(\frac{\partial^2 u}{\partial t^2}\right)_k^n + \frac{(\Delta x)^2}{2!}\left(\frac{\partial^2 u}{\partial x^2}\right)_k^n$$
$$- \frac{\Delta t}{1!}\frac{\Delta x}{1!}\left(\frac{\partial^2 u}{\partial t \partial x}\right)_k^n + \frac{(\Delta t)^3}{3!}\left(\frac{\partial^3 u}{\partial t^3}\right)_k^n - \frac{\Delta t^2}{2!}\frac{\Delta x}{1!}\left(\frac{\partial^3 u}{\partial t^2 \partial x}\right)_k^n$$
$$- + \frac{\Delta t}{1!}\frac{\Delta x^2}{2!}\left(\frac{\partial^3 u}{\partial t \partial x^2}\right)_k^n - \frac{(\Delta x)^3}{3!}\left(\frac{\partial^3 u}{\partial x^3}\right)_k^n$$

$$u_k^{n+1} = u_k^n + \Delta t\left(\frac{\partial u}{\partial t}\right)_k^n + \frac{(\Delta t)^2}{2!}\left(\frac{\partial^2 u}{\partial t^2}\right)_k^n + \frac{(\Delta t)^3}{3!}\left(\frac{\partial^3 u}{\partial t^3}\right)_k^n + \cdots$$

Substituting we get

$$\left(\frac{\partial u}{\partial t}\right)_k^n = D\left(\frac{\partial^2 u}{\partial x^2}\right)_k^n + HOT$$

where

$$HOT = -\frac{\Delta t}{2}\left(\frac{\partial^2 u}{\partial t^2}\right)_k^n + D\Delta t\left(\frac{\partial^3 u}{\partial t \partial x^2}\right)_k^n + \frac{D(\Delta x)^2}{12}\left(\frac{\partial^4 u}{\partial x^4}\right)_k^n.$$

Hence, the scheme has first-order accuracy with a TE of $O\left[\Delta t,\ (\Delta x)^2\right]$ and is consistent as $\Delta t \to 0$ and $\Delta x \to 0$.

4.2.2.2 The Crank-Nicolson Method [1947]

This method, shown in Figure 4.3, involves central difference approximations for both time and space derivatives at the grid point $(k\Delta x,\ n\Delta t + 1/2)$ which is not a grid point. The central difference approximation for this time derivative $\partial u/\partial t$ at this point in the solution domain is the same as the forward difference approximation of this derivative at grid point $(k,\ n)$. The spatial derivative $\partial^2 u/\partial x^2$ at $(k\Delta x,\ n\Delta t + 1/2)$ is approximated by the average of central difference approximation of these spatial derivatives at grid points $(k,\ n)$ and $(k,\ n+1)$.

The finite difference approximation is therefore

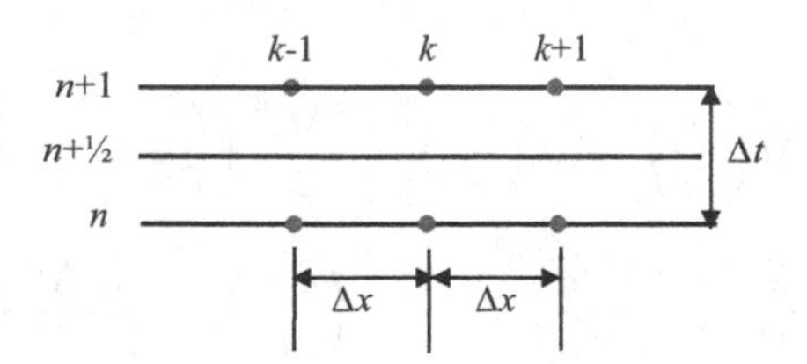

FIGURE 4.3 Schematic Diagram of the Crank-Nicolson Method.

$$\frac{u_k^{n+1} - u_k^n}{\Delta t} = \frac{1}{2} D \left[\frac{u_{k+1}^{n+1} - 2u_k^{n+1} + u_{k-1}^{n+1}}{\Delta x^2} + \frac{u_{k+1}^n - 2u_k^n + u_{k-1}^n}{\Delta x^2} \right] \tag{4.7a}$$

Rearranging we get

$$-\frac{1}{2}\lambda u_{k-1}^{n+1} + (1+\lambda)u_k^{n+1} - \frac{1}{2}\lambda u_{k+1}^{n+1} = \frac{1}{2}\lambda u_{k-1}^n + (1-\lambda)u_k^n + \frac{1}{2}\lambda u_{k+1}^n \tag{4.7b}$$

The truncation error is $O\left[(\Delta t)^2, (\Delta x)^2\right]$

Commencing with $n = 0$, then taking $n = 1, 2, 3\ldots$ in turn, the right-hand side of this equation is known for $k = 1, 2, 3\ldots, (k\text{-}1)$. Application of boundary conditions, which define the values of u_0^{n+1} and u_k^{n+1} then yields a set of simultaneous equations in the unknowns u_k^{n+1}, $k = 1, 2, \ldots. (k\text{-}1)$, i.e.

$$\begin{bmatrix} (1+\lambda)u_1^{n+1} & -{}^1\!/_2\lambda u_2^{n+1} & \cdots & \cdots & \cdots & \cdots \\ -{}^1\!/_2\lambda u_1^{n+1} & (1+\lambda)u_2^{n+1} & -{}^1\!/_2\lambda u_3^{n+1} & \cdots & \cdots & \cdots \\ \cdots & -{}^1\!/_2\lambda u_2^{n+1} & (1+\lambda)u_3^{n+1} & -{}^1\!/_2\lambda u_4^{n+1} & \cdots & \cdots \\ & & \vdots & & & \\ & & & \vdots & & \\ \cdots & & & -{}^1\!/_2\lambda u_{k-2}^{n+1} & (1+\lambda)u_{k-1}^{n+1} & \end{bmatrix} = \begin{bmatrix} d_1 \\ d_2 \\ \vdots \\ \vdots \\ \vdots \\ d_{k-1} \end{bmatrix} \tag{4.8a}$$

where

$$d_1 = \frac{\lambda}{2}u_0^n + (1-\lambda)u_1^n + \frac{\lambda}{2}u_2^n + \frac{\lambda}{2}u_0^{n+1};$$

$$d_k = \frac{\lambda}{2}u_{k-1}^n + (1-\lambda)u_k^n + \frac{\lambda}{2}u_{k+1}^n,\ k = 2, \ldots (k-2); \tag{4.8b}$$

$$d_{k-1} = \frac{\lambda}{2}u_{k-2}^n + (1-\lambda)u_{k-1}^n + \frac{\lambda}{2}u_k^n + \frac{\lambda}{2}u_k^{n+1}$$

are all known. This system can be solved by Thomas algorithm.

Application of Von Neumann's method yield the amplification factor ξ for Fourier component of the error distribution at any time level

$$\xi = \frac{1 - 2\lambda sin^2\left(\theta/2\right)}{1 + 2\lambda sin^2\left(\theta/2\right)}$$

and $|\xi| \leq 1$, for all positive values of λ and all θ. It is clear, that Crank-Nicolson method is unconditionally stable.

Consistency:

Expanding each term in the Taylor series about the grid point (k, n) yields

$$\left(\frac{\partial u}{\partial t}\right)_k^n = D\left(\frac{\partial^2 u}{\partial x^2}\right)_k^n + \frac{D(\Delta x)^2}{12}\left(\frac{\partial^4 u}{\partial x^4}\right)_k^n + \left[(\Delta t)^2, (\Delta x)^2\right]$$

In the limit as $\Delta x \to 0$, $\Delta t \to 0$, HOT $\to 0$ and hence the method is consistent.

Since the Crank-Nicolson method is second-order accurate both in time and space, a larger time increment can be used with the same degree of accuracy for this scheme.

The unconditionally stable Crank-Nicolson method may be considered a combination of a conditionally stable FTCS explicit method and an unconditionally stable classical implicit method. This is shown in Figure 4.4.

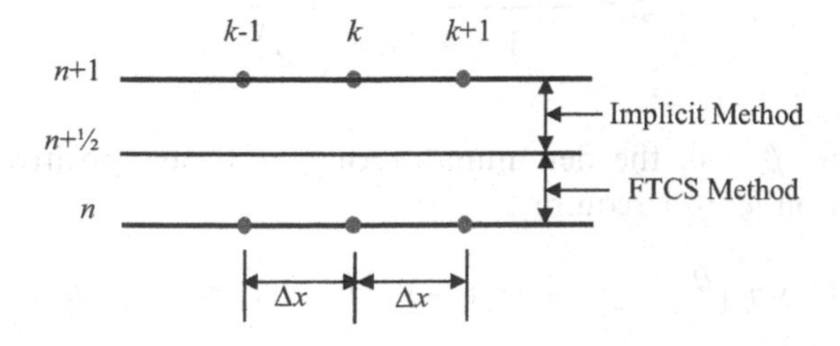

FIGURE 4.4 Reason for the Unconditional Stability of the Crank-Nicolson Method.

Applying FTCS method between n and n+½ we get

$$u_k^{n+1/2} = \frac{\lambda}{2}u_{k-1}^n + (1-\lambda)u_k^n + \frac{\lambda}{2}\, u_{k+1}^n$$

Applying a classical implicit method between n+½ and n+1, we get

$$-\frac{\lambda}{2}u_{k-1}^{n+1} + (1+\lambda)u_k^{n+1} - \frac{\lambda}{2}u_{k+1}^{n+1} = u_k^{n+1/2}.$$

Adding these two equations is the finite difference equation between time level n and $n+1$, which is identical to C-N scheme.

$$-\frac{1}{2}\lambda u_{k-1}^{n+1} + (1+\lambda)u_k^{n+1} - \frac{1}{2}\lambda u_{k+1}^{n+1} = \frac{1}{2}\lambda u_{k-1}^n + (1-\lambda)u_k^n + \frac{1}{2}\lambda u_{k+1}^n$$

4.2.2.3 The Method of Weighted Averages

Crandall [1956] used a weighted average instead of the C-N simple average of the central difference approximation to $\partial^2 u/\partial x^2$ at time level n and $(n+1)$. The finite difference approximation to diffusion equation then becomes

$$\frac{u_k^{n+1} - u_k^n}{\Delta t} = D\left[\beta\, \frac{u_{k+1}^{n+1} - 2u_k^{n+1} + u_{k-1}^{n+1}}{\Delta x^2} + (1-\beta)\frac{u_{k+1}^n - 2u_k^n + u_{k-1}^n}{\Delta x^2}\right] \tag{4.9a}$$

Where $0 \le \beta \le 1$. If $\beta = 1/2$, it is then C-N method; $\beta = 1$ this equation reduces to classical Implicit method and if $\beta = 0$, the classical FTCS scheme is defined.

Rearranging we get,

$$\begin{aligned}-\beta\lambda u_{k-1}^{n+1} + (1+2\beta\lambda)u_k^{n+1} - \beta\lambda u_{k+1}^{n+1} = \lambda(1-\beta)u_{k-1}^n + [1-2\lambda(1-\beta)]u_k^n \\ +\lambda(1-\beta)u_{k+1}^n\end{aligned} \tag{4.9b}$$

Application of Von Neumann's stability analysis yields

$$\xi = \frac{1 - 4\lambda(1-\beta)sin^2\left(\theta/2\right)}{1 + 4\lambda\beta sin^2\left(\theta/2\right)}$$

With $\lambda > 0$ and $\beta > 0$, the denominator of ξ is always positive, so that the stability condition $|\xi| \le 1$ require

$$-1 - 4\lambda\beta sin^2\left(\theta/2\right) \le 1 - 4\lambda(1-\beta)sin^2\left(\theta/2\right) \le 1 + 4\lambda\beta sin^2\left(\theta/2\right)$$

Considering the right-hand side inequality yields

$$4\lambda sin^2\left({}^{\theta}\!/_{2}\right) \geq 0,$$

which is always satisfied because $\lambda > 0$. Left hand side inequality leads to condition

$$2\lambda(1 - 2\beta)sin^2\left({}^{\theta}\!/_{2}\right) \leq 1 \tag{4.10}$$

If $\beta < ½$, this can be written as

$$\lambda \leq \frac{1}{2(1 - 2\beta)sin^2\left({}^{\theta}\!/_{2}\right)},$$

which is satisfied for all θ if

$$\lambda \leq \frac{1}{2(1 - 2\beta)}$$

If $\beta > ½$, since $\lambda > 0$, the left side of condition [Eq. (4.10)] is always negative or zero.

Hence, weighted average scheme defined by Eq. (4.9b) conditionally stable requiring

$$0 < \lambda \leq \frac{1}{2(1 - 2\beta)} \qquad for\ \beta < \frac{1}{2}$$

$$\lambda > 0 \qquad for\ \beta \geq \frac{1}{2}$$

The explicit method marches the solution outward from the initial data and calculates the unknown u at a point P without knowing the values in the boundaries as can be seen in Figure 4.5.

Since the parabolic heat equation has the characteristic t = constant, the value at point P depends on the left and right boundary conditions. Hence, the explicit scheme fails to represent the physical behavior of the parabolic equation. The implicit scheme will be a better method for solving the parabolic PDE.

4.3 ONE-DIMENSIONAL TRANSPORT EQUATIONS

PDE governing one-dimensional transport equation is given by:

$$\frac{\partial u}{\partial t} + c\frac{\partial u}{\partial x} = D\frac{\partial^2 u}{\partial x^2} \tag{4.11}$$

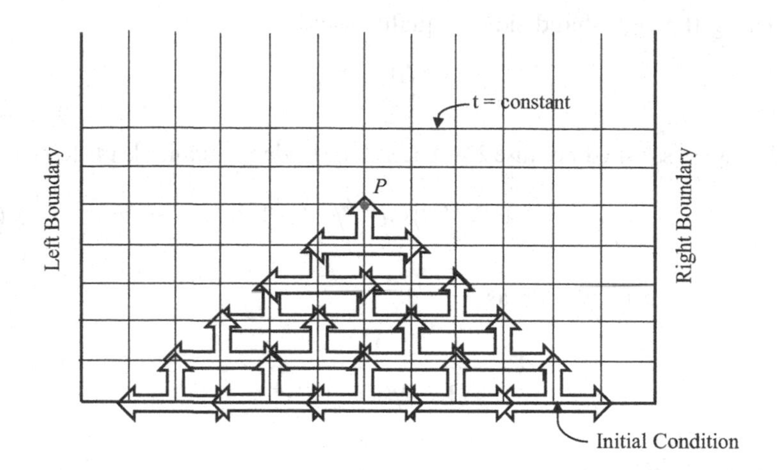

FIGURE 4.5 Stencil of the Explicit Scheme.

This describes the change of scalar property u associated with diffusion and convection in a fluid moving with a speed c parallel to X-axis. Already we have seen the discretization methods of the diffusion equation. Hence, we shall first concentrate on the discretization schemes of the convection equation before taking up the complete transport equation.

4.3.1 The Wave Equation

$$\frac{\partial u}{\partial t} + c\frac{\partial u}{\partial x} = 0 \tag{4.12}$$

This is a linear hyperbolic equation in which the wave propagates with a constant velocity c This equation also can be used to represent the non-linear equations governing inviscid flows which will be described afterwards. The general solution of this equation for constant c is $u = f(x - ct)$ where $f(x)$ is the initial distribution of x. This means that the initial distribution of u is translated without any change along the X-axis at constant speed c.

4.3.1.1 The FTCS Method

The schematic diagram of the FTCS method is shown in Figure 4.6.

Discretization of the Eq. (4.12) takes the form:

$$\frac{u_k^{n+1} - u_k^n}{\Delta t} + c\left[\frac{u_{k+1}^n - u_{k-1}^n}{2\Delta x}\right] = 0 \tag{4.13a}$$

or

$$u_k^{n+1} = u_k^n - \frac{c\ \Delta t}{2\ \Delta x}\left[u_{k+1}^n - u_{k-1}^n\right] \tag{4.13b}$$

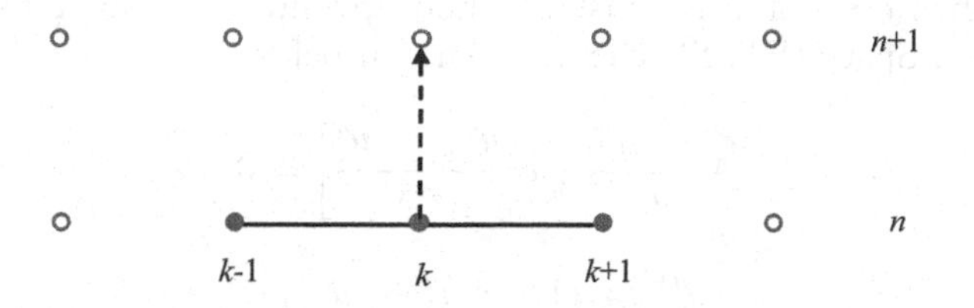

FIGURE 4.6 Schematic Diagram of the FTCS Method.

or

$$u_k^{n+1} = u_k^n - \frac{\lambda}{2}\left[u_{k+1}^n - u_{k-1}^n\right] \tag{4.13c}$$

where

$$\lambda = \frac{c\,\Delta t}{\Delta x} = \textit{Courant number}$$

If Eq. (4.13c) is analysed for stability using Von Neumann's spectral method, it can be found that the amplification factor is given by

$$\xi = 1 - i\lambda\, sin\theta$$

or

$$|\xi|^2 = 1 + \lambda^2 sin^2\theta$$

which is always ≥ 1 for all values of Δx and Δt. This means that for wave equation, *FTCS method is always unstable.*

4.3.1.2 Upwind Differencing

To overcome this instability problem, the upwind differencing scheme is used. In this scheme, a one-sided rather than central space differencing is used for the convective term, $\partial u/\partial x$, the direction of the difference being upwind, which is shown in Figure 4.7. That is when, $c > 0$, the backward difference scheme is used at grid point (k, n) and for $c < 0$, the forward difference is used.

This has the following physical basis. When determining the value of u_k^{n+1} from known values of u at the n-th time level, if $c > 0$ the fluid flow carries information from x_{k-1} to x_k. So, a backward difference approximation is chosen for the spatial derivative in the convective term $c\partial u/\partial x$. To use a forward difference approximation would be unrealistic, since information cannot be carried from position x_{k+1} to x_k when $c > 0$. However, this choice is appropriate if $c < 0$.

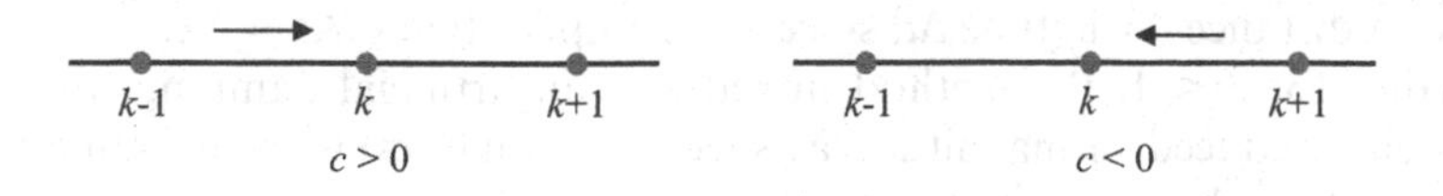

FIGURE 4.7 Upwind Difference Scheme.

Hence, when $c < 0$ and is constant, then upwind differencing uses Forward Time Forward Space (FTFS) scheme as shown below.

$$\frac{u_k^{n+1} - u_k^n}{\Delta t} + c\left[\frac{u_{k+1}^n - u_k^n}{\Delta x}\right] = 0 \tag{4.14a}$$

or

$$u_k^{n+1} = (1 - \lambda)u_k^n - \lambda u_{k+1}^n \tag{4.14b}$$

For $c > 0$, Forward Time Backward Space (FTBS) scheme approximation is used. That is

$$\frac{u_k^{n+1} - u_k^n}{\Delta t} + c\left[\frac{u_k^n - u_{k-1}^n}{\Delta x}\right] = 0 \tag{4.14c}$$

or

$$u_k^{n+1} = (1 - \lambda)u_k^n + \lambda u_{k-1}^n \tag{4.14d}$$

Both the finite difference form of above equations has a truncation error of $O[\Delta t, \Delta x]$.

Stability Analysis: When $c > 0$, the error propagates according to the relation

$$E_k^{n+1} = (1 - \lambda)E_k^n + \lambda E_{k-1}^n$$

Application of Von Neumann's stability analysis to this error equation yields the amplification factor

$$\xi = (1 - \lambda + \lambda \cos \theta) - i\lambda \sin\theta$$

or

$$|\xi|^2 = (1 - \lambda + \lambda \cos \theta)^2 + \lambda^2 \sin^2\theta$$

or

$$|\xi|^2 = 1 - 2\lambda(1 - \lambda)(1 - \cos \theta) \tag{4.15}$$

When $\lambda \leq 1$, $|\xi|^2$ has a maximum value of one when $\cos \theta = 1$, since the second term in the right-hand side of Eq. (4.15) is always positive.

When $\lambda > 1$, $|\xi|^2$ has a maximum value of $(2\lambda - 1)^2 > 1$ when $\cos \theta = -1$ and the method is unstable. The requirement of stability criterion is therefore $\lambda \leq 1$.

This relation is called *Courant-Friedrichs-Lewy (CFL)* condition. This condition generally applies to explicit finite difference approximations to hyperbolic PDE. It implies that $c\,\Delta t \leq \Delta x$, i.e. fluid should not travel more than one grid spacing in one time-step. If $\lambda = 1$, when $c > 0$, Eq. (4.14b) becomes $u_k^{n+1} = u_{k-1}^n$, which is the exact solution to the convective equation. The initial distribution of u moves a distance Δx in time Δt, since $\lambda = 1$ implies that $c\,\Delta t = \Delta x$.

Further for $\lambda < 1$, the method introduces an artificial damping, the values of u_k^n being reduced in magnitude at successive time levels in the same way as the errors E_k^n reduces, since the equations are analogues.

***Consistency*:**

Expanding each term of Eq. (4.14b) by Taylor's series we get

$$u_k^{n+1} = u_k^n + \Delta t\left(\frac{\partial u}{\partial t}\right)_k^n + \frac{(\Delta t)^2}{2!}\left(\frac{\partial^2 u}{\partial t^2}\right)_k^n + \frac{(\Delta t)^3}{3!}\left(\frac{\partial^3 u}{\partial t^3}\right)_k^n + \ldots$$

$$u_{k-1}^n = u_k^n - \Delta x\left(\frac{\partial u}{\partial x}\right)_k^n + \frac{(\Delta x)^2}{2!}\left(\frac{\partial^2 u}{\partial x^2}\right)_k^n - \frac{(\Delta x)^3}{3!}\left(\frac{\partial^3 u}{\partial x^3}\right)_k^n + \ldots$$

$$\therefore \quad u_k^n + \Delta t\left(\frac{\partial u}{\partial t}\right)_k^n + \frac{(\Delta t)^2}{2!}\left(\frac{\partial^2 u}{\partial t^2}\right)_k^n + O\left[(\Delta t)^3\right]$$

$$= (1-\lambda)u_k^n + \lambda u_k^n - \lambda\Delta x\left(\frac{\partial u}{\partial x}\right)_k^n + \lambda\frac{(\Delta x)^2}{2!}\left(\frac{\partial^2 u}{\partial x^2}\right)_k^n + O\left[(\Delta x)^3\right]$$

$$\left(\frac{\partial u}{\partial t}\right)_k^n + \lambda\frac{\Delta x}{\Delta t}\left(\frac{\partial u}{\partial x}\right)_k^n = -\frac{\Delta t}{2!}\left(\frac{\partial^2 u}{\partial t^2}\right)_k^n + \lambda\frac{(\Delta x)^2}{\Delta t 2!}\left(\frac{\partial^2 u}{\partial x^2}\right)_k^n + O\left[(\Delta t)^2 + (\Delta x)^2\right]$$

Now $\dfrac{\partial u}{\partial t} = -c\dfrac{\partial u}{\partial x}$; *or* $\dfrac{\partial^2 u}{\partial t^2} = -c\dfrac{\partial}{\partial x}\left(\dfrac{\partial u}{\partial t}\right) = -c\dfrac{\partial}{\partial x}\left(-c\dfrac{\partial u}{\partial x}\right) = c^2\dfrac{\partial^2 u}{\partial x^2}$

$$\therefore \quad \left(\frac{\partial u}{\partial t}\right)_k^n + c\left(\frac{\partial u}{\partial x}\right)_k^n = -\frac{\Delta t}{2!}\left(c^2\frac{\partial^2 u}{\partial x^2}\right)_k^n + c\frac{\Delta x}{2!}\left(\frac{\partial^2 u}{\partial x^2}\right)_k^n + O\left[(\Delta t)^2 + (\Delta x)^2\right]$$

$$\text{or} \quad \left(\frac{\partial u}{\partial t}\right)_k^n + c\left(\frac{\partial u}{\partial x}\right)_k^n = c\frac{\Delta x}{2}(1-\lambda)\left(\frac{\partial^2 u}{\partial x^2}\right)_k^n + O\left[(\Delta t)^2 + (\Delta x)^2\right]$$

$$= D'\left(\frac{\partial^2 u}{\partial x^2}\right)_k^n + O\left[(\Delta t)^2 + (\Delta x)^2\right]$$

where $D' = c\frac{\Delta x}{2}(1-\lambda)$. As $\Delta x \to 0$, $\Delta t \to 0$, the coefficient $D' \to 0$ as do the term $O\left[(\Delta t)^2 + (\Delta x)^2\right]$. Eq. (4.14b) is therefore consistent with the wave equation. The upwind differencing method will introduce an artificial diffusion because of non-physical coefficient D' of $\partial^2 u/\partial x^2$. For given $\lambda \neq 1$, larger the speed c and larger value of step Δx results in larger values of D' and, hence introducing considerable artificial diffusion. Similar results are obtained for $c < 0$.

Further Comments:

For upwind difference scheme ($c > 0$), we have seen

$$\left(\frac{\partial u}{\partial t}\right)_k^n + c\left(\frac{\partial u}{\partial x}\right)_k^n = c\frac{\Delta x}{2}(1-\lambda)\left(\frac{\partial^2 u}{\partial x^2}\right)_k^n + O\left[(\Delta t)^2 + (\Delta x)^2\right] \tag{4.16}$$

This equation is known as modified equation. Now if $\lambda = 1$, the right-hand side of the modified equation becomes zero and the wave equation is solved exactly.

In this case, the upwind difference scheme reduces to $u_k^{n+1} = u_{k-1}^n$ which is equivalent to solving the wave equation exactly. Finite difference algorithms which exhibit such behavior are said to satisfy *"shift"* condition. From Eq. (4.16) we can observe that first term of HOT contains $c\frac{\Delta x}{2}(1-\lambda)\left(\partial^2 u/\partial x^2\right)$ which is like the viscous term. For $\lambda < 1$, the method introduces an artificial damping, the values of u_k^n being reduced in magnitude at successive time levels thus introducing an "*artificial viscosity*" into the solution. The reduction of all gradients in the solution which is numerically induced is due to artificial viscosity and is due to even derivative term in the truncation error. This effect is known as *dissipation* and is explained in Figure 4.8.

A different effect arises due to odd derivative terms in the TE; this results in a distortion of phase relations between waves. This effect is known as *dispersion.* The combination of dissipation and dispersion is known as *diffusion* and results in sharp dividing lines in the computational domain.

4.3.1.3 The Lax Method (Lax-Friedrichs Method)

The FTCS method (Euler Method) can be made stable by replacing u_k^n in Eq. (4.13a) with the averaged term $\left(u_{k+1}^n + u_{k-1}^n\right)/2$. The resulting algorithm, known as the Lax method [1954] is given by

$$\frac{u_k^{n+1} - \left(u_{k+1}^n + u_{k-1}^n\right)/2}{\Delta t} + c\left[\frac{u_{k+1}^n - u_{k-1}^n}{2\Delta x}\right] = 0 \tag{4.17}$$

This explicit, one step scheme is first order accurate with truncation error of $O\left[\Delta t, (\Delta x)^2/\Delta t\right]$ and stable if $\lambda \leq 1$. A schematic diagram of the Lax method is shown in Figure 4.9.

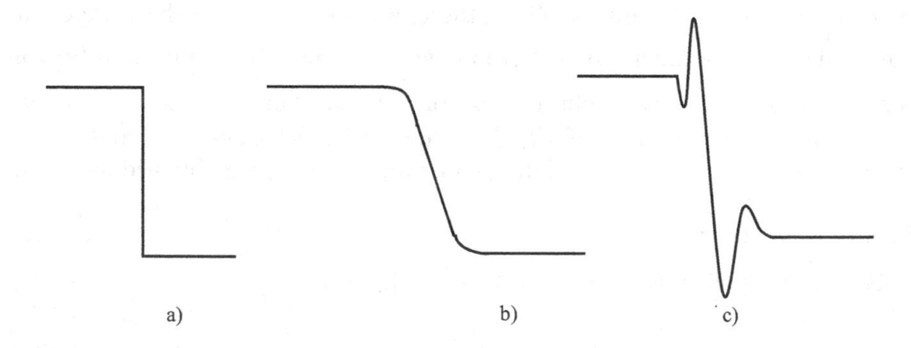

FIGURE 4.8 Numerical Dissipation and Dispersion a) Exact Solution; b) Numerical Solution Distorted by Dissipation Error; c) Numerical Solution Distorted by Dispersion Error.

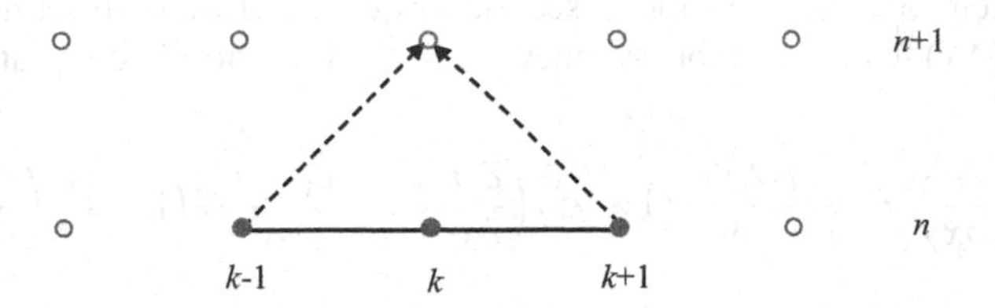

FIGURE 4.9 Schematic Diagram of the Lax Method.

The modified equation is

$$\left(\frac{\partial u}{\partial t}\right)_k^n + c\left(\frac{\partial u}{\partial x}\right)_k^n = c\frac{\Delta x}{2}\left(\frac{1}{\lambda} - \lambda\right)\left(\frac{\partial^2 u}{\partial x^2}\right)_k^n + \frac{c(\Delta x)^3}{3}(1-\lambda^2)\left(\frac{\partial^3 u}{\partial x^3}\right)_k^n + \ldots$$

Note that this method is not uniformly consistent since $(\Delta x)^2/\Delta t$ may not approach zero as Δt and Δx approach to zero. However, if λ is held constant as Δt and Δx approach to zero, the method is consistent. The Lax method is known for its large dissipation error when $\lambda \neq 1$. The amplification number is given by

$$|\xi|^2 = cos\ \theta - i\lambda\ sin\ \theta$$

4.3.1.4 The Lax-Wendroff Method [1960]

The Lax-Wendroff method can be obtained from a Taylor-series expansion in the following manner:

$$u_k^{n+1} = u_k^n + \Delta t\left(\frac{\partial u}{\partial t}\right)_k^n + \frac{(\Delta t)^2}{2!}\left(\frac{\partial^2 u}{\partial t^2}\right)_k^n + O\left[(\Delta t)^3\right] \tag{4.18a}$$

Using wave equation, we get

$$\frac{\partial u}{\partial t} = -c\frac{\partial u}{\partial x};\ or\ \frac{\partial^2 u}{\partial t^2} == c^2\frac{\partial^2 u}{\partial x^2} \tag{4.18b}$$

or

$$u_k^{n+1} = u_k^n - c\Delta t\left(\frac{\partial u}{\partial x}\right)_k^n + \frac{c^2(\Delta t)^2}{2}\left(\frac{\partial^2 u}{\partial x^2}\right)_k^n + O\left[(\Delta t)^3\right] \tag{4.18c}$$

Hence,

$$u_k^{n+1} = u_k^n - \frac{c\ \Delta t}{2\ \Delta x}\left[u_{k+1}^n - u_{k-1}^n\right] + \frac{c^2(\Delta t)^2}{2(\Delta x)^2}\left(u_{k+1}^n - 2u_k^n + u_{k-1}^n\right)$$

This explicit, one-step scheme is second order accurate with truncation error of $O\left[(\Delta t)^2, (\Delta x)^2\right]$ and is stable whenever $\lambda \leq 1$. The modified equation is

$$\left(\frac{\partial u}{\partial t}\right)_k^n + c\left(\frac{\partial u}{\partial x}\right)_k^n = -\frac{c(\Delta x)^2}{6}(1-\lambda^2)\left(\frac{\partial^3 u}{\partial x^3}\right)_k^n - \frac{c(\Delta x)^3}{8}\lambda(1-\lambda^2)\left(\frac{\partial^4 u}{\partial x^4}\right)_k^n + \ldots$$

The amplification factor is $|\xi|^2 = 1 - \lambda^2(1 - cos\ \theta) - i\lambda\ sin\ \theta$

The scheme normally has a large lagging phase error.

4.3.1.5 The Two-Step Lax-Wendroff Method

For non-linear equations, such as inviscid flow equations, (e.g. $\frac{\partial u}{\partial t} + u\frac{\partial u}{\partial x} = 0$), a two-step variation of the original Lax-Wendroff method can be used. This explicit, two-step three-time level method when applied to the wave equation, becomes

Step 1:

$$\frac{u_{k+1/2}^{n+1/2} - \left(u_{k+1}^n + u_k^n\right)/2}{\Delta t/2} + c\left[\frac{u_{k+1}^n - u_k^n}{\Delta x}\right] = 0 \tag{4.19a}$$

Step 2:

$$\frac{u_k^{n+1} - u_k^n}{\Delta t} + c\left[\frac{u_{k+1/2}^{n+1/2} - u_{k-1/2}^{n+1/2}}{\Delta x}\right] = 0 \tag{4.19b}$$

This scheme is second order accurate with truncation error of $O\left[(\Delta t)^2, (\Delta x)^2\right]$ and is stable whenever $\lambda \leq 1$. Step 1 is the Lax method applied at the mid-point $k+½$ for a half time-step and step 2 is the leap-frog method for the remaining half-time step. When applied to the linear wave equation, two-step Lax-Wendroff method is equivalent to the original Lax-Wendroff method. This can be readily shown by substituting Eq. (4.19a) into Eq. (4.19b). Since the two schemes are equivalent, it follows that the modified equation and the amplification factor are same for two methods.

4.3.1.6 The MacCormack Method

The MacCormack method is an explicit method for solving hyperbolic and hyperbolic-parabolic systems of equations. It is a variation of the two-step Lax-Wendroff method. When applied to a first-order wave equation, Eq. (4.11), MacCormack's predictor-corrector scheme [1969] becomes

Predictor step:

$$\overline{u_k^{n+1}} = u_k^n - \frac{c\Delta t}{\Delta x}\left(u_{k+1}^n - u_k^n\right) \tag{4.20a}$$

Corrector step:

$$u_k^{n+1} = \frac{1}{2}\left[u_k^n + \overline{u_k^{n+1}} - \frac{c\Delta t}{\Delta x}\left(\overline{u_k^{n+1}} - \overline{u_{k-1}^{n+1}}\right)\right] \tag{4.20b}$$

The predictor step gives an approximate value of u_k at time level $n+1$, which is denoted by $\overline{u_k^{n+1}}$. The corrector step then provides the final value of u_k at time level $n+1$. The present predictor equation utilizes a forward difference of spatial derivative, while the corrector equation utilizes a backward difference. This differencing can be reversed without affecting the formal accuracy of the method.

The explicit MacCormack method is second order accurate with a truncation error of $O\left[(\Delta t)^2, (\Delta x)^2\right]$ and is stable whenever $\lambda \le 1$. For linear wave equation, Eq. (4.20a) can be substituted in Eq. (4.20b) to give one-step scheme which is identical to Lax-Wendroff method.

$$u_k^{n+1} = u_k^n - \frac{c\ \Delta t}{2\ \Delta x}\left[u_{k+1}^n - u_{k-1}^n\right] + \frac{c^2(\Delta t)^2}{2(\Delta x)^2}\left(u_{k+1}^n - 2u_k^n + u_{k-1}^n\right)$$

The modified equation for this method is

$$\left(\frac{\partial u}{\partial t}\right)_k^n + c\left(\frac{\partial u}{\partial x}\right)_k^n = -\frac{c(\Delta x)^2}{6}(1-\lambda^2)\left(\frac{\partial^3 u}{\partial x^3}\right)_k^n - \frac{c(\Delta x)^3}{8}\lambda(1-\lambda^2)\left(\frac{\partial^4 u}{\partial x^4}\right)_k^n + \ldots$$

and the amplification factor becomes

$$|\xi|^2 = 1 - \lambda^2(1 - \cos\theta) - i\lambda \sin\theta$$

Note that both the modified equation and amplification factor is like Lax-Wendroff method.

4.3.1.7 The Beam-Warming Method

The Beam-Warming method, a second-order upwind method proposed by Warming and Beam [1975], is a variant of the MacCormack method. In this method, only backward difference is used for the predictor and corrector steps. For $c > 0$,

Predictor step:

$$\overline{u_k^{n+1}} = u_k^n - \frac{c\Delta t}{\Delta x}\left(u_k^n - u_{k-1}^n\right) \tag{4.20a}$$

Corrector step:

$$u_k^{n+1} = \frac{1}{2}\left[u_k^n + \overline{u_k^{n+1}} - \frac{c\Delta t}{\Delta x}\left(\overline{u_k^{n+1}} - \overline{u_{k-1}^{n+1}}\right) - \frac{c\Delta t}{\Delta x}\left(u_k^n - 2u_{k-1}^n + u_{k-2}^n\right)\right] \quad (4.21b)$$

If we substitute Eq. (4.21a) into Eq. (4.21b), we obtain one-step equation as follows

$$u_k^{n+1} = u_k^n - \frac{c\,\Delta t}{\Delta x}\left[u_k^n - u_{k-1}^n\right] + \frac{1}{2}\frac{c\Delta t}{\Delta x}\left[\frac{c\Delta t}{\Delta x} - 1\right]\left(u_k^n - 2u_{k-1}^n + u_{k-2}^n\right)$$

The method is second order accurate with a truncation error of $O\left[(\Delta t)^2, (\Delta t)(\Delta x),\ (\Delta x)^2\right]$ and is stable whenever $0 \le \lambda \le 2$.

For a wave equation, which is hyperbolic in nature, explicit schemes provide a more natural finite difference approximation because a hyperbolic PDE has a limited zone of influence. Implicit methods are more appropriate for solving a parabolic PDE.

4.3.1.8 The Implicit Method

For linear one-dimensional wave equations, the second-order Lax-Wendroff scheme produces excellent results with minimum computational effort. An implicit scheme is not an optimal choice, as we require smaller time steps in obtaining a solution. However, we shall be discussing the second-order Crank-Nicolson method.

The unsteady term of the wave equation can be expressed as

$$u_k^{n+1} = u_k^n + \left[\left(\frac{\partial u}{\partial t}\right)^n + \left(\frac{\partial u}{\partial t}\right)^{n+1}\right]_k + O(\Delta t^3)$$

Now

$$\frac{\partial u}{\partial t} = -c\frac{\partial u}{\partial x}$$

Hence, we get

$$u_k^{n+1} = u_k^n - \frac{c\Delta t}{2}\left[\left(\frac{\partial u}{\partial x}\right)^n + \left(\frac{\partial u}{\partial x}\right)^{n+1}\right]_k + O(\Delta t^3)$$

or

$$u_k^{n+1} = u_k^n - \frac{\lambda}{4}\left[u_{k+1}^n + u_k^n - u_{k+1}^{n+1} + u_k^{n+1}\right] \quad (4.22)$$

where

$$\lambda = \frac{c\Delta t}{\Delta x}$$

This method has a truncation error $0(\Delta t^2, \Delta x^2)$ and is unconditionally stable. The tridiagonal matrix must be solved at each time level. As mentioned earlier, explicit methods are recommended for solving the linear wave equation.

4.3.2 The Complete Transport Equation

The complete transport equation is given by

$$\frac{\partial u}{\partial t} + c\frac{\partial u}{\partial x} = D\frac{\partial^2 u}{\partial x^2} \tag{4.23}$$

4.3.2.1 Central Difference

Using central difference approximation for the spatial derivative in the convection term, which produces an unconditionally unstable method in the absence of a diffusion term, the Eq. (4.23) becomes

$$\frac{u_k^{n+1} - u_k^n}{\Delta t} + c\left[\frac{u_{k+1}^n - u_{k-1}^n}{2\Delta x}\right] = D\left[\frac{u_{k+1}^n - 2u_k^n + u_{k-1}^n}{(\Delta x)^2}\right] \tag{4.24a}$$

Rearranging we get,

$$u_k^{n+1} = \left(\frac{\lambda}{2} + s\right)u_{k-1}^n + (1-2s)u_k^n + \left(-\frac{\lambda}{2} + s\right)u_{k+1}^n \tag{4.24b}$$

where

$$\lambda = \frac{c\Delta t}{\Delta x} \quad and \quad s = \frac{D\Delta t}{(\Delta x)^2} \tag{4.24c}$$

Applying Von Neumann's stability analysis, the amplification factor for the above equation becomes

$$\xi = 1 - 2s(1 - cos\ \theta) - i\lambda\ sin\ \theta \tag{4.25}$$

The diffusion term in Eq. (4.24a) is responsible for the second term on the right-hand side of Eq. (4.25). It has a stabilizing effect, as we shall see later, by reducing the real part of ξ below unity.

Hence, $|\xi|^2 = 1 + 4s^2 + 4s^2 cos^2\theta - 4s + 4s cos\ \theta - 8s^2 \cos\theta + \lambda^2 sin^2\ \theta$

If $\lambda \le$ s, then

$$|\xi|^2 \le 1 - 4s\ (1 - 2s)\ (1 - \cos\theta)$$

Now as $cos\ \theta \le 1$, $|\xi|^2 \le 1$ provided $s \le ½$.
Hence, the stability condition is $\lambda \le 2s \le 1$.
Substituting the values of λ and s we get

$$r = \frac{c\Delta x}{D} \leq 2$$

where r can be considered as the Reynolds number based on the grid size. Later while dealing with the finite volume method we shall see that this term is also called as cell Peclet number.

4.3.2.2 The Richardson Method

Applying central time differencing with central space differencing for both convection and diffusion terms we get,

$$\frac{u_k^{n+1} - u_k^{n-1}}{2\Delta t} + c\left[\frac{u_{k+1}^n - u_{k-1}^n}{2\Delta x}\right] = D\left[\frac{u_{k+1}^n - 2u_k^n + u_{k-1}^n}{(\Delta x)^2}\right] \tag{4.26a}$$

Rearranging we get,

$$u_k^{n+1} = (1 + \lambda + 2s)u_{k-1}^n - 4s\, u_k^n + (2s - \lambda)u_{k+1}^n \tag{4.26b}$$

This scheme, like the pure diffusion equation, is unconditionally unstable.

4.3.2.3 The DuFort-Frankel Method

Replacing the value of u_k^n in the second order space derivative on the right-hand side of Eq. (4.26a) by its average at time levels (n+1) and (n-1), we get

$$\frac{u_k^{n+1} - u_k^{n-1}}{2\Delta t} + c\left[\frac{u_{k+1}^n - u_{k-1}^n}{2\Delta x}\right] = D\left[\frac{u_{k+1}^n - \left(u_k^{n+1} + u_k^{n-1}\right) + u_{k-1}^n}{(\Delta x)^2}\right] \tag{4.27a}$$

Rearranging

$$u_k^{n+1} = \left(\frac{1 - 2s}{1 + 2s}\right)u_{k-1}^n + \left(\frac{\lambda + 2s}{1 + 2s}\right)u_{k-1}^n - \left(\frac{\lambda - 2s}{1 + 2s}\right)u_{k+1}^n \tag{4.27b}$$

Von Neumann's stability analysis show that D-F method is stable for $\lambda \leq 1$. Using Taylor series expansion about grid point (n, k) produces the following equation which is hyperbolic in nature.

$$\left(\frac{\partial u}{\partial t}\right)_k^n + \mathrm{c}\left(\frac{\partial u}{\partial x}\right)_k^n + D\left(\frac{\Delta t}{\Delta x}\right)^2\left(\frac{\partial^2 u}{\partial x^2}\right)_k^n = D\left(\frac{\partial^2 u}{\partial x^2}\right)_k^n + \ldots\ldots\ldots \tag{4.28}$$

The stability of the method is due to the third term on the left-hand side of Eq. (4.28) which makes the equation hyperbolic in nature. The finite difference equation is consistent if ${}^{\Delta t}/_{\Delta x} \to 0$ *as* $\Delta x \to 0$ *and* $\Delta t \to 0$.

4.3.2.4 The Upwind Method

FTCS for the time derivative and diffusion term in Eq. (4.23), together with upwind differencing for the convective term, gives for $c > 0$

$$\frac{u_k^{n+1} - u_k^n}{\Delta t} + c\left[\frac{u_k^n - u_{k-1}^n}{\Delta x}\right] = D\left[\frac{u_{k+1}^n - 2u_k^n + u_{k-1}^n}{(\Delta x)^2}\right] \tag{4.29a}$$

or

$$u_k^{n+1} = (\lambda + s)u_{k-1}^n + (1 - \lambda - 2s)\, u_k^n + s\, u_{k+1}^n \tag{4.29b}$$

Carrying out Von Neumann's stability analysis, we get the stability criteria as $\lambda + 2s \leq 1$.

For checking the consistency, expanding each term of Eq. (4.29b) by Taylor series at grid point (n, k) we get

$$\left(\frac{\partial u}{\partial t}\right)_k^n + c\left(\frac{\partial u}{\partial x}\right)_k^n = (D + D')\left(\frac{\partial^2 u}{\partial x^2}\right)_k^n + O\left[(\Delta t)^2 + (\Delta x)^2\right]$$

where $D' = c\frac{\Delta x}{2}(1 - \lambda)$.

The artificial diffusion effect, D', will be minimized for λ as close to unity as possible. However, in a practical problem with c variable, this is difficult to achieve throughout the complete flow field and so some artificial diffusion will necessarily be present. Accurate solution cannot be achieved unless $D' \ll D$ which, in general, requires $^1/_2 c\Delta x \ll D$.

4.4 TWO-DIMENSIONAL DIFFUSION EQUATION

We are now ready for the discretization of a two-dimensional diffusion equation. It is a simple extension of the one-dimensional diffusion equation. The two-dimensional diffusion equation is given by

$$\frac{\partial u}{\partial t} = \frac{\partial}{\partial x}\left(D_x \frac{\partial u}{\partial x}\right) + \frac{\partial}{\partial y}\left(D_y \frac{\partial u}{\partial y}\right) + S \tag{4.30}$$

where $D_{x,y}$ is the diffusion coefficient in x and y direction and S is the source term.

For heat conduction equation

$$D_{x,y} = \frac{k_{x,y}}{\rho C_P} \text{ and } S = \frac{\dot{q}}{\rho C_P}$$

For isotropic material, $D_x = D_y = D$ and neglecting source term the Eq. (4.30) becomes

$$\frac{\partial u}{\partial t} = D\left(\frac{\partial^2 u}{\partial x^2} + \frac{\partial^2 u}{\partial y^2}\right) \tag{4.31}$$

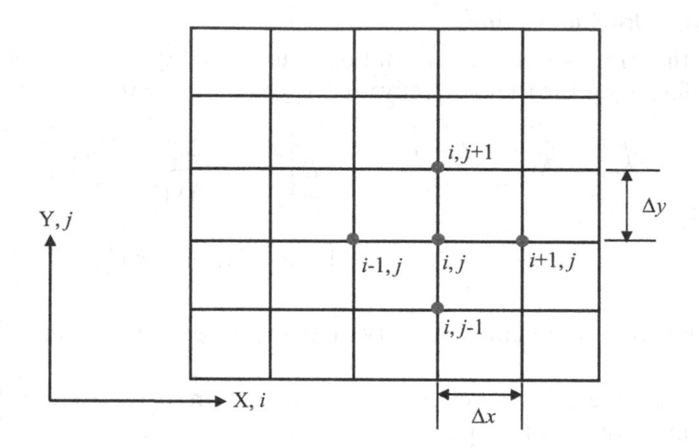

FIGURE 4.10 Finite Difference Representation in Two Dimensions.

The first step in establishing a finite difference procedure for solving PDE is to replace the continuous problem domain with a finite-difference mesh or grid as shown in Figure 4.10.

4.4.1 The Explicit Method

Applying the FTCS method explained earlier, we get

$$\frac{u_{i,j}^{n+1} - u_{i,j}^{n}}{\Delta t} = D\left[\frac{u_{i+1,j}^{n} - 2u_{i,j}^{n} + u_{i-1,j}^{n}}{(\Delta x)^2} + \frac{u_{i,j+1}^{n} - 2u_{i,j}^{n} + u_{i,j-1}^{n}}{(\Delta y)^2}\right] \tag{4.32a}$$

$$Let\ \lambda_x = \frac{D\Delta t}{(\Delta x)^2}\ and\ \lambda_y = \frac{D\Delta t}{(\Delta y)^2}$$

Then Eq. (4.32a) becomes

$$u_{i,j}^{n+1} = u_{i,j}^{n} + \lambda_x\left[u_{i+1,j}^{n} - 2u_{i,j}^{n} + u_{i-1,j}^{n}\right] + \lambda_y\left[u_{i,j+1}^{n} - 2u_{i,j}^{n} + u_{i,j-1}^{n}\right] \tag{4.32b}$$

which is of order $\left[\Delta t, (\Delta x)^2, (\Delta y)^2\right]$.

Stability analysis indicates the scheme is stable under the following condition:

$$\lambda_x + \lambda_y \leq \frac{1}{2}$$

If $\Delta x = \Delta y$, then $\lambda_x = \lambda_y = \lambda$. Then the stability criteria become $\lambda \leq 1/4$, which is twice as restrictive as one-dimensional case. Such a restriction on the time-step size makes the explicit scheme inefficient.

4.4.2 Implicit Methods

4.4.2.1 The Fully Implicit Method

If we employ a fully implicit unconditionally stable finite difference scheme to Eq. (4.31), we get

$$\frac{u_{i,j}^{n+1} - u_{i,j}^{n}}{\Delta t} = D\left[\frac{u_{i+1,j}^{n+1} - 2u_{i,j}^{n+1} + u_{i-1,j}^{n+1}}{(\Delta x)^2} + \frac{u_{i,j+1}^{n+1} - 2u_{i,j}^{n+1} + u_{i,j-1}^{n+1}}{(\Delta y)^2}\right] \tag{4.33a}$$

Rearranging we get

$$-\lambda_x u_{i-1,j}^{n+1} - \lambda_x u_{i+1,j}^{n+1} + 2(1 + \lambda_x + \lambda_y)\, u_{i,j}^{n+1} - \lambda_y u_{i,j-1}^{n+1} - \lambda_y u_{i,j+1}^{n+1} = u_{i,j}^{n} \tag{4.33b}$$

Hence, we obtain a penta-diagonal matrix which need to be solved simultaneously. The solution procedure for a penta-diagonal system of equation is also very time consuming. This scheme is of order $\left[\Delta t, (\Delta x)^2, (\Delta y)^2\right]$ like explicit scheme.

4.4.2.2 The Crank-Nicolson Method

The fully implicit method discussed earlier is only first-order accurate in time. The Crank-Nicolson method can be used and is second-order accurate in time. Finite difference approximation for the two-dimensional diffusion equation, Eq. (4.31), is given by

$$\frac{u_{i,j}^{n+1} - u_{i,j}^{n}}{\Delta t} = \frac{1}{2}D\left[\frac{u_{i+1,j}^{n+1} - 2u_{i,j}^{n+1} + u_{i-1,j}^{n+1}}{(\Delta x)^2} + \frac{u_{i+1,j}^{n} - 2u_{i,j}^{n} + u_{i-1,j}^{n}}{(\Delta x)^2} + \frac{u_{i,j+1}^{n+1} - 2u_{i,j}^{n+1} + u_{i,j-1}^{n+1}}{(\Delta y)^2} + \frac{u_{i,j+1}^{n} - 2u_{i,j}^{n} + u_{i,j-1}^{n}}{(\Delta y)^2}\right] \tag{4.34a}$$

$$-\frac{1}{2}\lambda_x u_{i-1,j}^{n+1} - \frac{1}{2}\lambda_x u_{i+1,j}^{n+1} + (1 + \lambda_x + \lambda_y)u_{i,j}^{n+1} - \frac{1}{2}\lambda_y u_{i,j-1}^{n+1} - \frac{1}{2}\lambda_y u_{i,j+1}^{n+1}$$

$$= \frac{1}{2}\lambda_x u_{i-1,j}^{n} + \frac{1}{2}\lambda_x u_{i+1,j}^{n} + (1 - \lambda_x - \lambda_y)u_{i,j}^{n} + \frac{1}{2}\lambda_y u_{i,j-1}^{n} + \frac{1}{2}\lambda_y u_{i,j+1}^{n} \tag{4.34b}$$

Crank-Nicolson method is unconditionally stable, and the truncation error is $O\left[(\Delta t)^2, (\Delta x)^2, (\Delta y)^2\right]$. Here also we obtain a penta-diagonal matrix which need to be solved simultaneously.

4.4.2.3 The Alternate Direction Implicit (ADI) Method

As discussed above, both the fully implicit scheme and the C-N scheme involve the solution of a penta-diagonal matrix simultaneously at each time step; this is very time consuming. The alternate direction implicit (ADI) method has been developed by Peaceman and Rachford [1955], to overcome this shortcoming and inefficiency. This scheme consists of solving two-dimensional diffusion equation implicitly in the x-direction and explicitly in the y-direction for time step n + ½ and then repeating the same for the n +1 time step with the y-direction as implicit and the x-direction as explicit. The algorithm produces two sets of tri-diagonal simultaneous equations to be solved in sequence. This is explained in Figure 4.11.

Finite difference equations of the ADI formulation are

$$\frac{u_{i,j}^{n+\frac{1}{2}} - u_{i,j}^{n}}{\left(\frac{\Delta t}{2}\right)} = D\left[\frac{u_{i+1,j}^{n+\frac{1}{2}} - 2u_{i,j}^{n+\frac{1}{2}} + u_{i-1,j}^{n+\frac{1}{2}}}{(\Delta x)^2} + \frac{u_{i,j+1}^{n} - 2u_{i,j}^{n} + u_{i,j-1}^{n}}{(\Delta y)^2}\right] \tag{4.35a}$$

and

$$\frac{u_{i,j}^{n+1} - u_{i,j}^{n}}{\left(\frac{\Delta t}{2}\right)} = D\left[\frac{u_{i+1,j}^{n+\frac{1}{2}} - 2u_{i,j}^{n+\frac{1}{2}} + u_{i-1,j}^{n+\frac{1}{2}}}{(\Delta x)^2} + \frac{u_{i,j+1}^{n+1} - 2u_{i,j}^{n+1} + u_{i,j-1}^{n+1}}{(\Delta y)^2}\right] \tag{4.35b}$$

Rearranging we get,

$$-\frac{1}{2}\lambda_x u_{i-1,j}^{n+\frac{1}{2}} + (1+\lambda_x)u_{i,j}^{n+\frac{1}{2}} - \frac{1}{2}\lambda_x u_{i+1,j}^{n+\frac{1}{2}} = \frac{1}{2}\lambda_y u_{i,j-1}^{n} + (1-\lambda_y)u_{i,j}^{n} + \frac{1}{2}\lambda_y u_{i,j+1}^{n} \tag{4.36a}$$

and

$$-\frac{1}{2}\lambda_y u_{i,j-1}^{n+1} + (1+\lambda_y)u_{i,j}^{n+1} - \frac{1}{2}\lambda_y u_{i,j+1}^{n+1} = \frac{1}{2}\lambda_x u_{i-1,j}^{n+\frac{1}{2}} + (1-\lambda_x)u_{i,j}^{n+\frac{1}{2}} + \frac{1}{2}\lambda_x u_{i+1,j}^{n+\frac{1}{2}} \tag{4.36b}$$

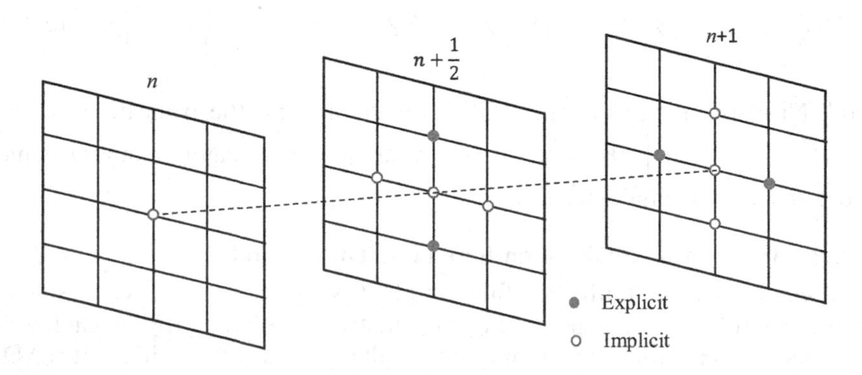

FIGURE 4.11 Alternate Direction Implicit (ADI) Scheme in Two Dimensions.

This scheme is unconditionally stable, and the truncation error is $O\left[(\Delta t)^2, (\Delta x)^2, (\Delta y)^2\right]$. Compared to other methods, ADI is fast and computationally efficient as both Eq. (4.36a) and (4.36b) can be solved by Thomas algorithm.

4.4.2.4 Comments on Diffusion Equations

Selecting the best method for solving diffusion equations is very difficult. However, implicit methods are more suitable than explicit methods. For a one-dimensional problem, the Crank-Nicolson method is highly recommended because of second-order accuracy in both time and space. For two- or three-dimensional cases, the ADI scheme may be used as this method allows for the use of the Thomas algorithm.

4.4.2.5 Further Comments on Conservative vs Non-Conservative Variables

As discussed in Section 2.3.1, the conservative form of the steady state continuity equation in one dimension is given by

$$\frac{\partial}{\partial x}(\rho u) = 0 \tag{4.37a}$$

The above equation in the non-conservative form can be written as

$$\rho \frac{\partial u}{\partial x} + u \frac{\partial \rho}{\partial x} = 0 \tag{4.37b}$$

The forward difference of Eq. (4.37a) is given by

$$\frac{(\rho u)_{i+1} - (\rho u)_i}{\Delta x} = 0 \tag{4.38a}$$

Hence, if we apply the difference equation in all points, as shown in Figure 4.12, we get

$$\frac{(\rho u)_1 - (\rho u)_0}{\Delta x} = 0; \quad \frac{(\rho u)_2 - (\rho u)_1}{\Delta x} = 0; \quad \frac{(\rho u)_3 - (\rho u)_2}{\Delta x} = 0 \tag{4.38b}$$

If we add, then we get

$$\frac{(\rho u)_3 - (\rho u)_0}{\Delta x} = 0 \tag{4.38c}$$

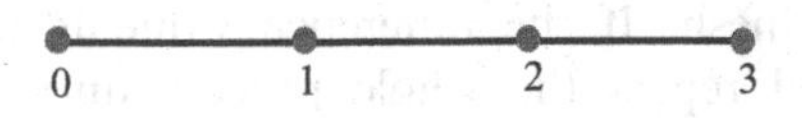

FIGURE 4.12 Discretization Points.

which is the discretised equation between 3 and 0. So we can see that by using the conservative form of variables we are able to preserve local as well as global conservation which is essential and desirable property of any discretisation scheme.

Now, let us see what happens when we use the non-conservative form of equation given in Eq. (4.37b). The discretized form is given by

$$\rho_i \frac{u_{i+1} - u_i}{\Delta x} + u_i \frac{\rho_{i+1} - \rho_i}{\Delta x} = 0 \tag{4.39a}$$

i.e.

$$\rho_0 \frac{u_1 - u_0}{\Delta x} + u_0 \frac{\rho_1 - \rho_0}{\Delta x} = 0; \; \rho_1 \frac{u_2 - u_1}{\Delta x} + u_1 \frac{\rho_2 - \rho_1}{\Delta x} = 0;$$

$$\rho_2 \frac{u_3 - u_2}{\Delta x} + u_2 \frac{\rho_3 - \rho_2}{\Delta x} = 0 \tag{4.39b}$$

The discretized equation between nodes 3 and 0 is

$$\rho_0 \frac{u_3 - u_0}{\Delta x} + u_0 \frac{\rho_3 - \rho_0}{\Delta x} = 0 \tag{4.39c}$$

which is totally different than adding all the equations between nodes 0 and 3.

Hence, it is necessary to formulate governing equations in the conservative form and integrate them using a conservative finite-difference procedure to guarantee that any discontinuity present in the flow will be processed correctly. That means that shock waves will be of correct intensity and in correct position. The use of non-conservative forms of equations can produce significant errors in the speed of the shock waves.

4.4.2.6 The Grid (Mesh) Independence Study

In the grid independence study, we make sure that the result (i.e. a monitored value) is independent of the number of meshes or grids. The monitored value may be the average outlet temperature, total drag force, etc. The grid independence study is different than convergence. In a convergence study, we continue iterating until we obtain a solution such that the residual is less than our prescribed value. In CFD, normally we start with coarse mesh (i.e. a smaller number of meshes) and find a converged solution. We then increase the number of meshes by around 20–50% of the previous value and find a new converged solution. The monitored value with increased number of meshes is then checked against the value obtained from the smaller number of grids. If the difference is negligible (normally within 1–3%), we choose the earlier number of meshes as the grid independent mesh. If the difference value is more, we increase the number of meshes and repeat the whole process until the difference is very small.

The computational time depends on the number of meshes used for computation. As the number of meshes increases, the computational time also increases.

Hence, to optimize computational time, the grid independence study is to be done at the beginning of any simulation. Once the grid independent mesh is obtained, further simulation is carried out with it. Remember that results cannot be accepted unless we get grid independent results.

4.5 BURGERS' EQUATION

So far, we have discussed the finite difference method applied to the solution of simple linear equations. However, governing equations associated with the fluid flow (Navier-Stokes equations) are non-linear. Burgers' equation can represent these problems. It has a convective term and a diffusion term in addition to the unsteady term.

The complete Burgers' equation can be expressed as

$$\frac{\partial u}{\partial t} + u\frac{\partial u}{\partial x} = \nu\frac{\partial^2 u}{\partial x^2} \tag{4.40}$$

with initial condition $u\,(x, 0) = u_0(x);\ a < x < b$

The Eq. (4.40) is similar to the transport Eq. (4.11). The difference is the wave velocity c which is constant is replaced by u which is not constant. The non-linearity arises because of $u^{\partial u}/_{\partial x}$ term. This equation is parabolic in nature.

Let us now discuss the characteristics of the linear transport equation and Burgers' equation. In the case of $\nu \rightarrow 0$, the viscous term drops out and Eq. (4.11) takes the form of the linear wave equation given by Eq. (4.12), discussed earlier. The solution to the linear wave equation becomes $u(x, t) = u_0(x - ct)$ where $u_0(x)$ is the initial distribution. So, the transport of the initial profile remains the same with respect to time and is shown in Figure 4.13.

For vanishing viscosity, Eq. (4.40) takes the form

$$\frac{\partial u}{\partial t} + u\frac{\partial u}{\partial x} = 0 \tag{4.41}$$

which is known as inviscid Burgers' equation. This is a non-linear wave equation in which each point of the wave front can move with a different

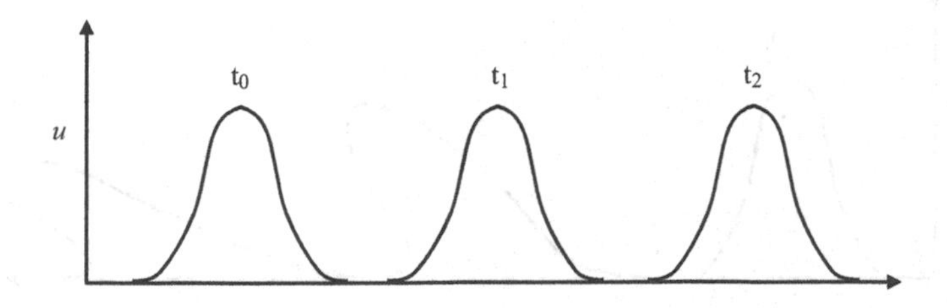

FIGURE 4.13 Transport of the Initial Profile by a Constant Velocity c.

speed. Thus, the characteristics combine, resulting in discontinuous solutions similar to the shock wave. The equation exhibits a wave-breaking phenomenon. The peak of the pulse moves faster because wave speed increases with increasing amplitude. An example of this phenomenon is the breaking of surface waves in shallow water on a beach. Propagation of such waves can be seen in Figure 4.14.

The presence of viscosity in the transport equation changes the profile. The effect of the viscosity is to dampen the amplitude. The resulting propagation of the wave front for linear transport and Burgers' equation are shown in Figure 4.15 and Figure 4.16, respectively.

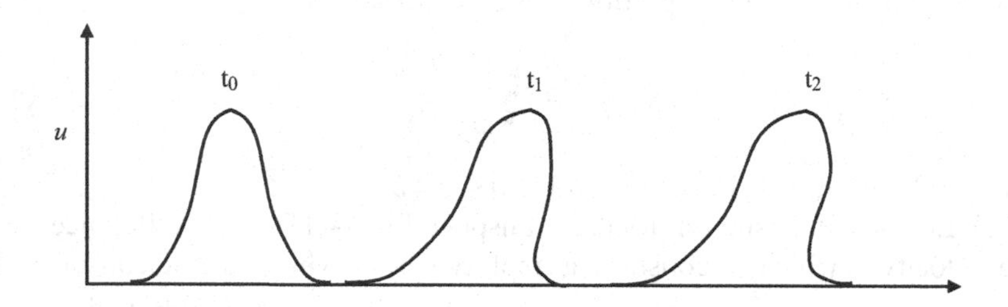

FIGURE 4.14 Propagation of a Wave Front for Inviscid Burgers' Equation.

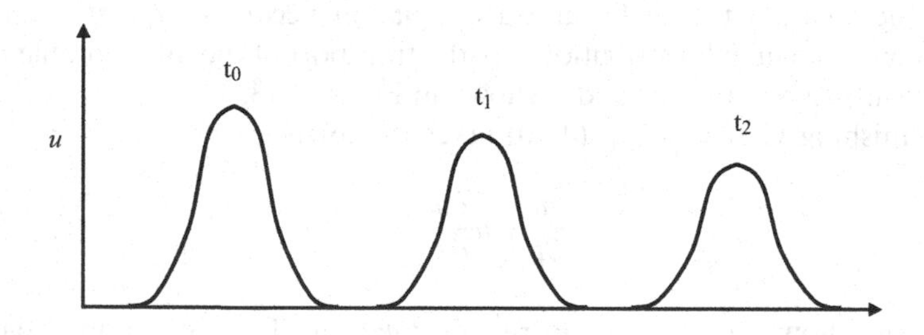

FIGURE 4.15 Propagation of the Wave Front for the Viscid Linear Transport Equation.

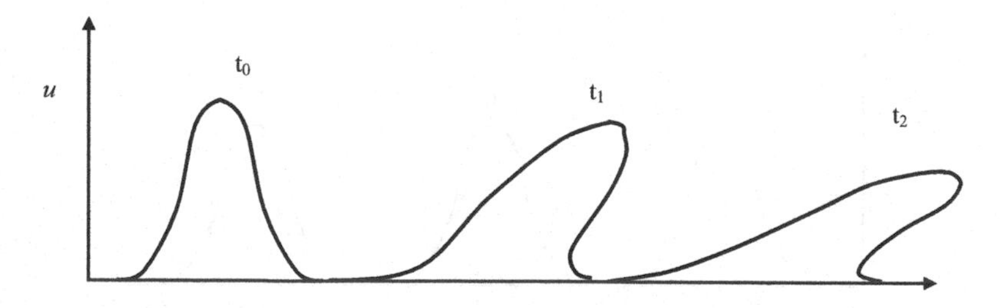

FIGURE 4.16 Propagation of the Wave Front for Viscid Burgers' Equation.

Now we shall discuss the finite difference discretization of the Burgers' equation. Initially, we shall take up the inviscid Burgers' equation and then the viscid Burgers' equation.

4.5.1 The Inviscid Burgers' Equation

As we have seen, the FTCS for the linear wave equation is not stable; the same is true for the inviscid Burgers' equation. Hence, we shall discuss other methods.

4.5.1.1 Upwind Differencing

For $u > 0$, Forward Time Backward Space (FTBS) scheme approximation is used. That is

$$\frac{u_k^{n+1} - u_k^n}{\Delta t} + u_k^n \left[\frac{u_k^n - u_{k-1}^n}{\Delta x} \right] = 0 \tag{4.42}$$

This is a non-conservative equation. Already in Section 4.4.2.5, we have seen that use of non-conservative form of equation may not converge for discontinuous solution. So, to prevent the method from converging to non-solution, we shall use the conservative from of Inviscid Burgers' equation which can be expressed as

$$\frac{\partial u}{\partial t} + \frac{\partial F}{\partial x} = 0; \ where\ F = \frac{u^2}{2} \tag{4.43}$$

Discretizing this equation we get,

$$\frac{u_k^{n+1} - u_k^n}{\Delta t} + \left[\frac{F_k^n - F_{k-1}^n}{\Delta x} \right] = 0 \tag{4.44a}$$

or

$$u_k^{n+1} = u_k^n - \frac{\Delta t}{\Delta x} \left[F_k^n - F_{k-1}^n \right] \tag{4.44b}$$

Both the finite difference form of above equations has a truncation error of $O[\Delta t, \Delta x]$.

4.5.1.2 The Lax Method (Lax-Friedrichs Method)

We know that

$$u(x, t + \Delta t) = u(x, t) + \Delta t \left(\frac{\partial u}{\partial t} \right)_{x,t} + \dots\dots\dots\dots$$

which we can write as

$$u(x, t + \Delta t) = u(x, t) - \Delta t \left(\frac{\partial F}{\partial x} \right)_{x,t} + \dots\dots\dots\dots$$

or

$$u_k^{n+1} = u_k^n - \frac{\Delta t}{\Delta x}\frac{1}{2}\left[F_{k+1}^n - F_{k-1}^n\right] \tag{4.45}$$

This is first order accurate. The amplification factor is given by

$$|\xi|^2 = cos\ \theta - i\frac{\Delta t}{\Delta x}\ u\ sin\ \theta$$

The stability condition is given by

$$\left|\frac{\Delta t}{\Delta x}u_{max}\right| \leq 1$$

This method is first-order accurate and captures discontinuity well, but the method is dissipative in nature (i.e. there is smearing of discontinuity over several mesh intervals). This becomes worse with the decrease in Courant number. The monotonic behavior of the solution is because of the first-order accuracy of the method.

4.5.1.3 The Lax-Wendroff Method

The Lax-Wendroff method can be obtained from a Taylor-series expansion in the following manner:

$$u_k^{n+1} = u_k^n + \Delta t\left(\frac{\partial u}{\partial t}\right)_k^n + \frac{(\Delta t)^2}{2!}\left(\frac{\partial^2 u}{\partial t^2}\right)_k^n + O\left[(\Delta t)^3\right]$$

Now

$$\frac{\partial u}{\partial t} = -\frac{\partial F}{\partial x};\ or\ \frac{\partial^2 u}{\partial t^2} = -\frac{\partial}{\partial t}\left(\frac{\partial F}{\partial x}\right) = -\frac{\partial}{\partial x}\left(\frac{\partial F}{\partial t}\right) = -\frac{\partial}{\partial x}\left(\frac{\partial F}{\partial u}\frac{\partial u}{\partial t}\right)$$

or

$$\frac{\partial^2 u}{\partial t^2} = -\frac{\partial}{\partial x}\left(A\frac{\partial u}{\partial t}\right) = \frac{\partial}{\partial x}\left(A\frac{\partial F}{\partial x}\right)$$

where

$$A(u) = \frac{\partial F}{\partial u}\ is\ a\ Jacobian\ matrix.$$

Then

$$u_k^{n+1} = u_k^n - \Delta t\left(\frac{\partial F}{\partial x}\right)_k^n + \frac{(\Delta t)^2}{2}\frac{\partial}{\partial x}\left(A\frac{\partial F}{\partial x}\right)_k^n$$

Hence,

$$u_k^{n+1} = u_k^n - \frac{\Delta t}{2\,\Delta x}\left[F_{k+1}^n - F_{k-1}^n\right] + \frac{(\Delta t)^2}{2(\Delta x)^2}\left[A_{k+\frac{1}{2}}^n\left(F_{k+1}^n - F_k^n\right) - A_{k-\frac{1}{2}}^n\left(F_k^n - F_{k-1}^n\right)\right] \quad (4.46)$$

where

$$A_{k+\frac{1}{2}}^n = A\left(\frac{u_k^n + u_{k+1}^n}{2}\right)$$

Since in inviscid Burgers' equation,

$$F = \frac{u^2}{2}, \quad A = u$$

Then

$$A_{k+\frac{1}{2}}^n = \left(\frac{u_k^n + u_{k+1}^n}{2}\right) \quad and \quad A_{k-\frac{1}{2}}^n = \left(\frac{u_k^n + u_{k-1}^n}{2}\right)$$

This explicit, one-step scheme is second order accurate with truncation error of $O\left[(\Delta t)^2, (\Delta x)^2\right]$ and is stable whenever

$$\frac{\Delta t}{\Delta x} u_{max} \leq 1$$

The amplification factor is

$$|\xi|^2 = 1 - 2\left(\frac{\Delta t}{\Delta x}u\right)^2 (1 - cos\ \theta) - 2i\frac{\Delta t}{\Delta x}u\ sin\ \theta$$

This method can predict the position of the discontinuity correctly and sharply. However, the method is dispersive in nature and hence produces oscillations near the discontinuity. The quality of the solution is degraded as the Courant number decreases below 1.0.

4.5.1.4 The MacCormack Method

The MacCormack method is explicitly for solving hyperbolic and hyperbolic-parabolic systems of equations. It is a variation of the two-step Lax-Wendroff method. When applied to the inviscid Burgers' equation, MacCormack's predictor-corrector scheme becomes

Predictor step:

$$\overline{u_k^{n+1}} = u_k^n - \frac{\Delta t}{\Delta x}\left(F_{k+1}^n - F_k^n\right) \quad (4.47a)$$

Corrector step:

$$u_k^{n+1} = \frac{1}{2}\left[u_k^n + \overline{u_k^{n+1}} - \frac{\Delta t}{\Delta x}\left(\overline{F_k^{n+1}} - \overline{F_{k-1}^{n+1}}\right)\right] \tag{4.47b}$$

The explicit MacCormack method is second order accurate with a truncation error of $O\left[(\Delta t)^2, (\Delta x)^2\right]$. The amplification and stability limit for this scheme is similar to Lax-Wendroff method. However, this scheme is easier to apply than Lax-Wendroff method as Jacobian does not appear in the equations.

4.5.1.5 Implicit Methods

The second-order Crank-Nicolson scheme for the inviscid Burgers' equation can be written as

$$u_k^{n+1} = u_k^n - \frac{\Delta t}{2}\left[\left(\frac{\partial F}{\partial x}\right)^n + \left(\frac{\partial F}{\partial x}\right)^{n+1}\right]$$

This is a non-linear problem. Hence, to solve the equation, we must linearize the equation. Beam and Warming [1976], suggested that

$$F^{n+1} \approx F^n + \left(\frac{\partial F}{\partial u}\right)^n \left(u^{n+1} - u^n\right) = F^n + A^n\left(u^{n+1} - u^n\right)$$

Hence, we get

$$u_k^{n+1} = u_k^n - \frac{\Delta t}{2}\left[2\left(\frac{\partial F}{\partial x}\right)^n + \frac{\partial}{\partial x}\left\{A^n\left(u_k^{n+1} - u_k^n\right)\right\}\right]_k \tag{4.48}$$

Using central difference for the x-derivatives, we get

$$-\frac{\Delta t A_{k-1}^n}{4\,\Delta x}u_{k-1}^{n+1} + u_k^{n+1} - \frac{\Delta t A_{k+1}^n}{4\,\Delta x}u_{k+1}^{n+1}$$

$$= -\frac{\Delta t}{\Delta x}\frac{F_{k+1}^n - F_{k-1}^n}{2} - \frac{\Delta t A_{k-1}^n}{4\,\Delta x}u_{k-1}^n + u_k^n - \frac{\Delta t A_{k+1}^n}{4\,\Delta x}u_{k+1}^n \tag{4.49}$$

Hence, we get a tri-diagonal system of equations that can be solved by the Thomas algorithm. This method is stable for any time step. However, some artificial damping is needed for obtaining an acceptable solution.

4.5.1.6 The Godunov Method

So far, we have discussed various finite difference methods for solving the inviscid Burgers' equation. The solutions for various methods applied to Burgers' equation are shown in Figure 4.17.

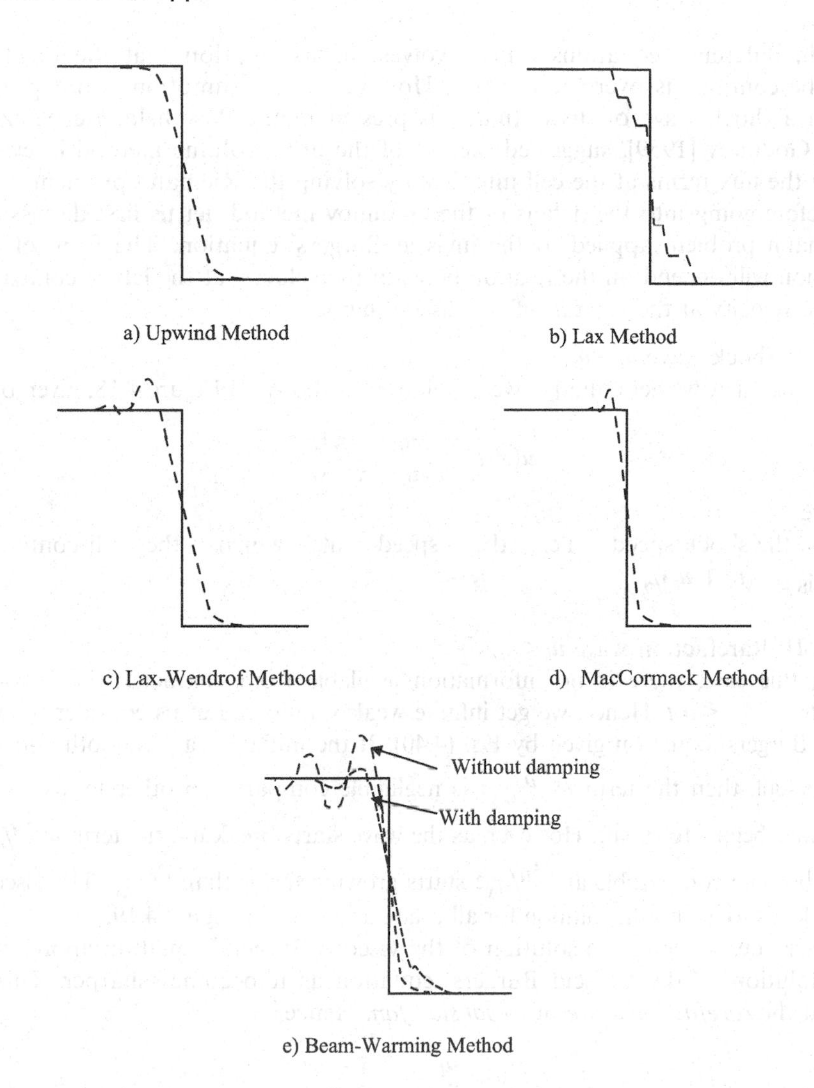

FIGURE 4.17 Typical Solutions of Burgers' Equation by Various Methods.

We can see that the solution of the inviscid Burgers' equation by explicit methods produces superior results compared to implicit methods, especially when there is discontinuity. The implicit scheme gives rise to more oscillations, which is not acceptable and requires damping. Also, the implicit method requires more computational effort as the time step is decided by the transient phenomenon. Hence, explicit methods are recommended for solving the inviscid Burgers' equation. However, as we can see in Figure 4.17, first-order methods lead to numerical diffusion, whereas second-order methods give rise to oscillations (i.e. dispersion), which may not be desirable. The reason for this is that the finite difference scheme uses the Taylor series expansion to obtain difference equations. This involves the assumption that the function will be continuous over the domain. However, this assumption is not proper when a shock wave or discontinuity is present in the flow field. Recognizing this, Godunov [1959], suggested the use of the finite volume method in evaluating the flux terms at the cell interface by solving the Riemann problem.

Before going into the details of the Godunov method, let us first discuss the Riemann problem applied to the inviscid Burgers' equation. The form of the solution will depend on the relation between the velocity at the left u_l compared to the velocity at the right u_r of the discontinuity.

Case I: Shock wave $u_l > u_r$

In this case, we get a unique weak solution as shown in Figure 4.18, given by

$$u(x,t) = \begin{matrix} u_l & x < st \\ u_r & x > st \end{matrix} \quad (4.50)$$

where

$s =$ the shock speed i.e. the speed at which the discontinuity travels $= {(u_l + u_r)}/{2}$

Case II: Rarefaction wave $u_l < u_r$

In this case, there is no information available from characteristics between points $u_l t < x < u_r t$. Hence, we get infinite weak solutions. Let us consider the viscous Burgers' equation given by Eq. (4.40). If the initial data is smooth and ν is very small, then the term $\nu {\partial^2 u}/{\partial x^2}$ is negligible compared to other terms before the wave begins to break. However, as the wave starts breaking, the term $\nu {\partial^2 u}/{\partial x^2}$ may become comparable as ${\partial^2 u}/{\partial x^2}$ starts growing faster than ${\partial u}/{\partial x}$. The viscous term leads to a smooth solution for all times, as shown in Figure 4.19.

As ν goes to zero, the solution of the viscous Burgers' equation approaches the solution of the inviscid Burgers' equation as it becomes sharper. This is called the *rarefaction wave* or *expansion fan*. Hence,

$$u(x,t) = \begin{matrix} u_l & x < u_l t \\ {x}/{t} & u_l t \leq x \leq u_r t \\ u_r & x > u_r t \end{matrix} \quad (4.51)$$

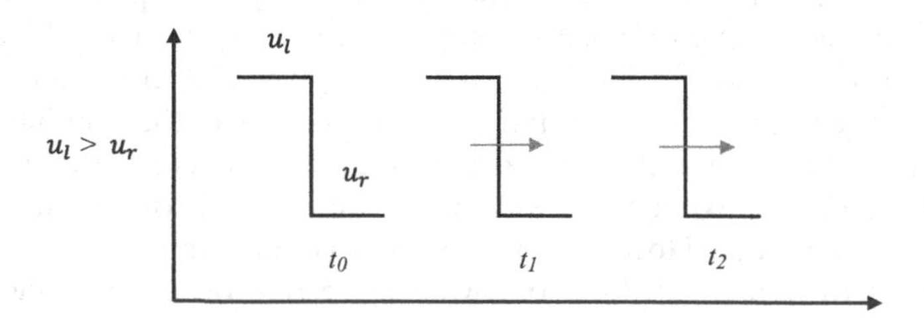

FIGURE 4.18 Propagation of the Shock Wave.

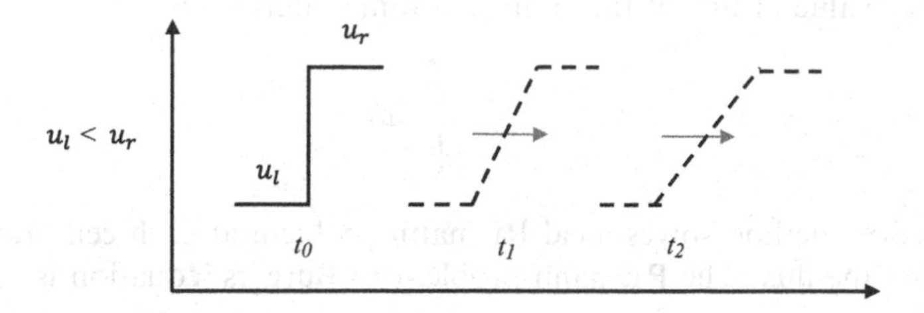

FIGURE 4.19 Propagation of the Rarefaction Wave.

So, shock waves and rarefaction waves are two possible solutions of the Riemann problem.

Let us now return to our discussion on the Godunov method for solving the inviscid Burgers' equation, Eq. (4.43). As discussed earlier, the Godunov method is an explicit finite volume method in which fluxes are evaluated at the cell face by solving the Riemann problem. Consider a control volume extending from t to t +Δt and from $x - {}^{\Delta x}/_{2}$ to $x + {}^{\Delta x}/_{2}$. Hence, the control volume is centered at $\left(k, n + {}^{1}/_{2}\right)$ as shown in Figure 4.20.

So, the discretized equation becomes

$$\bar{u}_k^{n+1} = \bar{u}_k^{n+1} - \frac{\Delta t}{\Delta x}\left[f\left(u_{k+\frac{1}{2}}\right) - f\left(u_{k-\frac{1}{2}}\right)\right] \tag{4.52}$$

where

$$\bar{u}_k = \text{average value of u over control volume} = \frac{1}{\Delta x}\int_{x-\frac{\Delta x}{2}}^{x+\frac{\Delta x}{2}} u(x,t)dt$$

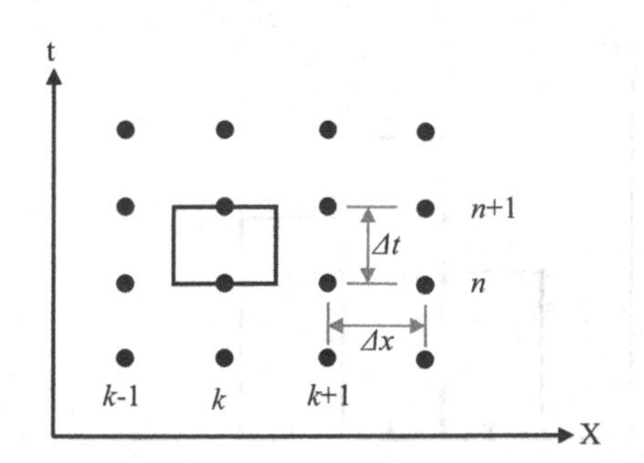

FIGURE 4.20 Control Volume for the Godunov Method.

The average value of flux at the control volume interface is

$$f = \frac{1}{\Delta t}\int_{t}^{t+\Delta t} f dt$$

The Godunov method solves local Riemann problem at each cell interface for determining the flux. The Riemann problem to Burgers' equation is

$$\frac{\partial u}{\partial t} + \frac{\partial}{\partial x}\left(\frac{u^2}{2}\right) = 0 \tag{4.53a}$$

with initial condition

$$u(x,0) = \begin{matrix} u_k & x \leq 0 \\ u_{k+1} & x > 0 \end{matrix} \tag{4.53b}$$

The distribution of average velocities for this problem are shown in Figure 4.21.

Defining shock speed as

$$s_{k+\frac{1}{2}} = \left(\frac{dx}{dt}\right)_{k+\frac{1}{2}} = \frac{u_k + u_{k+1}}{2}$$

the solutions are:

Case I: Shock wave $u_k > u_{k+1}$

$$u = \begin{matrix} u_k & x/t < s_{k+\frac{1}{2}} \\ u_{k+1} & x/t > s_{k+\frac{1}{2}} \end{matrix} \tag{4.54a}$$

$$f_{k+\frac{1}{2}} = \begin{matrix} \frac{1}{2}u_k^2 & s_{k+\frac{1}{2}} > 0 \\ \frac{1}{2}u_{k+1}^2 & s_{k+\frac{1}{2}} < 0 \end{matrix} \tag{4.54b}$$

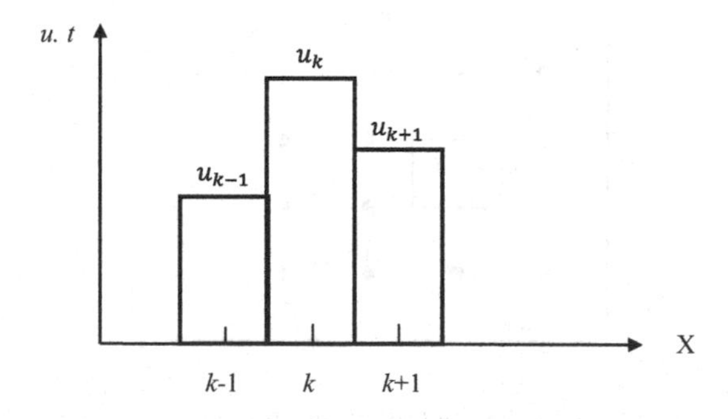

FIGURE 4.21 Wave Diagram of Godunov's Method.

Case II: Expansion wave $u_k < u_{k+1}$

$$u = \begin{cases} u_k & x/t < u_k \\ x/t & u_k < x/t < u_{k+1} \\ u_{k+1} & x/t > u_{k+1} \end{cases} \tag{4.55a}$$

$$f_{k+\frac{1}{2}} = \begin{cases} 0 & & u_k < 0 < u_{k+1} \\ \frac{1}{2}u_k^2 & s_{k+\frac{1}{2}} > 0 & u_{k+1} > u_k > 0 \\ \frac{1}{2}u_{k+1}^2 & s_{k+\frac{1}{2}} < 0 & u_k < u_{k+1} < 0 \end{cases} \tag{4.55b}$$

Here the assumption is the waves from the adjacent cells do not interact. This is possible only if the wave travels maximum of half the one cell distance. Hence, the stability criterion is given by

$$\left| u_{max} \frac{\Delta t}{\Delta x} \right| \leq \frac{1}{2}.$$

Now by evaluating the fluxes at each cell interface we get the solution of the Riemann problem by using Eq. (4.52). Figure 4.22 shows a typical solution of shock waves by Godunov's method. Thus, we can see that Godunov's method is superior in capturing the shock wave. For the expansion wave, this method is equally good.

4.5.1.7 The Roe Method

Godunov's method gives excellent results for Burgers' equation. However, the method involves the exact solution of the Riemann problem. For complicated equations governing fluid flow like Euler's equation, Godunov's method becomes computationally expensive as it involves inefficient iterative methods in dealing with non-linearity. Roe [1980,1981] observed that instead of an exact Riemann solution, an approximate Riemann problem does well for most cases.

Roe method involves finding a constant matrix $\tilde{A}(u_k, u_{k+1})$ such that

$$\frac{\partial u}{\partial t} + \tilde{A}\frac{\partial u}{\partial x} = 0 \tag{4.56}$$

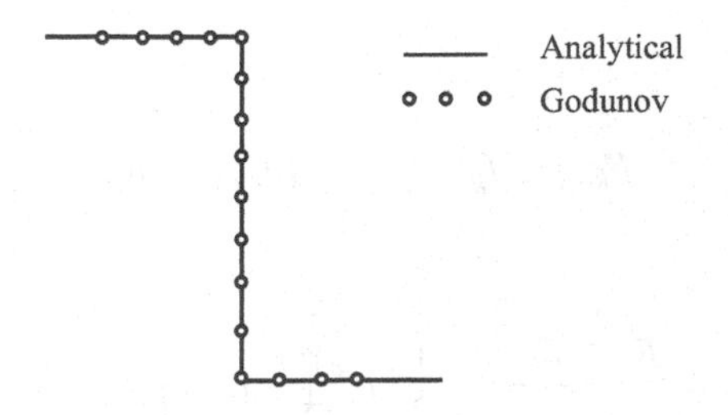

FIGURE 4.22 Typical Solution of Shock waves by Godunov's Method.

The matrix $\tilde{A}(u_k, u_{k+1})$ should satisfy

i) As $u_k \to u_{k+1} \to \mathbf{u}, \tilde{A}(u_k, u_{k+1}) \to A(\mathbf{u})$ where $A = {}^{\partial F}/_{\partial u}$
ii) For any $u_k, u_{k+1}, F_{k+1} - F_k = \tilde{A}(u_{k+1} - u_k)$..

The first condition gives exact value in the smooth region and the second condition guarantees the correct jump condition across a discontinuity.

Roe's method when applied to Burgers' equation, Eq. (4.53) we get

$$\frac{\partial u}{\partial t} + \bar{u}\frac{\partial u}{\partial x} = 0 \tag{4.57}$$

Where $\bar{u}$ is locally constant according to Roe and is average value of $\tilde{A}$. The value of $\bar{u}$ applied for cell between k and k+1 can be determined from the conditions mentioned above and this leads to

$$\bar{u} = \bar{u}_{k+\frac{1}{2}} = \frac{F_{k+1} - F_k}{u_{k+1} - u_k} \tag{4.58}$$

For Burgers' equation, Eq. (4.58) reduces to

$$\bar{u}_{k+\frac{1}{2}} = \begin{matrix} \frac{u_k + u_{k+1}}{2} & u_k \neq u_{k+1} \\ u_k & u_k = u_{k+1} \end{matrix} \tag{4.59}$$

The discretized Burgers' equation in terms of flux terms can be expressed as

$$\bar{u}_k^{n+1} = \bar{u}_k^{n+1} - \frac{\Delta t}{\Delta x}\left[f_{k+\frac{1}{2}} - f_{k-\frac{1}{2}}\right] \tag{4.60}$$

The Riemann problem is now solved for a simple linear wave equation with a constant wave speed. The single wave coming originating from cell interface will travel in either direction of the cell face depending on the value of $\bar{u}_{k+\frac{1}{2}} = {}^{dx}/_{dt}$.

Hence, we get

If $\bar{u}_{k+\frac{1}{2}} < 0$

$$f_{k+\frac{1}{2}} - F_k = \bar{u}^-_{k+\frac{1}{2}}(u_{k+1} - u_k)$$

If $\bar{u}_{k+\frac{1}{2}} > 0$

$$F_{k+1} - f_{k+\frac{1}{2}} = \bar{u}^+_{k+\frac{1}{2}}(u_{k+1} - u_k)$$

Hence, we get

$$f_{k+\frac{1}{2}} = \frac{F_k + F_{k+1}}{2} + \frac{1}{2}\left(\bar{u}^-_{k+\frac{1}{2}} - \bar{u}^+_{k+\frac{1}{2}}\right)(u_{k+1} - u_k)$$

The wave diagram of Roe's method applied to Burgers' equation is shown in Figure 4.23.

If we consider the contribution of wave travel individually either left or right, then the numerical flux can be calculated separately from

$$f_{k+\frac{1}{2}} = \frac{F_k + F_{k+1}}{2} - \frac{1}{2}\left|\bar{u}_{k+\frac{1}{2}}\right|(u_{k+1} - u_k) \tag{4.61}$$

Figure 4.24 shows the prediction of Roe's method in capturing the shock wave and expansion wave. We can clearly see that Roe's method captures the shock wave excellently. However, for the expansion wave, Roe's method gives rise to non-physical expansion shock, which violates entropy. This is because the method cannot distinguish between compression shock and expansion shock.

To avoid this non-physical expansion shock, Harten and Hyman [1983] proposed the following. Let

$$\varepsilon = max\left(0, \frac{u_{k+1} - u_k}{2}\right)$$

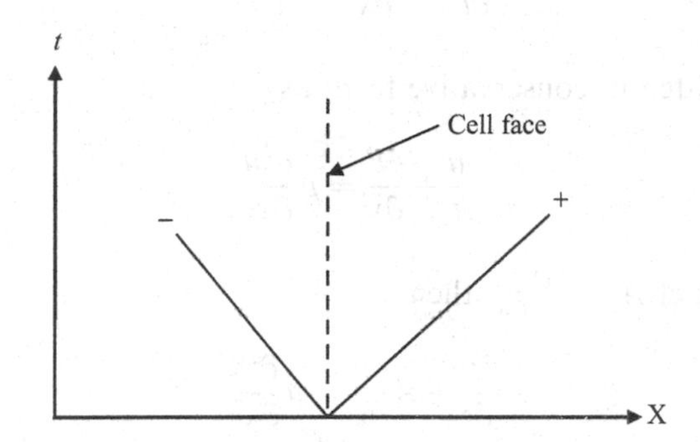

FIGURE 4.23 Wave Diagram of Roe's Method Applied to Burgers' Equation.

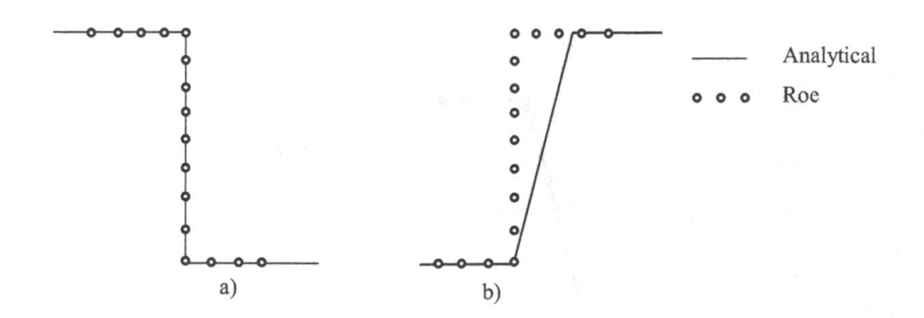

FIGURE 4.24 Typical Solution of Shock and Expansion Waves by the Roe Method: a) Shock Wave; b) Expansion Wave.

Then

$$\bar{u}_{k+\frac{1}{2}} = \begin{bmatrix} \bar{u}_{k+\frac{1}{2}}\varepsilon & \bar{u}_{k+\frac{1}{2}} \geq \varepsilon \\ \varepsilon & \bar{u}_{k+\frac{1}{2}} < \varepsilon \end{bmatrix} \tag{4.62}$$

where $\varepsilon = 0$ denotes the compression shock in which unaltered $\bar{u}_{k+\frac{1}{2}}$ is used. In the case of expansion wave $\left[\varepsilon = {(u_{k+1} - u_k)}/{2}\right]$ modification is required. Roe's method with an entropy fix for an expansion fan is shown in Figure 4.25.

Both Godunov's and Roe's method are first-order accurate schemes in which state variables are assumed to be constant in the control volume. In order to improve accuracy, higher-order convective schemes like monotone upstream-centered schemes for conservation laws and total variation diminishing schemes are developed. However, these schemes are not discussed here.

4.5.2 The Viscid Burgers' Equation

The complete Burgers' equation can be expressed as

$$\frac{\partial u}{\partial t} + u\frac{\partial u}{\partial x} = \mu\frac{\partial^2 u}{\partial x^2} \tag{4.63}$$

which can be written in conservative form as

$$\frac{\partial u}{\partial t} + \frac{\partial F}{\partial x} = \mu\frac{\partial^2 u}{\partial x^2} \tag{4.64}$$

where $F = {u^2}/{2}$. Let $A = {\partial F}/{\partial u}$, then

$$\frac{\partial u}{\partial t} + A\frac{\partial u}{\partial x} = \mu\frac{\partial^2 u}{\partial x^2} \tag{4.65}$$

where $A = u$ for non-linear Burgers' equation.

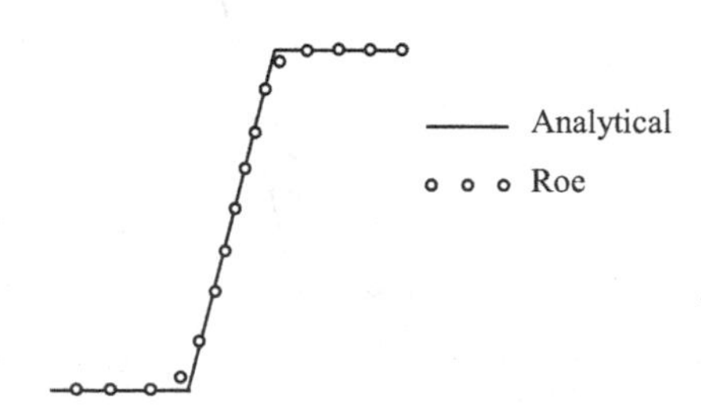

FIGURE 4.25 Typical Solution of an Expansion wave by Roe's Method with Entropy Fix.

4.5.2.1 The FTCS Method

Roache [1972] developed the FTCS method by applying forward time and central space difference to the linear Burgers' equation, Eq. (4.65) with $A = c$. As a result, the discretized equation becomes

$$\frac{u_k^{n+1} - u_k^n}{\Delta t} + c\left[\frac{u_{k+1}^n - u_{k-1}^n}{2\Delta x}\right] = D\left[\frac{u_{k+1}^n - 2u_k^n + u_{k-1}^n}{(\Delta x)^2}\right] \tag{4.66a}$$

Rearranging we get,

$$u_k^{n+1} = \left(\frac{\lambda}{2} + s\right)u_{k-1}^n + (1 - 2s)u_k^n + \left(-\frac{\lambda}{2} + s\right)u_{k+1}^n \tag{4.66b}$$

where

$$\lambda = \frac{c\Delta t}{\Delta x} \quad and \quad s = \frac{\mu \Delta t}{(\Delta x)^2} \tag{4.66c}$$

This is first-order explicit one-step method with a truncation error $O\left[\Delta t, (\Delta x)^2\right]$. The stability condition is $\lambda \leq 2s \leq 1$.

Substituting the values of λ and s we get

$$r = \frac{c\Delta x}{D} \leq 2$$

where r is cell Reynolds number. When $r > 2$, oscillations are produced in the solution which may lead the solution to diverge. To avoid the oscillations, upwind scheme is used for the convective term. Hence, we get for $c > 0$

$$\frac{u_k^{n+1} - u_k^n}{\Delta t} + c\left[\frac{u_k^n - u_{k-1}^n}{\Delta x}\right] = D\left[\frac{u_{k+1}^n - 2u_k^n + u_{k-1}^n}{(\Delta x)^2}\right] \tag{4.67a}$$

or

$$u_k^{n+1} = (\lambda + s)u_{k-1}^n + (1 - \lambda - 2s)\, u_k^n + s\, u_{k+1}^n \tag{4.67b}$$

The requirement for stability for this method becomes $\lambda + 2s \leq 1$. However, this scheme gives rise to large dissipation.

4.5.2.2 The DuFort-Frankel Method

The DuFort-Frankel method applied to Eq. (4.65) becomes

$$\frac{u_k^{n+1} - u_k^{n-1}}{2\Delta t} + A_k^n\left[\frac{u_{k+1}^n - u_{k-1}^n}{2\Delta x}\right] = \mu\left[\frac{u_{k+1}^n - \left(u_k^{n+1} + u_k^{n-1}\right) + u_{k-1}^n}{(\Delta x)^2}\right] \tag{4.68a}$$

Rearranging

$$u_k^{n+1} = \left(\frac{1-2s}{1+2s}\right)u_{k-1}^n + \left(\frac{\lambda+2s}{1+2s}\right)u_{k-1}^n - \left(\frac{\lambda-2s}{1+2s}\right)u_{k+1}^n \tag{4.68b}$$

Von Neumann's stability analysis show that for linear case (i.e. $A = c$), D-F method is stable for $\lambda \leq 1$.

This scheme is first-order accurate with a truncation error of $O\left[\left({}^{\Delta t}/_{\Delta x}\right)^2,\right.$ $\left.(\Delta t)^2, (\Delta x)^2\right]$. The finite difference equation is consistent if ${}^{\Delta t}/_{\Delta x} \to 0$ *as* $\Delta x \to 0$ *and* $\Delta t \to 0$ which requires a smaller time step. Hence, the method is better suited for steady solutions.

4.5.2.3 The Lax-Wendroff Method

Thomman [1966] proposed the following scheme for the two-step Lax-Wendroff method to complete Burgers' equation as

Step 1:

$$\begin{aligned} u_k^{n+\frac{1}{2}} =& \frac{1}{2}\left(u_{k+\frac{1}{2}}^n - u_{k-\frac{1}{2}}^n\right) - \frac{\Delta t}{\Delta x}\left(F_{k+\frac{1}{2}}^n - F_{k-\frac{1}{2}}^n\right) \\ &+ s\left[\left(u_{k-\frac{3}{2}}^n - 2u_{k-\frac{1}{2}}^n + u_{k+\frac{1}{2}}^n\right) + \left(u_{k+\frac{3}{2}}^n - 2u_{k+\frac{1}{2}}^n + u_{k-\frac{1}{2}}^n\right)\right] \end{aligned} \tag{4.69a}$$

Step 2:

$$u_k^{n+1} = u_k^n - \frac{\Delta t}{\Delta x}\left(F_{k+\frac{1}{2}}^{n+\frac{1}{2}} - F_{k-\frac{1}{2}}^{n+\frac{1}{2}}\right) + s(u_{k+1}^n - 2u_k^n + u_{k-1}^n) \tag{4.69b}$$

This scheme is first-order accurate method with a truncation error $O\left[\Delta t, (\Delta x)^2\right]$.

4.5.2.4 The MacCormack Method

The MacCormack method applied to the complete Burgers' equation is

Predictor step:

$$\overline{u_k^{n+1}} = u_k^n - \frac{\Delta t}{\Delta x}(F_{k+1}^n - F_k^n) + s(u_{k+1}^n - 2u_k^n + u_{k-1}^n) \tag{4.70a}$$

Corrector step:

$$u_k^{n+1} = \frac{1}{2}\left[u_k^n + \overline{u_k^{n+1}} - \frac{\Delta t}{\Delta x}\left(\overline{F_k^{n+1}} - \overline{F_{k-1}^{n+1}}\right) + s\left(\overline{u_{k+1}^{n+1}} - 2\overline{u_k^{n+1}} + \overline{u_{k-1}^{n+1}}\right)\right] \tag{4.70b}$$

This scheme is second-order accurate both in space and time. In the predictor step, a forward difference is used for ${}^{\partial F}/_{\partial x}$ whereas in the corrector step, backward difference is used. However, we can also use backward difference in predictor step together with forward difference in corrector step. The accuracy for both the versions remain same.

4.6 THE LAPLACE EQUATION

The Laplace equation is given by

$$\left(\frac{\partial^2 u}{\partial x^2}+\frac{\partial^2 u}{\partial y^2}\right)=0 \tag{4.71}$$

This equation is elliptic PDE and represents steady state heat equation in a solid, incompressible irrotational (potential) flow of a fluid. The steady incompressible N-S equations are also elliptic but are coupled, which will be explained in Chapter 6. Another important example of elliptic equation is the Poisson equation, which is given by

$$\left(\frac{\partial^2 u}{\partial x^2}+\frac{\partial^2 u}{\partial y^2}\right)=f(x,y) \tag{4.72}$$

where $f(x,y)$ is the source term which can be thought as the heat generation in the solid in heat conduction equation.

The finite difference discretization of Eq. (4.71) is given by

$$\frac{u_{i+1,j}-2u_{i,j}+u_{i-1,j}}{(\Delta x)^2}+\frac{u_{i,j+1}-2u_{i,j}+u_{i,j-1}}{(\Delta y)^2}=0 \tag{4.73a}$$

or

$$u_{i+1,j}-2u_{i,j}+u_{i-1,j}+\beta^2\left(u_{i,j+1}-2u_{i,j}+u_{i,j-1}\right)=0 \tag{4.73b}$$

where $\beta^2=\left(\Delta x/\Delta y\right)^2$.

The solution methods for Eq. (4.73b) are discussed in Chapter 8.

4.7 EXAMPLES

Example 4.1 A flat plate is 7m wide, 6m high and 0.3m long, as shown in Figure 4.26. The temperature of the left surface is maintained at 200^{0}C and the right surface at 100^{0}C. The conductivity of the plate material is 0.6 W/mk. Find the temperature distribution inside the plate, the flux through the plate and the total heat loss through it.

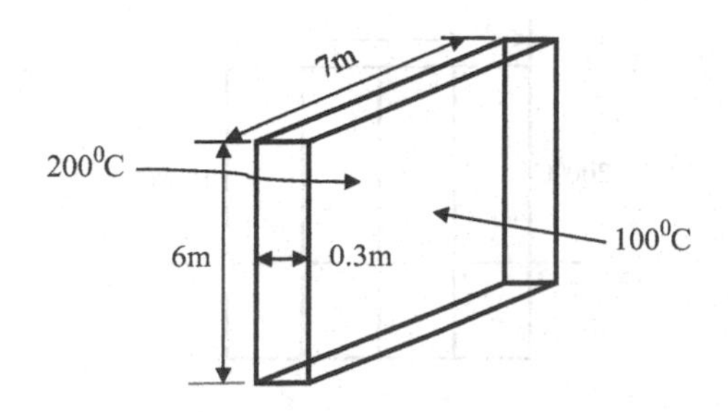

FIGURE 4.26 Problem Definition for Example 4.1.

Answer:
Since the length of the plate is very small compared to the width and height, it is steady state one-dimensional heat conduction. The corresponding governing equation is

$$\frac{d^2T}{dx^2} = 0$$

The corresponding FDE is

$$T_{i-1} - 2T_i + T_{i+1} = 0 \tag{4.74}$$

Let us divide the plate length into 4 equal divisions. Hence, we shall have 5 nodes as shown in Figure 4.27.

Given:

$$T_1 = 200^0\text{C};\ T_5 = 100^0\text{C};\ k = 0.6\ \text{W/mK};\ L = 0.3\text{m}$$

We shall find the temperatures at points T_2, T_3 and T_4 which are interior nodes.
Now Eq. (4.74) for interior nodes can be written as

Node 2:

$$T_1 - 2T_2 + T_3 = 0$$

or

$$2T_2 - T_3 = 200 \tag{4.75a}$$

Node 3:

$$T_2 - 2T_3 + T_4 = 0 \tag{4.75b}$$

Node 4:

$$T_3 - 2T_4 + T_5 = 0$$

or

$$-T_3 + 2T_4 = 100 \tag{4.75c}$$

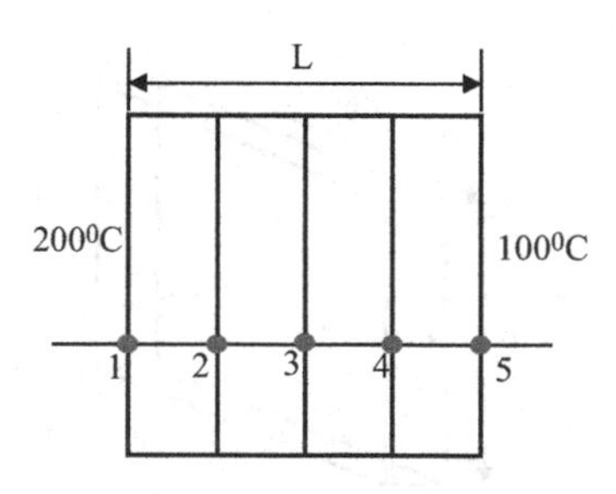

FIGURE 4.27 Grid Points for Example 4.1.

By solving the Eqs. (4.75a), (4.75b) and (4.75c), we can find the temperature distribution across the plate.

Multiplying the Eq. (4.75b) by 2 and then subtracting from Eq. (4.75a) we get

$$3T_3 - 2T_4 = 200 \tag{4.75b'}$$

Now multiply Eq. (4.75c) by 3 and then adding with Eq. (4.75b'), we get $T_4 = 125$.
Substituting value of T_4 in Eq. (4.75b') we get $T_3 = 150$
Similarly, we get from Eq. (4.75a) $T_2 = 175$.
Hence, temperature distribution across the plate is

$$T_1 = 200^0C;\ T_2 = 175^0C;\ T_3 = 150^0C;\ T_4 = 125^0C\ and\ T_5 = 100^0C$$

Now the heat flux through the plate is

$$q = -k\frac{dT}{dx} = -k\frac{T_5 - T_1}{L} = -0.6\frac{100 - 200}{0.3} = 200\ W/m^2$$

And total heat loss $Q = 200 \times 7 \times 6 = 8400\ W$.

Please note that we have used Gauss elimination in solving these equations. This method is suitable for a small number of equations. However, when the number of nodes (i.e. equations) are large, it is better to use the Thomas algorithm for solving algebraic equations. The solution procedures are discussed in Chapter 8.

Example 4.2 A large uranium plate of thickness L=0.3m and thermal conductivity k = 28 W/mK is uniformly generating heat at a constant rate of 5 kW/m^3. One side of the plate is maintained at 200^{0}C and other side at 100^{0}C. Find the temperature distribution inside the plate and compare with the analytical results. Take 5 nodes spaced equally.

Answer:

Given: L = 0.3m; k = 28 W/mK; $\dot{q}$ = 5000 W/m^3; nx = 5; T_1 = 200^{0}C; T_5 = 100^{0}C

Hence, $\Delta x = L/(nx - 1) = 0.075$.

The governing equation is

$$\frac{d^2T}{dx^2} + \frac{\dot{q}}{k} = 0 \tag{4.76a}$$

Corresponding analytical solution is given by

$$T(x) = T_1 + \left[\frac{\dot{q}(L - x)}{2k} + \frac{T_2 - T_1}{L}\right]x \tag{4.76b}$$

The corresponding FDE is

$$T_{i-1} - 2T_i + T_{i+1} + \frac{\dot{q}\Delta x^2}{k} = 0 \tag{4.76c}$$

Since the values of T_1 and T_5 are given, we must find the temperatures at points T_2, T_3 and T_4, which are interior nodes (see Figure 4.27).

Now,

$$\frac{\dot{q}\Delta x^2}{k} = \frac{5000 \times (0.075)^2}{28} = 1.004$$

Hence, $$T_{i-1} - 2T_i + T_{i+1} + 1.004 = 0$$

For Node 2: $$2T_2 - T_3 = 201.004$$

For Node 3: $$-T_2 + 2T_3 - T_4 = 1.004$$

For Node 4: $$-T_3 + 2T_4 = 101.004$$

The above equations can be solved by Gauss elimination and the results are:

$$T_2 = 176.504^0 C;\ T_3 = 152.004^0 C;\ T_4 = 126.504^0 C$$

We shall see how the above equations can be solved by Thomas algorithm.

All the above equations can be expressed as

$$b_i T_{i-1} + d_i T_i + a_i T_{i+1} = c_i$$

Hence, we get the matrix like:

$$\begin{bmatrix} d_2 & a_2 & \ldots \\ b_3 & d_3 & a_3 \\ \ldots & b_4 & a_4 \end{bmatrix} \begin{bmatrix} T_2 \\ T_3 \\ T_4 \end{bmatrix} = \begin{bmatrix} c_2 \\ c_3 \\ c_4 \end{bmatrix}$$

So, for IP = 3

$$d(3) = d(3) - \frac{b(3)}{d(2)} \times a(2) = 2 - \left[\frac{-1}{2} \times (-1)\right] = 1.5$$

$$c(3) = c(3) - \frac{b(3)}{d(2)} \times c(2) = 1.004 - \left[\frac{-1}{2} \times (201.004)\right] = 101.506$$

For IP = 4

$$d(4) = d(4) - \frac{b(4)}{d(3)} \times a(3) = 2 - \left[\frac{-1}{1.5} \times (-1)\right] = 1.333$$

$$c(4) = c(4) - \frac{b(4)}{d(3)} \times c(3) = 101.004 - \left[\frac{-1}{1.5} \times (101.506)\right] = 168.67$$

So

$$T_4 = \frac{c(4)}{d(4)} = \frac{168.67}{1.333} = 126.54$$

$$T_3 = \frac{c(3) - a(3) \times T_4}{d(3)} = \frac{101.506 - (-1) \times 126.54}{1.5} = 152.03$$

$$T_2 = \frac{c(2) - a(2) \times T_3}{d(2)} = \frac{201.004 - (-1) \times 152.03}{2} = 176.52$$

A comparison of the results of analytical, Gauss elimination and Thomas algorithm solutions is shown in Table 4.1.

Let IMIN is the subscript of first equation = 2
IMAX is the subscript of last equation = Nx = 4

Algorithm:

Compute

$IP = IMIN + 1$

$d_k = d_k - \frac{b_k}{d_{k-1}} a_{k-1}$; *for* $k = IP, IP + 1, \ldots, \ldots, IMAX$

and new c_k by

$c_k = c_k - {b_k c_{k-1}}/{d_{k-1}}$; *for* $k = IP, IP + 1, \ldots, \ldots, IMAX$

Then we compute unknowns by back substitution according to

$T_{IMAX} = {c_{IMX}}/{d_{NIMX}}$

$T_k = {[c_k - a_k T_{k+1}]}/{d_k}; k = IMAX - 1, IMAX - 2, \ldots\ldots\ldots IMIN$

TABLE 4.1
Comparison of the Results

	Analytical	FDE	
		Gauss Elimination	Thomas Algorithm
T_1	200.00	-	-
T_2	176.504	176.51	176.52
T_3	152.004	152.01	152.03
T4	126.504	126.51	126.54
T5	100.00	-	-

Example 4.3 A turbine blade of length 5cm with cross-sectional area 4.5cm^2 and perimeter 12cm is exposed to combustion gases at 900^{0}C. The blade is made of high alloy steel with a thermal conductivity of 25W/mK. The heat-transfer coefficient between the blade and combustion gas is 500W/m^{2}K. The temperature at the blade attachment point is kept at 500^{0}C. This is shown in Figure 4.28. Take 5 equal divisions and find the temperature distribution in the blade. Compare with the analytical results.

Answer:

Governing equation:

$$\frac{d^2T}{dx^2} + \frac{hP}{kA}(T_\infty - T) = 0 \tag{4.77a}$$

Corresponding analytical solution:

$$\frac{T - T_\infty}{T_b - T_\infty} = \frac{\cosh m(L - x)}{\cosh mL} \tag{4.77b}$$

where

$$m^2 = \frac{hP}{kA}$$

Given:

$$L = 0.05m; A = 4.5 \times 10^{-4}m^2; P = 0.12m;\ T_b = T_1 = 500^0C;\ T_\infty = 900^0C$$

$$k = 25W/mK; h = 500W/m^2K; ndiv = 5$$

Hence,

$$nx = number\ of\ nodes = 6; \Delta x = {}^{0.05}/_5 = 0.01m$$

$$m^2 = \frac{hP}{kA} = 5333.33$$

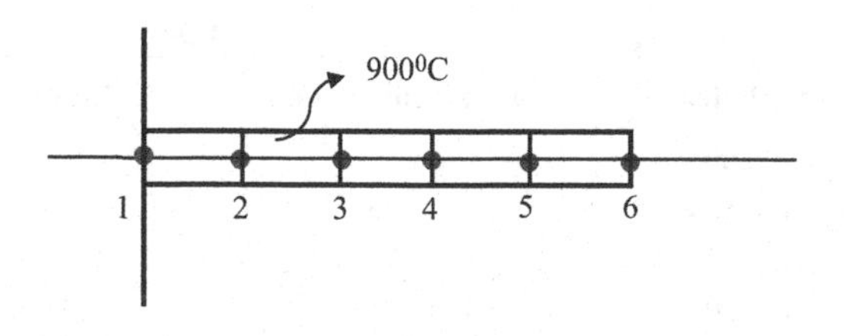

FIGURE 4.28 Problem Definition for Example 4.3.

Since the temperature of node 1 is fixed, we need to find the temperature of nodes 2 to 6.

The FDE for nodes 2 to 5 is given by

$$-\frac{1}{\Delta x^2}T_{i-1}+\left(\frac{2}{\Delta x^2}+\frac{hP}{kA}\right)T_i-\frac{1}{\Delta x^2}T_{i+1}=\frac{hP}{kA}T_\infty \tag{4.78}$$

Putting the values and simplifying we get

$$-T_{i-1}+2.533T_i-T_{i+1}=480$$

Node 2:

$$-T_1+2.533T_2-T_3=480 \; or \; 2.533T_2-T_3=980 \tag{4.79a}$$

Node 3:

$$-T_2+2.533T_3-T_4=480 \tag{4.79b}$$

Node 4:

$$-T_3+2.533T_4-T_5=480 \tag{4.79c}$$

Node 5:

$$-T_4+2.533T_5-T_6=480 \tag{4.79d}$$

Node 6:

The tip of the blade can be assumed to be insulated. Hence, the FDE for node 6 becomes

$$-\frac{2}{\Delta x^2}T_5+\left(\frac{2}{\Delta x^2}+\frac{hP}{kA}\right)T_6=\frac{hP}{kA}T_\infty$$

or

$$-T_5+1.267T_6=240 \tag{4.79e}$$

Solving Eqs. (4.79a) to (4.79e) by Thomas algorithm we get

$T_2=704.16^0C$; $T_3=803.64^0C$; $T_4=851.08^0C$; $T_5=871.69^0C$ and $T_6=877.36^0C$

A comparison of the results is shown in Table 4.2.

From Table 4.2 we can see that FDE gives satisfactory results. This prediction can be improved if we increase the number of nodes. However, hand calculation becomes increasingly difficult if we continue to increase the number of nodes. We can write a small program to solve the problem; it will be shown below.

TABLE 4.2

A Comparison of Analytical and Computational Results

	Distance, x	Analytical	FDE
T_1	0	500	500
T_2	0.01	706.94	704.16
T_3	0.02	806.09	803.64
T_4	0.03	852.90	851.08
T_5	0.04	873.47	871.69
T_6	0.05	879.25	877.36

Eq. (4.78) can be written in general form as

$$b_i T_{i-1} + d_i T_i + a_i T_{i+1} = c_i$$

Program:

Comments: Specify dimensions of b, d, a, c and T. Dimension should be at least equal to the number of nodes. Arbitrarily, let us assume 100.

```
Dimension b (100), d (100), a (100), c (100), T (100)
```

Comments: Read the specified values.

```
Read L, A, P, Tb, Tf, k, h, ndiv
```

Comments: Calculate distance between nodes (delx), m^2 and total number of nodes

```
delx = L/ndiv
m2 = (h*P)/(k*A)
nx = ndiv+1
```

Comments: IMIN is subscript of first equation (2 in this example)
Comments: IMAX is the subscript of last equation (nx in this case)

```
IMAX = nx
T(1)= Tb
```

Comments: Calculate the values of b, d, a, c

```
i = imin
do while (i ≤ IMAX-1)
{
  b (i) = -1.0/(delx*delx)
  d (i) = ((2.0/(delx*delx)) + m2)
  a (i) = -1.0/(delx*delx)
  c (i) = m2*Tf
```

```
    i++
  }
  i = IMAX
  b(i) = -2.0/(delx*delx)
  d(i) = ((2.0/(delx*delx)) + m2)
  c(i) = m2*Tf
```

Comments: Solution by the Thomas algorithm

```
  c(IMIN) = c(IMIN) – b(IMIN)*T(IMIN-1)
  i = IMIN+1
  do while (i ≤ IMAX)
  {
    r = b(i)/d(i-1)
    d(i) = d(i) – r*a(i-1)
    c(i) = c(i) – r*c(i-1)
    i++
  }
```

Comments: Back substitution

```
  T(IMAX) = c(IMAX)/d(IMAX)
  i = imin+1
  do while (i ≤ IMAX)
  {
    j = IMAX -i +IMIN
    T(j) = (c(j)- a(j)*T(j+1))/d(j)
    i++
  }
  end
```

Example 4.4 A large uranium plate of thickness L= 8cm with thermal conductivity k = 28 W/mK and thermal diffusivity $\alpha = 12.5 \times 10^{-6}$ m^2/s is initially at temperature 100^0C. The plate is uniformly generating heat at a constant rate of $1000kW/m^3$. At time t = 0, the left side of the plate is insulated, and right side is subjected to convection with a fluid of 20^0C with h = 35W/ m^2K as shown in Figure 4.29. Take Δx = 2cm. Find the temperature distribution inside the plate after 3 minutes using the explicit method.

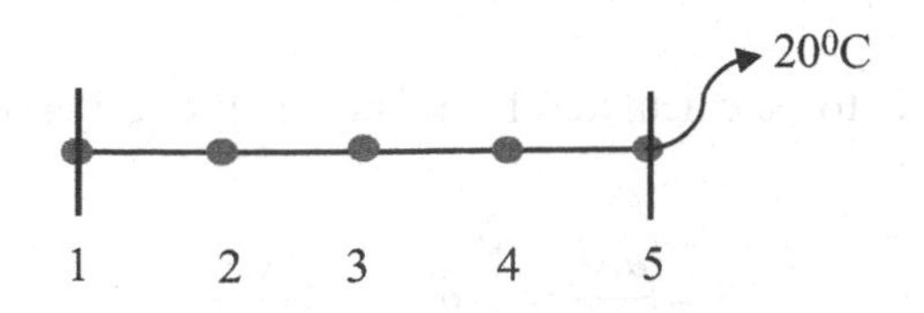

FIGURE 4.29 Problem Definition for Example 4.4.

Answer:

Given: L = 0.08m; k = 28 W/mK; $\alpha = 12.5 \times 10^{-6}$ m^2/s, $\dot{q} = 10^6$ W/m^3; T_0 = 100^{0}C; T_∞ = 20^{0}C;

Δx = 0.02m; h = 35W/m^{2}K; t = 180s.

Number of nodes = 0.08/0.02 = 4. Number of nodes = 4 + 1 = 5.

Here node 1 is insulated and node 5 is convective boundary.

The governing equation is

$$\frac{d^2T}{dx^2} + \frac{\dot{q}}{k} = \frac{\rho C_p}{k}\frac{\partial T}{\partial t} \tag{4.80a}$$

FDE for interior nodes 2,3,4

$$T_i^{n+1} = \lambda T_{i-1}^n + (1 - 2\lambda)T_i^n + \lambda T_{i+1}^n + \lambda\frac{\dot{q}\Delta x^2}{k} \tag{4.80b}$$

Node 2

$$T_2^{n+1} = \lambda T_1^n + (1 - 2\lambda)T_2^n + \lambda T_3^n + \lambda\frac{\dot{q}\Delta x^2}{k} \tag{4.81a}$$

Node 3

$$T_3^{n+1} = \lambda T_2^n + (1 - 2\lambda)T_3^n + \lambda T_4^n + \lambda\frac{\dot{q}\Delta x^2}{k} \tag{4.81b}$$

Node 4

$$T_4^{n+1} = \lambda T_3^n + (1 - 2\lambda)T_4^n + \lambda T_5^n + \lambda\frac{\dot{q}\Delta x^2}{k} \tag{4.81c}$$

Node 1 is insulated. Hence, FDE for this node

$$T_1^{n+1} = (1 - 2\lambda)T_1^n + 2\lambda T_2^n + \lambda\frac{\dot{q}\Delta x^2}{k} \tag{4.81d}$$

Node 5 is convective boundary. FDE for this node is

$$T_5^{n+1} = \left(1 - 2\lambda - 2\lambda\frac{h\Delta x}{k}\right)T_5^n + 2\lambda T_4^n + 2\lambda\frac{h\Delta x}{k}T_\infty + \lambda\frac{\dot{q}\Delta x^2}{k} \tag{4.81e}$$

where

$$\lambda = \frac{\alpha\Delta t}{\Delta x^2}$$

Not the time-step is to be calculated from the stability criteria.
For internal nodes:

$$\lambda = \frac{\alpha\Delta t}{\Delta x^2} \leq \frac{1}{2} \ or \ \Delta t \leq \frac{\Delta x^2}{2\alpha}$$

Putting the values, we get $\Delta t \leq 16\ s$.

For node 6,

$$1 - 2\lambda - 2\lambda \frac{h\Delta x}{k} \geq 0 \ or \ \lambda \leq \frac{1}{2\left(1 + \frac{h\Delta x}{k}\right)} \ or \ \Delta t \leq \frac{\Delta x^2}{2\alpha\left(1 + \frac{h\Delta x}{k}\right)}$$

From which we get $\Delta t \leq 15.6\ s$.
Now the time step should be lesser of the two. Let $\Delta t = 15\ s$.
Hence,

$$\lambda = \frac{\alpha \Delta t}{\Delta x^2} = 0.469$$

Table 4.3 shows the temperature at various nodes after solving Eq. (4.81a) to Eq. (4.81e).

Example 4.5 A property ∅ is transported by means of convection and diffusion through the one-dimensional domain as shown in Figure 4.30.

i) Write down the transport equation for steady state
ii) Using 5 equal division, write down the discretized equations for each node using central difference scheme for both convection and diffusion terms and find the value of ∅ at each node.

TABLE 4.3
Temperature Distribution at Various Nodes

	Temperatures (^{0}C)				
Time (s)	**T1**	**T2**	**T3**	**T4**	**T5**
0	100	100	100	100	100
15	106.7	106.75	106.75	106.75	104.969
30	113.4469	113.4799	113.5034	112.6681	111.4984
45	120.1779	120.2322	119.8574	119.2676	117.311
60	126.9288	126.791	126.5165	125.3862	123.7334
75	133.4996	133.4903	132.8784	131.9038	129.7298
90	140.1908	139.9744	139.4747	138.1072	136.0829
105	146.688	146.6116	145.8375	144.5682	142.1559
120	153.3163	153.0577	152.3782	150.8044	148.459
135	159.7738	159.6368	158.7349	157.2179	154.5607
150	166.3454	166.0579	165.2258	163.4618	160.8205
165	172.7759	172.5855	171.5713	169.8321	166.9276
180	179.2974	178.9854	178.0171	176.0705	173.1471

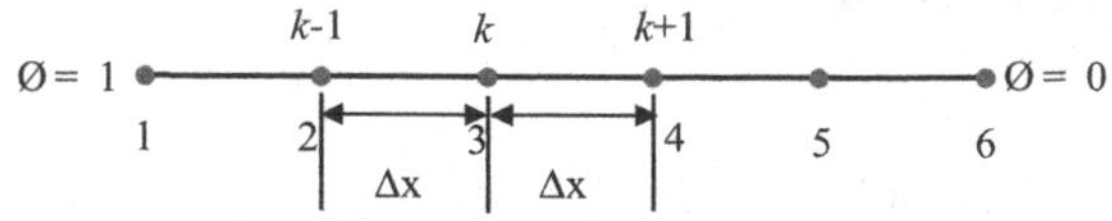

FIGURE 4.30 Problem Definition for Example 4.5.

Take u = 0.1 m/s, Length, L = 1.0 m, ρ = 1.0 kg/m^3, Γ = 0.1 kg/m/s, $\emptyset|_{x=0} = 1$ and $\emptyset|_{x=L} = 0$

Answer:

Governing equation:

$$\frac{d}{dx}(\rho u \emptyset) = \frac{d}{dx}\left(\Gamma \frac{d\emptyset}{dx}\right) + S \tag{4.82}$$

The values in node 1 and node 6 are known. The discretized equations for internal nodes

$$\rho u \left[\frac{\emptyset_{k+1} - \emptyset_{k-1}}{2\Delta x}\right] = \Gamma \left[\frac{\emptyset_{k+1} - 2\emptyset_k + \emptyset_{k-1}}{(\Delta x)^2}\right]$$

Let $F = \rho u$ *and* $D = {}^{\Gamma}/_{\Delta x}$

Hence, we get

$$-\left(\frac{F}{2} + D\right)\emptyset_{k-1} + 2D\emptyset_k - \left(-\frac{F}{2} + D\right)\emptyset_{k+1} = 0$$

Now, $F = 0.1$, $\Delta x = 1/5 = 0.2$ and $D = 0.5$

So

$$-0.55\ \emptyset_{k-1} + \emptyset_k - 0.45\ \emptyset_{k+1} = 0$$

Node 2:

$$-0.55\emptyset_1 + \emptyset_2 - 0.45\emptyset_3 = 0$$

Since $\emptyset_1$ is known, the above equation becomes

$$\emptyset_2 - 0.45\emptyset_3 = 0.55$$

Node 3: $-0.55\emptyset_2 + \emptyset_3 - 0.45\emptyset_4 = 0$

Node 4: $-0.55\emptyset_3 + \emptyset_4 - 0.45\emptyset_5 = 0$

Node 5: $$-0.55\emptyset_4 + \emptyset_5 - 0.45\emptyset_6 = 0$$

Now $\emptyset_6$ is known, the above equation becomes

$$-0.55T_4 + \emptyset_5 = 0$$

Solving the above equations, we get

$$\emptyset_2 = 0.871; \emptyset_3 = 0.714; \emptyset_4 = 0.522; \emptyset_5 = 0.287$$

4.8 SUMMARY

- For diffusion equations, central differencing is used for the discretization of the diffusion term.
- An explicit scheme fails to represent the physical behavior of the parabolic equation. The implicit scheme will be a better method for solving the parabolic PDE.
- Upwind differencing is used for discretization of convective terms. However, using the first-order upwind scheme gives rise to numerical diffusion.
- For a wave equation, which is hyperbolic in nature, explicit schemes provide a more natural finite difference approximation as hyperbolic PDE has a limited zone of influence.
- The selection of the best method for solving a diffusion equation is very difficult. However, implicit methods are more suitable than explicit methods. For one-dimensional problems, the Crank-Nicolson method is highly recommended because of second-order accuracy in time and space. For two- or three-dimensional cases, the ADI scheme may be used as this method allows for the use of the Thomas algorithm.
- The discretization of the conservative form of equation guarantees proper processing of any discontinuity present in the flow.
- A grid independence study is a must for any simulation.
- The solution of the inviscid Burgers' equation by explicit methods produces superior results compared to implicit methods specially when there is discontinuity. The implicit scheme requires more computational effort as the time step is decided by the transient phenomenon. Hence, explicit methods are recommended for solving the inviscid Burgers' equation.
- Godunov's method is superior in capturing the shock wave. For the expansion wave, this method is equally good. However, for Euler's equation, Godunov's method becomes computationally expensive as it involves inefficient iterative methods in dealing with non-linearity.

QUESTIONS

1) Distinguish between *explicit* and *implicit* methods for solving transient heat conduction equations.

2) What do you understand about the grid independence test?
3) What is meant by the Courant number?
4) Distinguish between the grid independence test and convergence.
5) What are sufficient and necessary conditions for the convergence of an iterative scheme?
6) What is numerical diffusion and how can it be minimized?
7) What is an upwind scheme? Why it is needed?
8) Is a higher-order differencing scheme always better than a scheme with lower order? Consider robustness, accuracy, CPU time and memory demand.
9) Among the different schemes (forward, backward and central difference), which is most accurate and how? What is the necessity of the other methods?
10) Consider the one-dimensional diffusion equation

$$\frac{\partial u}{\partial t} = \alpha \frac{\partial^2 u}{\partial x^2}$$

Write the finite difference formulation using
i) explicit; ii) implicit and iii) the Crank-Nicolson methods.
State the accuracy and stability criteria for each method.

11) Find the finite difference formulation of the DuFort-Frankel method and discuss the accuracy and consistency for the following equation:

$$\frac{\partial T}{\partial t} = \frac{k}{\rho C_P} \frac{\partial^2 T}{\partial x^2}$$

12) Consider the wave equation

$$\frac{\partial u}{\partial t} + c \frac{\partial u}{\partial x} = 0$$

where c is a constant
a. Find the finite difference discretization using FTCS.
b. State whether the scheme is stable with an explanation. If not, what procedure you will adopt to make the FDE stable?

13) What do you understand about artificial viscosity? Explain.
14) Consider the two-dimensional heat-conduction equation

$$\frac{\partial u}{\partial t} = \alpha \left(\frac{\partial^2 u}{\partial x^2} + \frac{\partial^2 u}{\partial y^2} \right)$$

Write the finite difference formulation using
i) explicit; ii) Crank-Nicolson and iii) ADI methods.
State the accuracy and stability criteria for each method.

15) An insulated rod of length 1.0m and 1.0cm and 1.0cm square cross-section has its temperature fixed at 50^0C at one end and 550^0C at the other end. Assuming steady-state heat conduction in one dimension, find the temperature at each node by the finite difference method and compare with the analytical solution. Divide the rod in five equal parts and take conductivity of the material k = 1000W/mK.
16) A large uranium plate of thickness L= 5cm and thermal conductivity k = 28W/mK is uniformly generating heat at a constant rate of 5×10^6 W/m^3. One side of the plate is insulated while the other side is subjected to convection to an environment at 30°C with a heat transfer coefficient of h = 50W/m^{2}K. Taking 6 equally spaced nodes, find the temperature distribution inside the plate.
17) A rectangular fin of uniform cross-section of area A and length 1m is exposed to the atmosphere at 25°C. The fin base is maintained at a constant temperature of 200°C. The fin tip is insulated. Assume 1 D heat transfer use hp/kA = 25m^{-2} (kA is constant).
    a) Write the governing differential equation and boundary conditions.
    b) Using five equidistant nodes, formulate a finite difference algorithm and assemble the solution matrix.
18) Solve the problem given in Example 4.4 by the implicit method using the same data. Use a time step of 15 seconds and compare with the results obtained by the explicit method.
19) Consider the two-dimensional steady state heat conduction equation

$$\frac{\partial^2 T}{\partial x^2} + \frac{\partial^2 T}{\partial y^2} = 0$$

Write the finite difference equation for this equation and discuss the various methods for solving the discretized equations.
20) Refer to Figure 4.31. An infinite plate in X-Z plane is initially at rest. The fluid (water) column above the plate is also at rest. Suddenly the plate is moved with a velocity U m/s at t = 0 s.

The Governing equation assuming no pressure gradient becomes

$$\frac{\partial u}{\partial t} = \nu \frac{\partial^2 u}{\partial y^2}$$

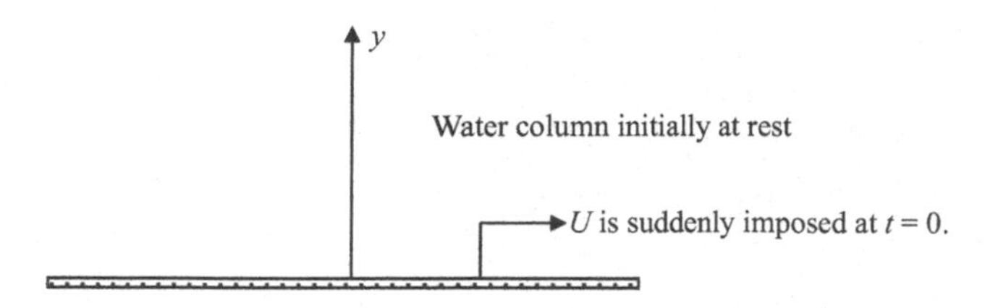

FIGURE 4.31 Problem Definition for Exercise 20.

with Initial and Boundary conditions:

$$u(y,\ 0) = 0 \ for \ y \geq 0$$

$$u(0, t) = U \ at \ t > 0$$

$$= 0 \ at \ t \leq 0$$

$$u(\infty, t) = 0$$

The expected analytical solution is

$$\frac{u}{U} = 1 - erf\left(\frac{y}{2\sqrt{\nu t}}\right)$$

Write the finite difference formulation using i) Explicit; ii) Implicit and iii) Crank-Nicolson Method and find *u(y)* with respect to increasing time. Compare with the analytical results.

5 Finite Volume Method

5.1 INTRODUCTION

In the previous chapter, we discussed finite difference approximation of governing equations. Initial development of CFD involved use of finite differences to approximate governing equations describing fluid flow. As we have seen, the partial spatial and temporal derivatives appearing in the equations are approximated through the Taylor series in the finite difference method. However, finite differences are mainly employed on Cartesian geometries. Since most practical engineering problems are associated with complex geometries, the use of the finite difference method becomes limited.

Finite element methods (FEMs) are very flexible in terms of geometry and mesh elements. The domain is discretized by elements. A simple functional form is assumed at each element to approximate the solution. The FEM aims to minimize the difference between the exact solution and the collection of base functions by various methods such as the Galerkin method, which is very popular for solid-mechanics problems.

However, fluid flow problems are generally governed by *local* as well as *global* conservation. For instance, the continuity equation dictates the local conservation of mass. Since in the FEM the difference between the base functions and the exact solution is minimized globally, the local conservation may not be obtained. Research is directed towards obtaining local conservation of the transport property of advection terms. Therefore, the method is not very popular in CFD.

The principle of the finite volume method (FVM) is local conservation; this is the key reason for its popularity in CFD. To solve the equations numerically with the FVM, the entire computational domain is divided into "small" subvolumes, called *cells*. It then balances the fluxes crossing the faces of a cell, thus enforcing the conservation rigorously and directly related to the physical quantities (e.g. mass flux). The FVM ensures local and global flux balance. This method has geometrical flexibility similar to that of the FEM. Hence, all commercial CFD software is based on FVM. As mentioned in Chapter 1, FVM uses integral form of the governing equation for discretization as it directly relates to the conservation of fluxes across a cell. This chapter discusses the finite volume discretization of governing equations.

The general scalar-transport equation for any transported physical quantity $\emptyset$ in conservative form is given by

$$\frac{\partial}{\partial t}(\rho\emptyset) + \frac{\partial}{\partial x_j}(\rho u_j \emptyset) = \frac{\partial}{\partial x_j}\left(\Gamma \frac{\partial \emptyset}{\partial x_j}\right) + s \quad (5.1)$$

Where Γ is the diffusion coefficient and s is the source term.

First, we shall address the most fundamental aspects of numerical procedures for solving diffusion problems, for example in heat conduction.

5.2 THE DIFFUSION EQUATION

General Diffusion Equation for an arbitrary variable $\emptyset$ is given by

$$\frac{\partial}{\partial t}(\rho\emptyset) = \frac{\partial}{\partial x_j}\left(\Gamma\frac{\partial\emptyset}{\partial x_j}\right) + s \tag{5.2}$$

For heat conduction equation

$$\Gamma = \frac{k}{C_P} \text{ and } s = \frac{\dot{q}}{C_P}$$

As mentioned earlier, the principle of the FVM is based on local conservation. First, the entire computational domain is divided into "small" sub-volumes, called *cells.* A typical control volume in three dimensions is shown in Figure 5.1. A typical control volume/cell in two dimensions is shown in Figure 5.2.

Normally, the cell centre is denoted P. With respect to point P, adjacent nodes are denoted *E, W, N, S, T* and *B* for coordinate directions east, west, north, south, top and bottom, respectively. Since, the flux crosses through the cell faces, e, w, n, s, t and b are used to denote the cell faces. Corresponding cell-face areas are denoted A_e, A_w, A_n, A_s, A_t, and A_b respectively and cell volumes V.

If we compare with FDM, the nodal values in FVM can be expressed in corresponding *ijk* notations as $\emptyset_P = \emptyset_{i,j,k}$; $\emptyset_E = \emptyset_{i+1,j,k}$ etc.

We already defined *cell.* Each cell is surrounded by its *faces.* These faces form a grid pattern throughout the domain. FVM can use both structured and unstructured grids, thus providing meshing flexibility similar to FEM. This aspect of the

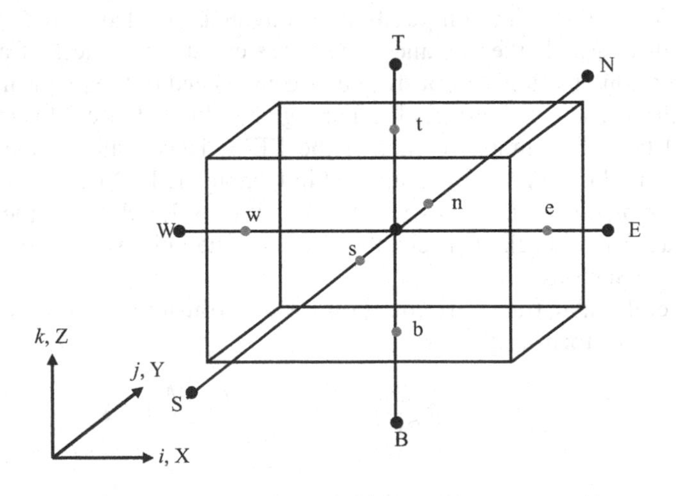

FIGURE 5.1 Typical Control Volume in Three Dimensions.

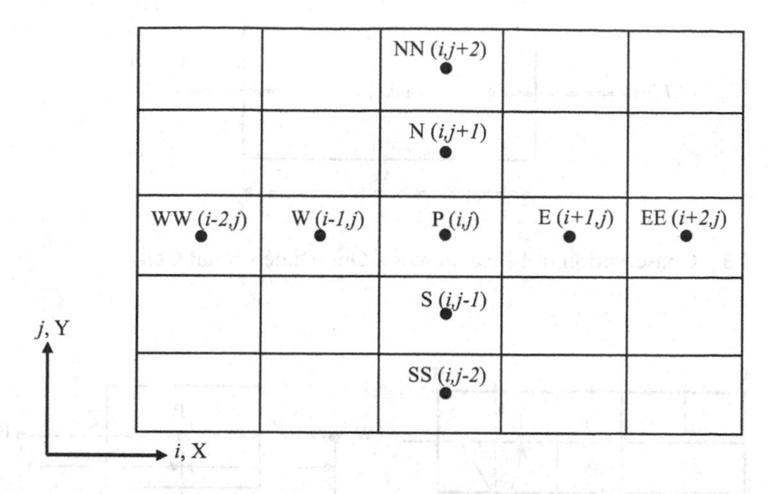

FIGURE 5.2 Typical Control Volume/Cell in Two Dimensions.

grid will be dealt with more details in the grid generation chapter (Chapter 10). In this chapter, we shall discuss the application of FVM to Cartesian grids. The application of FVM for non-orthogonal grids will be discussed in Chapter 7.

5.2.1 The Steady-State One-Dimensional Diffusion Equation

Let us consider first the steady-state, one-dimensional diffusion equation, which

- helps us simplify analysis and is easily understandable;
- in most cases provides analytical results;
- allows discretized equations to be solved by hand;
- provides straightforward extension of the method to two and three dimensions.

The one-dimensional diffusion equation for steady state is given by

$$\frac{d}{dx}\left(\Gamma\frac{d\emptyset}{dx}\right)+s=0 \tag{5.3}$$

In FVM, the fluxes are conserved across the cell as shown in Figure 5.3.

Integrating the control volume as shown in Figure 5.4, we get:

$$\int_V \frac{d}{dx}\left(\Gamma\frac{d\emptyset}{dx}\right)dV+\int_V s\,dV=0$$

$$\int_A \left(\Gamma\frac{d\emptyset}{dx}\right)dA+\int_V s\,dV=0$$

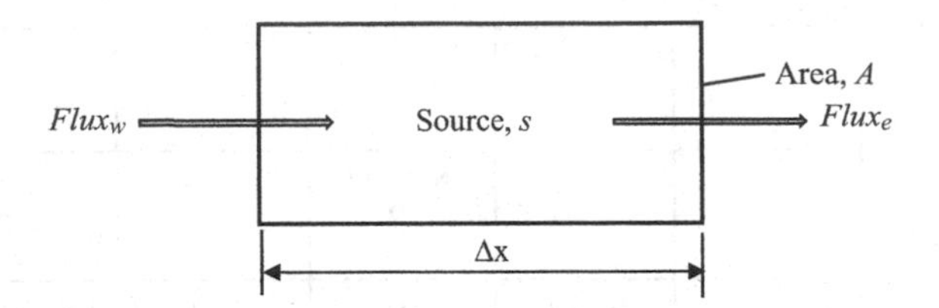

FIGURE 5.3 Conservation of Fluxes across a One-Dimensional Cell.

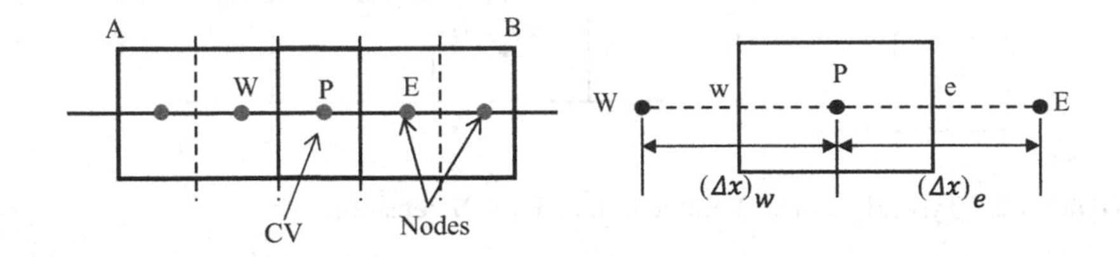

FIGURE 5.4 Typical Control Volume and Nomenclature.

or

$$\sum_{faces}\left(\Gamma\frac{d\emptyset}{dx}\right)A + sV = 0$$

i.e.

$$Flux_e - Flux_w + S = 0$$

where

$$Flux_e = \left(\Gamma A\frac{d\emptyset}{dx}\right)_e \text{ and } Flux_w = \left(\Gamma A\frac{d\emptyset}{dx}\right)_w$$

Hence,

$$\left(\Gamma A\frac{d\emptyset}{dx}\right)_e - \left(\Gamma A\frac{d\emptyset}{dx}\right)_w + S = 0 \tag{5.4}$$

Normally, central differencing is used for the discretization of the gradient in a diffusion equation. Let us consider Figure 5.4. We get:

$$\left(\Gamma A\frac{d\emptyset}{dx}\right)_e = \left(\frac{\Gamma A}{\Delta x}\right)_e(\emptyset_E - \emptyset_P) = D_e(\emptyset_E - \emptyset_P) \tag{5.5a}$$

where

$$D_e = \left(\frac{\Gamma A}{\Delta x}\right)_e$$

Similarly,

$$\left(\Gamma A \frac{d\emptyset}{dx}\right)_w = \left(\frac{\Gamma A}{\Delta x}\right)_w (\emptyset_P - \emptyset_W) = D_w(\emptyset_P - \emptyset_W) \quad (5.5b)$$

Note:
Normally, fluxes cross the boundary through *cell faces, not nodes.* Since, diffusion acts equally in all directions, equal weighting from the nodes either side of each face can be applied while approximating the diffusion term $(d\emptyset)/dx$ leading to central differencing scheme. Hence, it is second-order accurate in Δx.

5.2.2 Discretization of the Source Term

The total source strength for the cell is given by

Source = (source per unit volume) × (volume) or $S = s \times V$.

In one-dimension problem, *V is the cell length,* Δx.

The source, S may depend on the solution $\emptyset$ or may be independent of the solution $\emptyset$. Hence, the source term can be conveniently expressed as

$$Source, S = b_P + S_P\emptyset_P; \text{ where } S_P \leq 0$$

Here, b_P is the part of the source which is independent of $\emptyset$, and S_P is dependent on $\emptyset$. The reason for this form will be discussed later.

Hence, discretized form of *the steady-state* diffusion problem can be as follows:

$$Flux_e - Flux_w + Source = 0$$

$$D_e(\emptyset_E - \emptyset_P) - D_w(\emptyset_P - \emptyset_W) + b_P + S_P\emptyset_P = 0 \quad (5.6a)$$

or

$$D_w\emptyset_W - (D_w + D_e - S_P)\emptyset_P + D_e\emptyset_E + b_P = 0 \quad (5.6b)$$

or

$$-a_w\emptyset_W + a_P\emptyset_P - a_e\emptyset_E = b_P \quad (5.7a)$$

or

$$a_P\emptyset_P - \sum_{nb} a_{nb}\emptyset_{nb} = b_p \quad (5.7b)$$

where, *nb* denotes the neighboring or adjacent nodes and

$$a_w = D_w;\ a_e = D_e \ and \ a_P = \ D_w + D_e - S_P \tag{5.7c}$$

This equation is applicable for interior nodes. Now let us find out the equations applicable for boundary nodes.

5.2.3 Discretized Equation at Boundaries

5.2.3.1 For Given Value at the Boundaries (Dirichlet Boundary Conditions)

A typical boundary cell is shown in Figure 5.5.

We know
$$Flux_e - Flux_w + Source = 0$$
or
$$D_e(\emptyset_e - \emptyset_P) - \ D_w(\emptyset_P - \emptyset_W) + b_P + S_P\emptyset_P = 0$$
or
$$-D_w\emptyset_W + (D_w + D_e - S_P)\emptyset_P - D_e\emptyset_e = \ b_P$$
or
$$-a_w\emptyset_W + a_P\emptyset_P - a_e\emptyset_e = b_P$$

As the value of $\emptyset_e$ is known at the east boundary, the coefficients become

$$D_e = \left(\frac{\Gamma A}{\Delta x}\right)_e;\ D_w = \left(\frac{\Gamma A}{\Delta x}\right)_w$$

$$a_w = D_w;\ a_e = 0;\ a_P = \ a_w + D_e - S_P\ ;\ b_P = \ D_e\emptyset_e$$

5.2.3.2 Insulated Boundary

Again
$$Flux_e - Flux_w + Source = 0.$$

Since no flux crossing the right boundary as it is insulated,

$$-Flux_w + Source = 0$$
or
$$D_w(\emptyset_P - \emptyset_W) + b_P + S_P\emptyset_P = 0$$
or
$$-a_w\emptyset_W + a_P\emptyset_P - a_e\emptyset_e = b_P$$

where

$$D_e = \ 0;\ D_w = \left(\frac{\Gamma A}{\Delta x}\right)_w \text{ and } b_P = \ 0$$

$$a_w = D_w;\ a_e = 0 \ and \ a_P = \ D_w + D_e - S_P$$

If boundary flux is specified, then $b_P = Flux_e$.

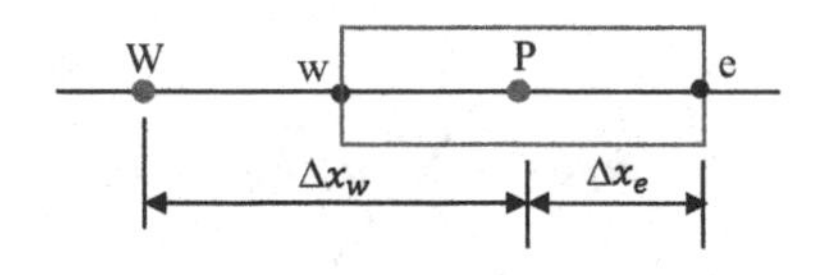

FIGURE 5.5 Boundary Cell in Which Value Is Specified at the Boundary.

5.2.3.3 Mixed Boundary Conditions

$$Flux_e - Flux_w + Source = 0$$

$$D_e(\emptyset_e - \emptyset_P) - D_w(\emptyset_P - \emptyset_W) + hA(\emptyset_\infty - \emptyset_e) = 0$$

$$-D_w\emptyset_W + (D_w + D_e - S_P)\emptyset_P - D_e\emptyset_e = b_P$$

$$D_e = \left(\frac{\Gamma A}{\Delta x} + hA\right)_e ; \; D_w = \left(\frac{\Gamma A}{\Delta x}\right)_w$$

$$a_w = D_w; \; a_e = 0; \; a_P = a_w + D_e - S_P ; \; b_P = hA\emptyset_\infty$$

Value of $\emptyset_e$ can be found out from

$$\emptyset_e = \frac{\left(\frac{\Gamma A}{\Delta x}\right)_e \emptyset_P + hA\emptyset_\infty}{\left(\frac{\Gamma A}{\Delta x}\right)_e + hA}.$$

5.2.4 Assembling the Algebraic Equations

Already we have seen that the discretized equation for one cell takes the form

$$a_P\emptyset_P - \sum_{nb} a_{nb}\emptyset_{nb} = b_p$$

If we assemble the discretized equation for each node, then we get a set of algebraic equations. These equations can be expressed in the matrix form $A\emptyset = b$

or

$$\underbrace{\begin{bmatrix} a_p & -a_s & 0 & 0 & 0 & 0 & 0 \\ -a_w & a_p & -a_e & 0 & 0 & 0 & 0 \\ 0 & -a_w & a_p & -a_e & 0 & 0 & 0 \\ 0 & 0 & -a_w & a_p & -a_e & 0 & 0 \\ 0 & 0 & 0 & -a_w & a_p & -a_e & 0 \\ 0 & 0 & 0 & 0 & -a_w & a_p & -a_e \\ 0 & 0 & 0 & 0 & 0 & -a_w & a_p \end{bmatrix}}_{A} \underbrace{\begin{bmatrix} \vdots \\ \vdots \\ \vdots \\ \phi_P \\ \vdots \\ \vdots \\ \vdots \end{bmatrix}}_{\emptyset} = \underbrace{\begin{bmatrix} \vdots \\ \vdots \\ \vdots \\ b_P \\ \vdots \\ \vdots \\ \vdots \end{bmatrix}}_{b} \tag{5.8}$$

The matrices derived from partial differential equations either by finite difference or FVMs lead to sparse matrices with the non-zero elements of the matrices lie on a small number of well-defined diagonals. For one-dimensional cases, we obtain tri-diagonal matrices. For constant coefficients, the matrix can be solved directly by Gaussian elimination or very efficiently by the tri-diagonal matrix algorithm known as the Thomas algorithm. If the elements of the matrix are dependent on the solution, then it must be solved iteratively.

5.2.5 Extension to Two Dimensions

The general conservation diffusion equation in two dimensions can be written

$$\frac{\partial}{\partial x}\left(\Gamma\frac{\partial\emptyset}{\partial x}\right)+\frac{\partial}{\partial y}\left(\Gamma\frac{\partial\emptyset}{\partial y}\right)+s=0 \tag{5.9}$$

Integrating over the control volume shown in Figure 5.6, we get

$$\int_V \frac{\partial}{\partial x}\left(\Gamma\frac{\partial\emptyset}{\partial x}\right)dV+\int_V \frac{\partial}{\partial y}\left(\Gamma\frac{\partial\emptyset}{\partial y}\right)dV+\int_V S\,dV=0$$

or

$$\int_A \left(\Gamma\frac{\partial\emptyset}{\partial x}\right)dA+\int_A \left(\Gamma\frac{\partial\emptyset}{\partial y}\right)dA+\int_V S\,dV=0$$

i.e. $$(Flux_e - Flux_w) + (Flux_n - Flux_s) + Source = 0$$

Like earlier,

$$Flux_e=\left(\Gamma A\frac{\partial\emptyset}{\partial x}\right)_e \text{ and } Flux_w=\left(\Gamma A\frac{\partial\emptyset}{\partial x}\right)_w$$

Similarly,

$$Flux_n=\left(\Gamma A\frac{\partial\emptyset}{\partial y}\right)_n \text{ and } Flux_s=\left(\Gamma A\frac{\partial\emptyset}{\partial y}\right)_s$$

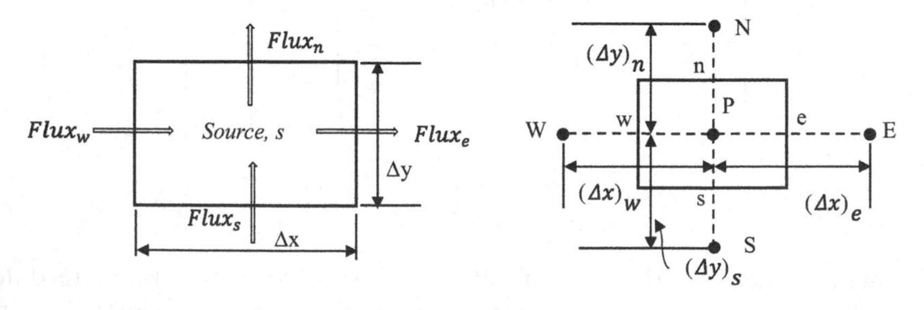

FIGURE 5.6 Conservation of Fluxes across a Two-Dimensional Cell.

Substituting, we get

$$-a_s\emptyset_S - a_w\emptyset_W + a_P\emptyset_P - a_e\emptyset_E - a_n\emptyset_N = b_P, \tag{5.10}$$

where

$$D_w = \left(\frac{\Gamma A}{\Delta x}\right)_w;\ D_e = \left(\frac{\Gamma A}{\Delta x}\right)_e;\ D_s = \left(\frac{\Gamma A}{\Delta y}\right)_s; D_n = \left(\frac{\Gamma A}{\Delta y}\right)_n \tag{5.11a}$$

$$a_w = D_w;\ a_e = D_e;\ a_s = D_s;\ a_n = D_n \text{ and } a_P = D_w + D_e + D_s + D_n - S_P \tag{5.11b}$$

This equation is for interior nodes. For boundary nodes, the treatment is similar to a one-dimensional case. It can be seen from Eq. (5.10) that we need to solve a penta-diagonal matrix. This is true for Cartesian/orthogonal meshes. For non-orthogonal meshes, **A** matrix becomes nine diagonals. This will be clear when we deal with the application of FVM for non-orthogonal meshes in Chapter 7.

5.2.6 Extension to Three Dimensions

The governing equation in three dimensions is given by

$$\frac{\partial}{\partial x}\left(\Gamma\frac{\partial\emptyset}{\partial x}\right) + \frac{\partial}{\partial y}\left(\Gamma\frac{\partial\emptyset}{\partial y}\right) + \frac{\partial}{\partial z}\left(\Gamma\frac{\partial\emptyset}{\partial z}\right) + s = 0 \tag{5.12}$$

The discretized equation becomes

$$-a_b\emptyset_b - a_s\emptyset_S - a_w\emptyset_W + a_P\emptyset_P - a_e\emptyset_E - a_n\emptyset_N - a_t\emptyset_t = b_P \tag{5.13}$$

where

$$D_w = \left(\frac{\Gamma A}{\Delta x}\right)_w; D_e = \left(\frac{\Gamma A}{\Delta x}\right)_e; D_s = \left(\frac{\Gamma A}{\Delta y}\right)_s; D_n = \left(\frac{\Gamma A}{\Delta y}\right)_n; D_b = \left(\frac{\Gamma A}{\Delta z}\right)_b; D_t = \left(\frac{\Gamma A}{\Delta z}\right)_t$$

$$a_w = D_w;\ a_e = D_e;\ a_s = D_s;\ a_n = D_n;\ a_b = D_b;\ a_t = D_t$$

and

$$a_P = D_b + D_s + D_w + D_e + D_n + D_t - S_P$$

Here we can see from Eq. (5.13) that for Cartesian/orthogonal meshes we obtain seven diagonal matrices, which become 19 diagonals for non-orthogonal meshes.

5.2.7 Desirable Properties of a Discretization Scheme

As discussed in Chapter 1, in discretization, certain rules to be followed to obtain realistic numerical results. These are stated again here for clarity.

Consistency
As the mesh size approaches to zero, if the resulting difference equations become equivalent to the original partial differential equations, then it is consistent.

$$(\emptyset_E - \emptyset_P)/_{\Delta x} \text{ is consistent approximation of } {}^{\partial\emptyset}/_{\partial x}.$$

Conservativeness
When the net flux crossing the control volume is zero (i.e. the outgoing flux from one cell becomes the incoming flux for the adjacent cell) it is conservative. We must remember that fluxes are associated with faces, not nodes. FVM automatically ensures this property.

Transportiveness
In a fluid problem, the fluid carries information from upstream to downstream (i.e. the advection scheme involves directional effects). This is known as the transportive property and appropriate weightage must be provided in the advection scheme.

Boundedness
In an advection-diffusion problem without sources, the solution of flow variables of a node must be between the values of the flow variables of the surrounding nodes.

Stability
Stability indicates whether we can obtain a solution. As we saw in Chapter 3, it is only possible if small errors do not grow during the solution procedure.

Hence, boundedness and stability lead to the following constraints:

- *Consistency at control volume face.*
- *Positive coefficients:* $a_F \geq 0$ *for all F where F denotes the faces.*
- *Negative slope of the source term:*

$$\text{Source, } S = b_P + S_P\emptyset_P; \text{ where } S_P \leq 0$$

- *Sum of neighboring coefficients:*

$$a_P \geq \sum_{nb} |a_{nb}|$$

This is only possible if $S_P \leq 0$

5.2.8 Further Comments on Interface Diffusion Coefficients

Γ_e represent the value of Γ pertaining to the control volume face "e". Similar is true for Γ_w etc. This require the values of Γ at faces. However, we only know the values at the nodes i.e. at W, P, E etc. Hence, how to find interface values?

One way to find this is by linear interpolation, i.e. $\Gamma_e = f_e\Gamma_P + (1 - f_e)\Gamma_E$. *where*

$$f_e = \frac{\Delta x_{e^+}}{\Delta x_e}$$

The notations are illustrated in Figure 5.7.

For e, midway between P and E, $f_e = 0.5$ *and* Γ_e is the arithmetic mean of Γ_P *and* Γ_E. This approach cannot handle the abrupt changes that may occur in some cases like dealing with composite materials. A better approach is

$$\Gamma_e = \left(\frac{1 - f_e}{\Gamma_P} + \frac{f_e}{\Gamma_E}\right)^{-1}.$$

For e midway between P and E

$$\Gamma_e = \frac{2\Gamma_P\Gamma_E}{\Gamma_P + \Gamma_E}.$$

For more explanation on the desirable properties on discretization, readers are requested to refer to Patankar [1980]. Other property values like density can be similarly calculated.

5.3 THE CONVECTION-DIFFUSION EQUATION

This section will discuss the most fundamental aspects of numerical procedures for solving convection-diffusion problems for incompressible flow.

General Convection-Diffusion Equation for an arbitrary variable ∅ is given by

$$\frac{\partial}{\partial t}(\rho\emptyset) + \frac{\partial}{\partial x_j}(\rho u_j\emptyset) = \frac{\partial}{\partial x_j}\left(\Gamma\frac{\partial\emptyset}{\partial x_j}\right) + S \tag{5.14a}$$

i.e.

$$\begin{aligned} &\frac{\partial}{\partial t}(\rho\emptyset) + \frac{\partial}{\partial x}(\rho u\emptyset) + \frac{\partial}{\partial y}(\rho v\emptyset) + \frac{\partial}{\partial z}(\rho w\emptyset) \\ &= \frac{\partial}{\partial x}\left(\Gamma\frac{\partial\emptyset}{\partial x}\right) + \frac{\partial}{\partial y}\left(\Gamma\frac{\partial\emptyset}{\partial y}\right) + \frac{\partial}{\partial z}\left(\Gamma\frac{\partial\emptyset}{\partial z}\right) + S \end{aligned} \tag{5.14b}$$

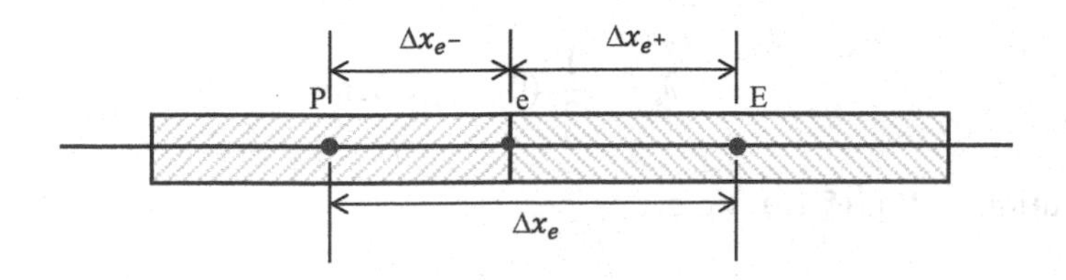

FIGURE 5.7 Interface Diffusion Coefficients.

5.3.1 The Steady-State One-Dimensional Advection-Diffusion Equation

Like earlier, we shall first deal with steady one-dimensional advection-diffusion equation. The one-dimensional advection-diffusion equation for steady state becomes

$$\frac{\partial}{\partial x}(\rho u \emptyset) = \frac{\partial}{x}\left(\Gamma \frac{\partial \emptyset}{\partial x}\right) + S \tag{5.15}$$

or

$$\frac{\partial}{\partial x}\left[(\rho u \emptyset) - \left(\Gamma \frac{\partial \emptyset}{\partial x}\right)\right] = S$$

The total flux consists of convective flux $(\rho u \emptyset)$ and diffusive flux $\left(\Gamma^{\partial \emptyset}/_{\partial x}\right)$.

Integrating the fluxes over the control volume as shown in Figure 5.3, we get

$$(flux_e - flux_w) = source$$

or

$$\left(\rho A u \emptyset - \Gamma A \frac{\partial \emptyset}{\partial x}\right)_e - \left(\rho A u \emptyset - \Gamma A \frac{\partial \emptyset}{\partial x}\right)_w = S\,\Delta x$$

Let $F = \rho A u$.

Then the above equation becomes

$$[F_e \emptyset_e - D_e(\emptyset_E - \emptyset_P)] - [F_w \emptyset_w - D_w(\emptyset_P - \emptyset_W)] = b_P + S_P \emptyset_P. \tag{5.16}$$

In the above equation, the values of $\emptyset_w$ *and* $\emptyset_e$ in the convective flux are not known and they must be approximated to the adjacent nodal values W, E etc. The approximation of these face values in terms of the values at adjacent nodes need special attention for convection-diffusion problem which we shall discuss now. The method of specifying these face values is called *convective scheme or convection-differencing scheme.*

5.3.1.1 The Central Differencing Scheme (CDS)

In the central differencing scheme (CDS) shown in Figure 5.8, we approximate the cell-face value by the average of the values at the adjacent nodes, that is

$$\emptyset_e = \frac{1}{2}(\emptyset_E + \emptyset_P)\text{etc.}$$

Substituting in Eq. (5.16), we get

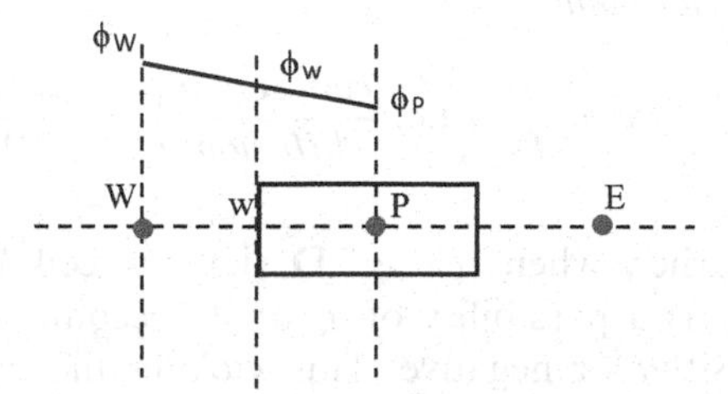

FIGURE 5.8 CDS Scheme.

$$\frac{1}{2}F_e(\emptyset_E+\emptyset_P)-D_e(\emptyset_E-\emptyset_P)-\frac{1}{2}F_w(\emptyset_W+\emptyset_P)+D_w(\emptyset_P-\emptyset_W)+b_P+S_P\emptyset_P=0$$

or

$$-\left(\frac{1}{2}F_w+D_w\right)\emptyset_W+\left(-\frac{1}{2}F_w+D_w+\frac{1}{2}F_e+D_e-S_P\right)\emptyset_P - \left(-\frac{1}{2}F_e+D_e\right)\emptyset_E=b_P \tag{5.17}$$

i.e.

$$a_P\emptyset_P-\sum_{nb}a_{nb}\emptyset_{nb}=b_P \tag{5.18a}$$

where

$$a_w={}^1\!/_2F_w+D_w;\, a_e=-{}^1\!/_2F_e+D_e;\, a_p=a_w+a_e-S_P+(F_e-F_w) \tag{5.18b}$$

Let us now see the implication of central differencing for a convective-diffusive equation. Let us assume $D_w = D_e = 1$; $F_e = F_w = 4$. In the absence of any sources, $a_w = 3$, $a_e = -1$ and $a_P = 2$.

Now, if $\emptyset_E = 200$ and $\emptyset_W = 100$, we get $\emptyset_P = 50$ which is unrealistic as it violates the boundedness criteria. Monotonicity of $\emptyset$ is not preserved and negative coefficients give unrealistic values at cell centre.

When $|F| > 2D$, depending on whether F is positive or negative, there is a possibility of a_e or a_w becoming negative which violates one of the basic rules. Negative coefficient also implies that $a_P \le \sum|a_{nb}|$ and this fails to satisfy the Scarborough Criteria. Satisfaction of Scarborough Criterion ensures numerical stability and realistic results.

Scarborough criterion is given by: $a_P = \sum a_{nb} \ge \sum|a_{nb}|$

This problem of instability with central difference scheme for convective term has already been discussed in earlier chapter dealing with finite difference. *So Central Difference scheme is to be restricted to low Reynolds Number i.e. to low value of F/D.*

Let us define *cell Peclet number Pe by*

$$Pe = \frac{|F|}{D}\left(i.e.\ \frac{convection}{diffusion}\right) = \frac{\rho u \Delta x}{D}$$

As we have seen earlier, when $|F| > 2D$ (i.e. the cell Peclet number Pe is greater than 2, there is a possibility of a_e or a_w becoming negative depending on whether F is positive or negative. This violates the boundedness criterion and we obtain unrealistic results. Physically, the fluid flow process is directional in nature (i.e. upwind properties influence downstream nodes). In the case of central differencing, we assign equal weightage to both upwind and downwind nodes. Hence, we obtain unrealistic results as discussed. Figures 5.9 and 5.10 provide clarity.

Let us consider an incense stick burning in the middle of a room. There is no air flowing inside the room. The aroma of the stick will fill the room after some time; the strength will depend on the distance from the burning stick. This is a pure diffusion process; the condition at point P is affected equally by upstream and downstream conditions as shown in Figure 5.9 a. Let us now consider that the burning incense stick is placed between a person and a pedestal fan and that air is blowing from right to left as shown in Figure 5.10a. The person standing on the left side will get a strong aroma as the smoke plume is blowing towards him. Now let the fan be placed between the person and the incense stick and blow from left to right as shown in Figure 5.10b. The air will blow the smoke away from the person and he will not get any aroma. The reason behind this phenomenon is the directional behavior of the blowing air, which takes information from upstream to downstream. This is also shown in Figure 5.9 b.

5.3.1.2 The Upwind Differencing Scheme (UDS)

One major problem associated with the central differencing scheme is that it gives equal weightage of both upstream and downstream nodes and does not take care

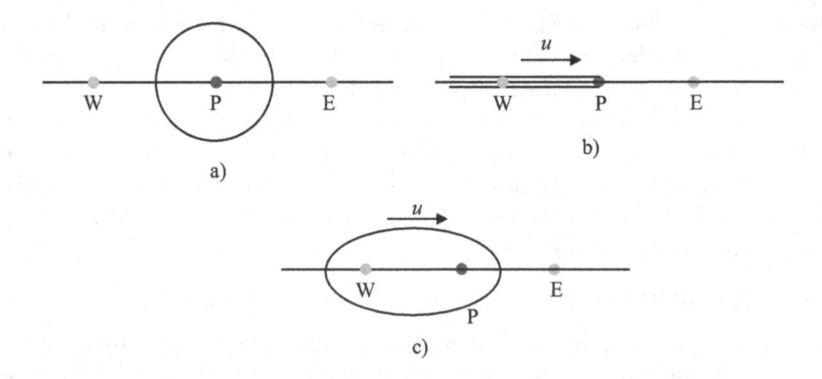

FIGURE 5.9 Zone of Influence: a) Pure Diffusion; b) Pure Convection and c) Convection + Diffusion.

FIGURE 5.10 Convective Scheme: a) Flow towards the Person; b) Flow away from the Person.

of the flow direction. For a strong convective flow from west-to-east, west cell face, w, should receive much stronger influence from node W than from node P. In *upwind differencing scheme which is also known as "donor-cell" differencing scheme*, face value $\emptyset_f$ is taken as the value at whichever is the upwind node i.e. $\emptyset_f = \emptyset_U$. Here f stands for the face, while U and D is upstream and downstream node respectively as shown in Figure 5.11.

For example:

$$\emptyset_e = \emptyset_P \quad if\ F_e > 0;\ \emptyset_e = \emptyset_E \quad if\ F_e < 0 \tag{5.19a}$$

$$\emptyset_w = \emptyset_W \quad if\ F_w > 0;\ \emptyset_w = \emptyset_P \quad if\ F_w < 0 \tag{5.19b}$$

This is known as an UDS and is explained in Figure 5.12.

When the flow is in the positive direction, the discretized equation becomes:

$$[F_e\emptyset_P - D_e(\emptyset_E - \emptyset_P)] - [F_w\emptyset_W - D_w(\emptyset_P - \emptyset_W)] = b_P + S_P\emptyset_P$$

or

$$-(F_w + D_w)\emptyset_w + (D_w + F_e + D_e - S_P)\emptyset_P - D_e\emptyset_E = b_P$$

or

$$-a_w\emptyset_W + a_P\emptyset_P - a_e\emptyset_E = b_P \tag{5.20}$$

where

$$a_w = F_w + D_w;\ a_e = D_e;\ a_P = a_w + a_e - S_p + (F_e - F_w) \tag{5.21a}$$

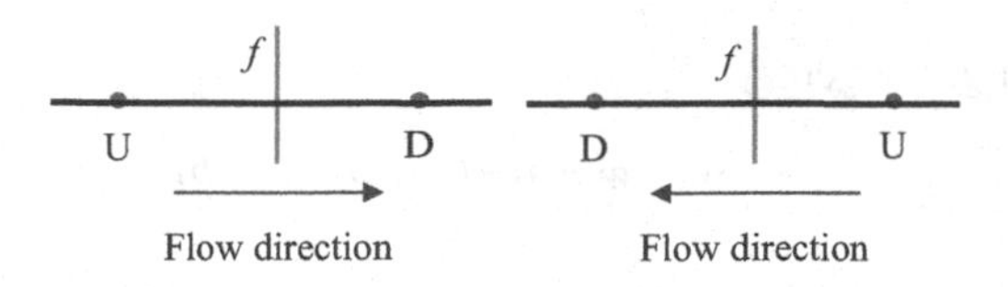

FIGURE 5.11 Definition of Upstream and Downstream.

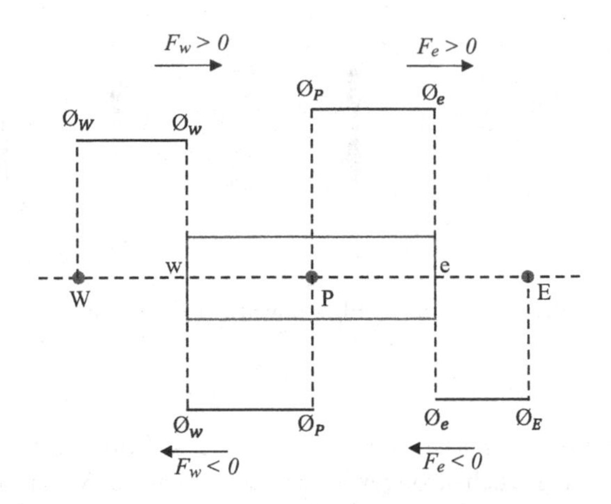

FIGURE 5.12 The Upwind Differencing Scheme (UDS).

Similarly, when the flow is in the negative direction, the discretized equation becomes:

$$-D_w \emptyset_w + (-F_w + D_w + D_e - S_P)\emptyset_P - (-F_e + D_e)\emptyset_E = b_P$$

or

$$-a_w \emptyset_W + a_P \emptyset_P - a_e \emptyset_E = b_P.$$

where

$$a_w = D_w;\ a_e = -F_e + D_e;\ a_P = a_w + a_e - S_p + (F_e - F_w) \quad (5.21b)$$

To take care of the flow directions, Eq. (5.19) can be combined and expressed as

$$F_e \emptyset_e = \emptyset_P [\![F_e, 0]\!] - \emptyset_E [\![-F_e, 0]\!]$$

$$F_w \emptyset_w = \emptyset_W [\![F_w, 0]\!] - \emptyset_P [\![-F_w, 0]\!]$$

Here, the symbol $[\![A, B]\!]$ is defined to denote greater of A and B.

Putting these values in Eq. (5.16), we get

$$\emptyset_P [\![F_e, 0]\!] - \emptyset_E [\![-F_e, 0]\!] - \emptyset_W [\![F_w, 0]\!] + \emptyset_P [\![-F_w, 0]\!] - D_e(\emptyset_E - \emptyset_P) + D_w(\emptyset_P - \emptyset_W)$$
$$= b_P + S_P \emptyset_P$$

After rearranging, we get

$$-a_w \emptyset_W + a_P \emptyset_P - a_e \emptyset_E = b_P$$

where,

$$a_w = [\![F_w, 0]\!] + D_w;\ a_e = [\![-F_e, 0]\!] + D_e \tag{5.22}$$

$$a_P = a_w + a_e - S_p + (F_e - F_w)$$

Assessment of Upwind Differencing Scheme (UDS)

Conservativeness:
In UDS, we use consistent expressions to balance the fluxes through cell faces. Hence, it is conservative.

Boundedness:
UDS always leads to positive coefficients and satisfies the requirement of boundedness. In the absence of source

$$a_P = a_w + a_e + (F_e - F_w)$$

Hence, a_P becomes equal to $a_w + a_e$, which is always $\geq \sum|a_{nb}|$ thereby satisfying the Scarborough criterion, when flow satisfies continuity, $(F_e - F_w) = 0$. This condition is essential for stability.

Transportiveness:
The scheme automatically takes care of the directional effect of flow and hence transportiveness.

Accuracy:
The scheme is first order accurate.

As discussed in Chapter 4, UDS, which is first-order accurate, leads to significant numerical diffusion. Hence, higher-order schemes are developed to improve the accuracy and reduction in numerical errors with a smaller number of grids. However, higher-order schemes demand more computational resources. In the following sections, we shall discuss various higher-order convective schemes and their advantages and disadvantages.

5.3.1.3 Exact Solution

For the convective-diffusive equation without any source, if Γ is constant, the equation can be solved exactly.

Let $0 \leq x \leq L$ and

B.C. $\quad x = 0 \quad \emptyset = \emptyset_0$

$\qquad x = L \quad \emptyset = \emptyset_L.$

The solution is

$$\frac{\emptyset - \emptyset_0}{\emptyset_L - \emptyset_0} = \frac{exp\left(\frac{Pex}{L}\right) - 1}{exp(Pe) - 1} \tag{5.23}$$

where Pe = Peclet Number = Ratio of strengths of convection and diffusion = F/D. $Pe = 0$ represents pure diffusion.

When flow is in a positive x-direction (i.e. Pe is positive), the value of $\emptyset$ is influenced by the upstream value $\emptyset_0$. For large positive values of Pe, the $\emptyset$ value of remains very close to the upstream value of $\emptyset_0$ as shown in Figure 5.13.

5.3.1.4 The Exponential Scheme

Let total flux

$$J = \rho u\emptyset - \Gamma\frac{d\emptyset}{dx}.$$

Hence, the convective-diffusive equation without any source becomes

$$\frac{dJ}{dx} = 0 \ or \ J_e = J_w$$

Putting in the exact solution, we get

$$J_e = F_e\left[\emptyset_P + \frac{\emptyset_P - \emptyset_E}{exp(Pe) - 1}\right]$$

Where

$$Pe = \frac{(\rho u)_e(\Delta x)_e}{\Gamma_e} = \frac{F_e}{D_e}$$

Hence,

$$F_e\left[\emptyset_P + \frac{\emptyset_P - \emptyset_E}{exp(Pe) - 1}\right] - F_w\left[\emptyset_W + \frac{\emptyset_W - \emptyset_P}{exp(Pw) - 1}\right] = 0$$

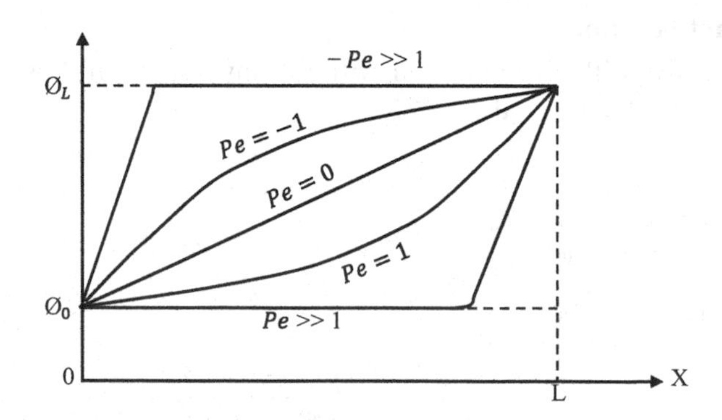

FIGURE 5.13 The Influence of the Cell Peclet Number.

or

$$-a_w\emptyset_W + a_P\emptyset_P - a_e\emptyset_E = b_P$$

where

$$a_w = \frac{F_w \exp(Pw)}{\exp(Pw) - 1};\ a_e = \frac{F_e}{\exp(Pe) - 1};\ a_P = a_w + a_e + (F_e - F_w) \tag{5.24}$$

Assessment

The method is conservative and satisfies the transportive and boundedness property required for stability. In the absence of source, for constant velocity and diffusivity, this scheme gives the exact solution. However, for two- or three-dimensional flow the scheme does not provide exact solution. The same is true when u or Γ vary for one-dimensional flow. The scheme is computationally highly expensive and hence not favored.

5.3.1.5 The Hybrid Differencing Scheme

The hybrid differencing scheme was developed by Spalding [1972] and is based on a combination of CDS and UDS. When the Peclet number is low (i.e. $Pe < 2$), the central differencing scheme can be employed as it gives second-order accuracy. However, for higher Peclet number (i.e. $Pe > 2$) upwind scheme is employed to account for transportiveness. For example, the convective flux at west face $\emptyset_w$ is given by

$$\emptyset_w = \left[\frac{1}{2}\left(1 + \frac{2}{Pw}\right)\emptyset_W + \frac{1}{2}\left(1 - \frac{2}{Pw}\right)\emptyset_P\right] \qquad for\ -2 < Pw < 2$$

$$\emptyset_w = \emptyset_W \qquad for\ Pw > 2$$

$$\emptyset_w = \emptyset_P \qquad for\ Pw < -2$$

The general form of the discretized equation is

$$-a_w\emptyset_W + a_P\emptyset_P - a_e\emptyset_E = b_P \tag{5.25a}$$

with

$$a_P = a_w + a_e + (F_e - F_w) \tag{5.25b}$$

$$a_w = \left[\!\left[F_w, \left(D_w + \frac{F_w}{2}\right), 0\right]\!\right];\ a_e = \left[\!\left[-F_e, \left(D_e - \frac{F_e}{2}\right), 0\right]\!\right] \tag{5.25c}$$

Assessment

The scheme is fully conservative, and it satisfies the transportiveness and boundedness requirement. The hybrid scheme remained popular for a long time in earlier commercial codes because of its inherent stability and robustness. However, for flow situations encountered in practical situation involve high advection/low diffusion regime leading to application of first order upwinding under this scheme.

Patankar [1980] developed a power-law approximation, in place of the exponential scheme, to reduce the computational time. However, this scheme suffers from numerical diffusion for three-dimensional cases.

5.3.1.6 The Second Order Upwind (SOU) Scheme

In the second order upwind (SOU) scheme, a linear profile is used with an upwind-biased stencil. The face value $\emptyset$ is extrapolated by a linear profile passing through two upstream nodes as shown in Figure 5.14. Here U is upstream, UU is far upstream and D is a downstream node with respect to the face, f.

Hence, for SOU scheme

$$\emptyset_f = \frac{3}{2}\emptyset_U - \frac{1}{2}\emptyset_{UU}$$

i.e.

$$\emptyset_w = \frac{3}{2}\emptyset_W - \frac{1}{2}\emptyset_{WW} \; F_w > 0; \; \emptyset_w = \frac{3}{2}\emptyset_P - \frac{1}{2}\emptyset_E \qquad F_w < 0$$

$$\emptyset_e = \frac{3}{2}\emptyset_P - \frac{1}{2}\emptyset_W \; F_e > 0; \; \emptyset_e = \frac{3}{2}\emptyset_E - \frac{1}{2}\emptyset_{EE} \qquad F_e < 0$$

This is explained in Figure 5.15.

Combining for the flow directions, we get

$$F_e\emptyset_e = \left(\frac{3}{2}\emptyset_P - \frac{1}{2}\emptyset_W\right)[\![F_e, 0]\!] - \left(\frac{3}{2}\emptyset_E - \frac{1}{2}\emptyset_{EE}\right)[\![-F_e, 0]\!] \tag{5.26a}$$

$$F_w\emptyset_w = \left(\frac{3}{2}\emptyset_W - \frac{1}{2}\emptyset_{WW}\right)[\![F_w, 0]\!] - \left(\frac{3}{2}\emptyset_P - \frac{1}{2}\emptyset_E\right)[\![-F_w, 0]\!] \tag{5.26b}$$

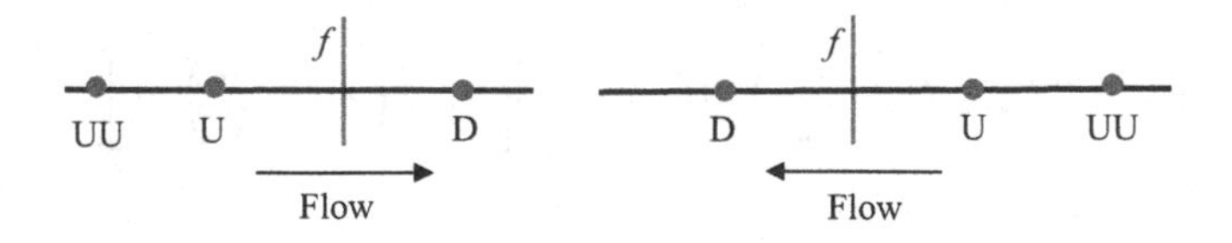

FIGURE 5.14 Notations for a Higher-Order Scheme.

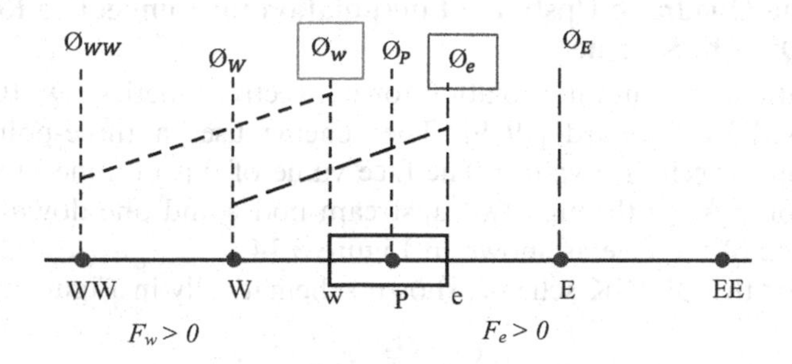

FIGURE 5.15 The SOU Scheme.

Hence, putting these values in Eq. (5.16) and rearranging we get

$$\left[\frac{1}{2}[\![F_w,0]\!]\right]\emptyset_{WW} + \left[-\frac{3}{2}[\![F_w,0]\!] - \frac{1}{2}[\![F_e,0]\!] - D_w\right]\emptyset_W$$
$$+\left[\frac{3}{2}[\![F_e,0]\!] + \frac{3}{2}[\![-F_w,0]\!] + D_w + D_e - S_P\right]\emptyset_P$$
$$+\left[-\frac{3}{2}[\![-F_e,0]\!] - \frac{1}{2}[\![-F_w,0]\!] - D_e\right]\emptyset_E + \left[\frac{1}{2}[\![-F_e,0]\!]\right]\emptyset_{EE} = b_P$$

which can be written as

$$-a_{ww}\emptyset_{WW} - a_w\emptyset_W + a_P\emptyset_P - a_e\emptyset_E - a_{ee}\emptyset_{EE} = b_P \tag{5.27}$$

where

$$a_w = \frac{3}{2}[\![F_w,0]\!] + \frac{1}{2}[\![F_e,0]\!] + D_w \ ; \quad a_e = \frac{3}{2}[\![-F_e,0]\!] + \frac{1}{2}[\![-F_w,0]\!] + D_e \tag{5.27b}$$

$$a_{ww} = -\frac{1}{2}[\![F_w,0]\!]; \quad a_{ee} = -\frac{1}{2}[\![-F_e,0]\!] \tag{5.27c}$$

$$a_P = a_{ww} + a_w + a_e + a_{ee} + (F_e - F_w) - S_P \tag{5.27d}$$

Assessment

The scheme is fully conservative, and it satisfies the transportiveness and boundedness requirement. The scheme is second-order accurate and is stable. However, this scheme as such cannot be solved by Tri-Diagonal Matrix Algorithm because of the presence of coefficients a_{ww} and a_{ee}.

5.3.1.7 The Quadratic Upstream Interpolation for Convective Kinetics (QUICK) Scheme

The quadratic upstream interpolation for convective kinetics (QUICK) scheme was proposed by Leonard [1979]. The scheme uses a three-point weighted interpolation for cell face values. The face value of $\emptyset$ is obtained from a quadratic function passing through two upstream nodes and one downstream node on either side of the face as shown in Figure 5.14.

Hence, for the QUICK scheme, shown schematically in Figure 5.16, we get

$$\emptyset_f = \frac{6}{8}\emptyset_U + \frac{3}{8}\emptyset_D - \frac{1}{8}\emptyset_{UU}$$

so

$$\emptyset_w = \frac{6}{8}\emptyset_W + \frac{3}{8}\emptyset_P - \frac{1}{8}\emptyset_{WW} \qquad \textit{for } F_w > 0$$

$$\emptyset_e = \frac{6}{8}\emptyset_P + \frac{3}{8}\emptyset_E - \frac{1}{8}\emptyset_W \qquad \textit{for } F_e > 0$$

$$\emptyset_w = \frac{6}{8}\emptyset_P + \frac{3}{8}\emptyset_W - \frac{1}{8}\emptyset_E \qquad \textit{for } F_w < 0$$

$$\emptyset_e = \frac{6}{8}\emptyset_E + \frac{3}{8}\emptyset_P - \frac{1}{8}\emptyset_{EE} \qquad \textit{for } F_e < 0.$$

Combining for the flow directions, we get

$$F_e\emptyset_e = \left(\frac{6}{8}\emptyset_P + \frac{3}{8}\emptyset_E - \frac{1}{8}\emptyset_W\right)[\![F_e, 0]\!] - \left(\frac{6}{8}\emptyset_E + \frac{3}{8}\emptyset_P - \frac{1}{8}\emptyset_{EE}\right)[\![-F_e, 0]\!] \tag{5.28a}$$

$$F_w\emptyset_w = \left(\frac{6}{8}\emptyset_W + \frac{3}{8}\emptyset_P - \frac{1}{8}\emptyset_{WW}\right)[\![F_w, 0]\!] - \left(\frac{6}{8}\emptyset_P + \frac{3}{8}\emptyset_W - \frac{1}{8}\emptyset_E\right)[\![-F_w, 0]\!] \tag{5.28b}$$

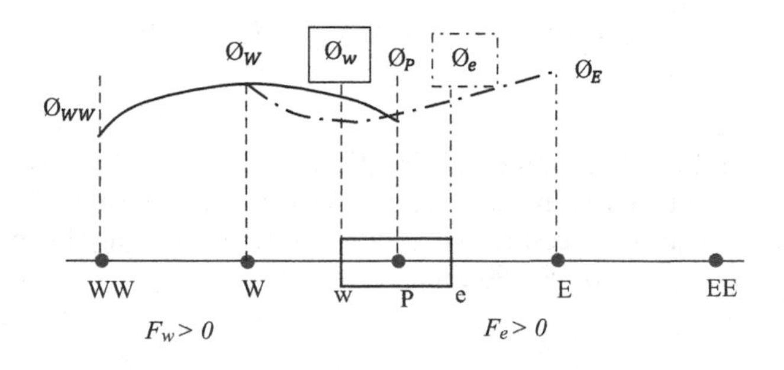

FIGURE 5.16 The QUICK Scheme.

Substituting these values in Eq. (5.16) and rearranging, we get

$$\left[\frac{1}{8}[\![F_w,0]\!]\right]\emptyset_{WW} + \left[-\frac{6}{8}[\![F_w,0]\!] + \frac{3}{8} - [\![F_w,0]\!] - \frac{1}{8}[\![F_e,0]\!] - D_w\right]\emptyset_W$$

$$+\left[\frac{6}{8}[\![F_e,0]\!] - \frac{3}{8}[\![-F_e,0]\!] - \frac{3}{8}[\![F_w,0]\!] + \frac{6}{8}[\![-F_w,0]\!] + D_w + D_e - S_P\right]\emptyset_P$$

$$+\left[-\frac{6}{8}[\![-F_e,0]\!] + \frac{3}{8}[\![F_e,0]\!] - \frac{1}{8}[\![-F_w,0]\!] - D_e\right]\emptyset_E + \left[\frac{1}{8}[\![-F_e,0]\!]\right]\emptyset_{EE} = b_P,$$

which can be written as

$$-a_{ww}\emptyset_{WW} - a_w\emptyset_W + a_P\emptyset_P - a_e\emptyset_E - a_{ee}\emptyset_{EE} = b_P, \tag{5.29}$$

where

$$\begin{aligned}
a_w &= \frac{6}{8}[\![F_w,0]\!] - \frac{3}{8}[\![-F_w,0]\!] + \frac{1}{8}[\![F_e,0]\!] + D_w;\ a_{ww} = -\frac{1}{8}[\![F_w,0]\!] \\
a_e &= \frac{6}{8}[\![-F_e,0]\!] - \frac{3}{8}[\![F_e,0]\!] + \frac{1}{8}[\![-F_w,0]\!] + D_e;\ a_{ee} = -\frac{1}{8}[\![-F_e,0]\!] \\
a_P &= a_{ww} + a_w + a_e + a_{ee} + (F_e - F_w) - S_P
\end{aligned} \tag{5.29b}$$

Assessment

QUICK Scheme uses consistent quadratic profiles for the cell face values of fluxes and hence conservative. On a uniform mesh, its accuracy in terms of the Taylor series truncation error is third order.

The transportiveness property is built into the scheme as the scheme uses two upstream and one downstream nodal value.

Boundedness:

- For modest Peclet number i.e. $Pe < 8/3$, if the flow field satisfies continuity, the coefficient $a_P$equals the sum of all neighbor coefficient which leads to satisfying boundedness criteria.
- For Peclet number (Pe) > 8/3, the coefficients a_e and a_{ww} or a_w and a_{ee} may not be positive depending on the flow direction. This will give rise to stability problems and unbounded solutions under certain flow conditions.

Since the discretized equations also involve a_{ww} and a_{ee}, Tri-diagonal matrix solution (TDMA) methods are not directly applicable for the QUICK scheme.

5.3.1.8 The FROMM Scheme

In the Fromm [1968] scheme, a linear profile is used with an upwind biased stencil. The face value $\emptyset$ is interpolated by a linear profile passing through

one far upstream (UU) and one downstream (D) node as shown in Figure 5.14.
Hence, for the FROMM scheme,

$$\emptyset_f = \emptyset_U + \frac{1}{4}\emptyset_D - \frac{1}{4}\emptyset_{UU}$$

Hence, from Figure 5.17, we get

$$\emptyset_w = \emptyset_W + \frac{1}{4}\emptyset_P - \frac{1}{4}\emptyset_{WW} \qquad \textit{for } F_w > 0$$

$$\emptyset_e = \emptyset_P + \frac{1}{4}\emptyset_E - \frac{1}{4}\emptyset_W \qquad \textit{for } F_e > 0$$

$$\emptyset_w = \emptyset_P + \frac{1}{4}\emptyset_W - \frac{1}{4}\emptyset_E \qquad \textit{for } F_w < 0$$

$$\emptyset_e = \emptyset_E + \frac{1}{4}\emptyset_P - \frac{1}{4}\emptyset_{EE} \qquad \textit{for } F_e < 0$$

Combining for the flow direction, we get

$$F_e\emptyset_e = \left(\emptyset_P + \frac{1}{4}\emptyset_E - \frac{1}{4}\emptyset_W\right)[\![F_e, 0]\!] - \left(\emptyset_E + \frac{1}{4}\emptyset_P - \frac{1}{4}\emptyset_{EE}\right)[\![-F_e, 0]\!] \tag{5.30a}$$

$$F_w\emptyset_w = \left(\emptyset_W + \frac{1}{4}\emptyset_P - \frac{1}{4}\emptyset_{WW}\right)[\![F_w, 0]\!] - \left(\emptyset_P + \frac{1}{4}\emptyset_W - \frac{1}{4}\emptyset_E\right)[\![-F_w, 0]\!] \tag{5.30b}$$

Substituting these values in Eq. (5.16) and rearranging we get

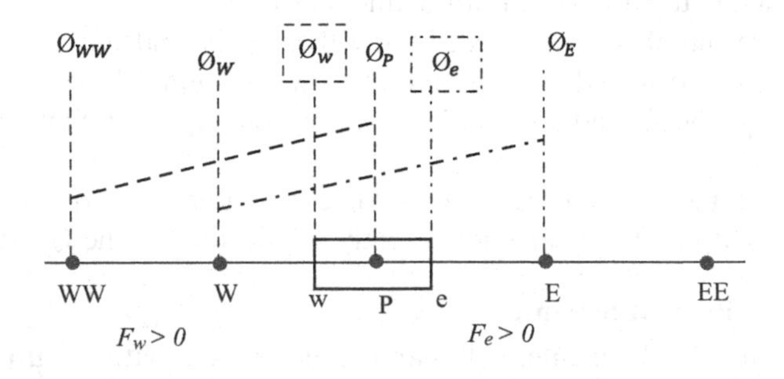

FIGURE 5.17 The FROMM Scheme.

$$\left[\frac{1}{4}[\![F_w,0]\!]\right]\emptyset_{WW}+\left[-[\![F,0]\!]+\frac{1}{4}[\![-F_w,0]\!]-\frac{1}{4}[\![F_e,0]\!]-D_w\right]\emptyset_W$$
$$+\left[[\![F_e,0]\!]-\frac{1}{4}[\![-F_e,0]\!]-\frac{1}{4}[\![F_w,0]\!]+[\![-F_w,0]\!]+D_w+D_e-S_P\right]\emptyset_P$$
$$+\left[-[\![-F_e,0]\!]+\frac{1}{4}[\![F_e,0]\!]-\frac{1}{4}[\![-F_w,0]\!]-D_e\right]\emptyset_E+\left[\frac{1}{4}[\![-F_e,0]\!]\right]\emptyset_{EE}=b_P$$

This can be written as

$$-a_{ww}\emptyset_{WW}-a_w\emptyset_W+a_P\emptyset_P-a_e\emptyset_E-a_{ee}\emptyset_{EE}=b_P, \tag{5.31}$$

where

$$a_w=[\![F_w,0]\!]-\frac{1}{4}[\![-F_w,0]\!]+\frac{1}{4}[\![F_e,0]\!]+D_w;\ a_{ww}=-\frac{1}{4}[\![F_w,0]\!] \tag{5.31b}$$

$$a_e=-[\![F_e,0]\!]-\frac{1}{4}[\![F_e,0]\!]+\frac{1}{4}[\![-F_w,0]\!]+D_e;\ a_{ee}=-\frac{1}{4}[\![-F_e,0]\!] \tag{5.31c}$$

$$a_P=a_{ww}+a_w+a_e+a_{ee}+(F_e-F_w)-S_P. \tag{5.31d}$$

Assessment
This scheme is fully conservative, and it satisfies the transportiveness and boundedness requirement. The scheme is second-order accurate and is stable. However, this scheme like SOU and QUICK scheme cannot be solved by Tri-Diagonal Matrix Algorithm because of the presence of coefficients a_{ww} and a_{ee}.

5.3.1.9 Advantages and Disadvantages of Various Convective Schemes

The advantages and disadvantages of the various convective discretization schemes are shown in Table 5.1.

5.3.2 Deferred Correction Approach

Because of the higher accuracy of higher-order schemes like SOU, QUICK and FROMM, they are widely used with some modifications to correct boundedness problems. We have seen that these schemes can be unstable due to the appearance of negative main coefficients. This can be corrected by modifying the scheme in such a way that the troublesome negative coefficients are placed in source terms. Positive main coefficients are thus retained. We treat the main coefficients implicitly and source term explicitly, known as *deferred correction*. The main idea of the deferred correction approach is to calculate the face value with an upwind term and higher-order terms. The upwind term is treated implicitly, whereas the higher-order terms are taken to the source term and treated explicitly. In this way, the schemes become stable and fast converging variant while

TABLE 5.1

Advantages and Disadvantages of Various Convective Discretization Schemes

Scheme	Comments
Central Differencing (CDS)	This scheme is second-order accurate and works well when diffusion dominates. However, for convective dominated flows, it produces solutions that oscillate. Hence, bounded variants are recommended for LES simulations.
First-order upwind (UDS)	This scheme is first-order accurate and behaves nicely when convection dominates. This scheme is bounded and robust and is a good scheme to start with. It introduces numerical diffusion, which helps make the solution stable at higher Peclet numbers. It should be replaced with higher-order schemes for the final calculations because of lower accuracy.
Second-Order Upwind (SOU)	This scheme is second-order accurate and stable for all Peclet numbers, but it is unbounded and not as robust as first-order upwind.
Power Law	This scheme is good for intermediate values of the Peclet number (Pe < 10).
QUICK	This scheme is good for all Peclet numbers. At low Peclet numbers, the accuracy of this scheme is comparable with CDS. At higher Peclet numbers, it produces oscillations of smaller magnitude. Also, it is unbounded. It has better accuracy than the second-order scheme for rotating or swirling flows.
FROMM	This scheme is second-order accurate. At high Peclet numbers, it produces small oscillations. This scheme is less accurate than QUICK and SOU.

retaining higher-order accuracy. We shall now discuss the deferred correction method as it is applied to various schemes.

5.3.2.1 CDS

$$\emptyset_f = \frac{1}{2}\left(\emptyset_U + \emptyset_D\right)$$

This can be written in deferred correction as

$$\emptyset_f = \emptyset_U + \frac{1}{2}\left(\emptyset_D - \emptyset_U\right)$$

so, we get

$$\emptyset_e = \emptyset_P + \frac{1}{2}\left(\emptyset_E - \emptyset_P\right) \qquad \text{if } F_e > 0$$

$$\emptyset_e = \emptyset_E + \frac{1}{2}\left(\emptyset_P - \emptyset_E\right) \qquad \text{if } F_e < 0$$

$$\emptyset_w = \emptyset_W + \frac{1}{2}\left(\emptyset_P - \emptyset_W\right) \qquad \text{if } F_w > 0$$

$$\emptyset_w = \emptyset_P + \frac{1}{2}\,(\,\emptyset_W - \emptyset_P) \qquad \textit{if } F_w < 0.$$

Combining the flow directions, we get

$$F_e\emptyset_e = \left[\emptyset_P + \frac{1}{2}\,(\,\emptyset_E - \emptyset_P)\right][\![F_e, 0]\!] - \left[\emptyset_E + \frac{1}{2}\,(\emptyset_P - \emptyset_E)\right][\![-F_e, 0]\!] \tag{5.32a}$$

$$F_w\emptyset_w = \left[\emptyset_W + \frac{1}{2}\,(\,\emptyset_P - \emptyset_W)\right][\![F_w, 0]\!] - \left[\emptyset_P + \frac{1}{2}\,(\,\emptyset_W - \emptyset_P)\right][\![-F_w, 0]\!] \tag{5.32b}$$

Substituting these values in Eq. (5.16) and rearranging we get

$$\begin{aligned} -a_w\emptyset_W + a_P\emptyset_P - a_e\emptyset_E &= \bar{b}_P \\ a_P &= a_w + a_e + (F_e - F_w) - S_P \\ a_w = [\![F_w, 0]\!] + D_w;\; a_e &= [\![-F_e, 0]\!] + D_e \\ a_P &= a_w + a_e + (F_e - F_w) - S_P \\ \bar{b}_P &= b_P + b_{CD} \end{aligned} \tag{5.33a}$$

$$\begin{aligned} b_{CD} = &- \left[\frac{1}{2}\,(\,\emptyset_E - \emptyset_P)\right][\![F_e, 0]\!] + \left[\frac{1}{2}\,(\,\emptyset_P - \emptyset_E)\right][\![-F_e, 0]\!] \\ &+ \left[\frac{1}{2}(\emptyset_P - \emptyset_W)\right][\![F_w, 0]\!] - \left[\frac{1}{2}\,(\emptyset_W - \emptyset_P)\right][\![-F_w, 0]\!] \end{aligned} \tag{5.33b}$$

Here the term b_{CD} is treated explicitly.

5.3.2.2 SOU

For SOU the face value is given by

$$\emptyset_f = \frac{3}{2}\emptyset_U - \frac{1}{2}\emptyset_{UU} = \emptyset_U + \frac{1}{2}(\emptyset_U - \emptyset_{UU})$$

or

$$\emptyset_e = \emptyset_P + \frac{1}{2}(\emptyset_P - \emptyset_W) \qquad F_e > 0$$

$$\emptyset_e = \emptyset_E + \frac{1}{2}(\emptyset_E - \emptyset_{EE}\,) \qquad F_e < 0$$

$$\emptyset_w = \emptyset_W + \frac{1}{2}(\emptyset_W - \emptyset_{WW}) \qquad F_w > 0$$

$$\emptyset_w = \emptyset_P + \frac{1}{2}(\emptyset_P - \emptyset_E) \qquad F_w < 0$$

Combining for the flow directions, we get

$$F_e\emptyset_e = \left[\emptyset_P + \frac{1}{2}(\emptyset_P - \emptyset)\right][\![F_e, 0]\!] - \left[\emptyset_E + \frac{1}{2}(\emptyset_E - \emptyset_{EE})\right][\![-F_e, 0]\!] \quad (5.34a)$$

$$F_w\emptyset_w = \left[\emptyset_W + \frac{1}{2}(\emptyset_W - \emptyset_{WW})\right][\![F_w, 0]\!] - \left[\emptyset_P + \frac{1}{2}(\emptyset_P - \emptyset_E)\right][\![-F_w, 0]\!] \quad (5.34b)$$

Hence, putting these values in Eq. (5.16) and rearranging, we get

$$\begin{aligned} -a_w\emptyset_W + a_P\emptyset_P - a_e\emptyset_E &= \bar{b}_P \\ a_P &= a_w + a_e + (F_e - F_w) - S_P \\ a_w = [\![F_w, 0]\!] + D_w;\ a_e &= [\![-F_e, 0]\!] + D_e \\ a_P &= a_w + a_e + (F_e - F_w) - S_P \\ \bar{b}_P &= b_P + b_{SOU} \end{aligned} \quad (5.35a)$$

$$\begin{aligned} b_{SOU} = &-\left[\frac{1}{2}(\emptyset_P - \emptyset_W)\right][\![F_e, 0]\!] + \left[\frac{1}{2}(\emptyset_E - \emptyset_{EE})\right][\![-F_e, 0]\!] \\ &+\left[\frac{1}{2}(\emptyset_W - \emptyset_{WW})\right][\![F_w, 0]\!] - \left[\frac{1}{2}(\emptyset_P - \emptyset_E)\right][\![-F_w, 0]\!] \end{aligned} \quad (5.35b)$$

5.3.2.3 QUICK

Hayase et al. [1992] proposed the following modification for the QUICK scheme:

$$\emptyset_f = \emptyset_U + \frac{1}{8}(3\emptyset_D - 2\emptyset_U - \emptyset_{UU})$$

or

$$\emptyset_w = \emptyset_W + \frac{1}{8}(3\emptyset_P - 2\emptyset_W - \emptyset_{WW}) \qquad \textit{for } F_w > 0$$

$$\emptyset_w = \emptyset_P + \frac{1}{8}(3\emptyset_W - 2\emptyset_P - \emptyset_E) \qquad \textit{for } F_w < 0$$

$$\emptyset_e = \emptyset_P + \frac{1}{8}(3\emptyset_E - 2\emptyset_P - \emptyset_W) \qquad \textit{for } F_e > 0$$

$$\emptyset_e = \emptyset_E + \frac{1}{8}(3\emptyset_P - 2\emptyset_E - \emptyset_{EE}) \qquad \textit{for } F_e < 0$$

Combining the flow directions, we get

$$F_e\emptyset_e = \left[\emptyset_P + \frac{1}{8}(3\emptyset_E - 2\emptyset_P - \emptyset_W)\right][\![F_e, 0]\!] - \left[\emptyset_E + \frac{1}{8}(3\emptyset_P - 2\emptyset_E - \emptyset_{EE})\right][\![-F_e, 0]\!] \quad (5.36a)$$

$$F_w\emptyset_w = \left[\emptyset_W + \frac{1}{8}(3\emptyset_P - 2\emptyset_W - \emptyset_{WW})\right][\![F_w, 0]\!] - \left[\emptyset_P + \frac{1}{8}(3\emptyset_W - 2\emptyset_P - \emptyset_E)\right][\![-F_w, 0]\!] \quad (5.36b)$$

Combining the discretized equation becomes:

$$-a_w\emptyset_W + a_P\emptyset_P - a_e\emptyset_E = \bar{b}_P$$

$$a_P = a_w + a_e + (F_e - F_w) - S_P$$

$$a_w = [\![F_w, 0]\!] + D_w; \quad a_e = [\![-F_e, 0]\!] + D_e \quad (5.37a)$$

$$a_P = a_w + a_e + (F_e - F_w) - S_P$$

$$\bar{b}_P = b_P + b_{QU}$$

$$b_{QU} = -\left[\frac{1}{8}(3\emptyset_E - 2\emptyset_P - \emptyset_W)\right][\![F_e, 0]\!] + \left[\frac{1}{8}(3\emptyset_P - 2\emptyset_E - \emptyset_{EE})\right][\![-F_e, 0]\!] + \left[\frac{1}{8}(3\emptyset_P - 2\emptyset_W - \emptyset_{WW})\right][\![F_w, 0]\!] - \left[\frac{1}{8}(3\emptyset_W - 2\emptyset_P - \emptyset_E)\right][\![-F_w, 0]\!] \quad (5.37b)$$

5.3.2.4 FROMM

For FROMM scheme, the face value is given as

$$\emptyset_f = \emptyset_U + \frac{1}{4}\emptyset_D - \frac{1}{4}\emptyset_{UU} = \emptyset_U + \frac{1}{4}(\emptyset_D - \emptyset_{UU})$$

i.e.

$$\emptyset_e = \emptyset_P + \frac{1}{4}(\emptyset_E - \emptyset_W) \qquad for\ F_e > 0$$

$$\emptyset_e = \emptyset_E + \frac{1}{4}(\emptyset_P - \emptyset_{EE}) \qquad for\ F_e < 0$$

$$\emptyset_w = \emptyset_W + \frac{1}{4}(\emptyset_P - \emptyset_{WW}) \qquad for\ F_w > 0$$

$$\emptyset_w = \emptyset_P + \frac{1}{4}(\emptyset_W - \emptyset_E)\ for \qquad F_w < 0$$

Combining for the flow direction, we get

$$F_e\emptyset_e = \left[\emptyset_P + \frac{1}{4}(\emptyset_E - \emptyset_W)\right][\![F_e, 0]\!] - \left[\emptyset_E + \frac{1}{4}(\emptyset_P - \emptyset_{EE})\right][\![-F_e, 0]\!] \quad (5.38a)$$

$$F_w\emptyset_w = \left[\emptyset_W + \frac{1}{4}(\emptyset_P - \emptyset_{WW})\right][\![F_w, 0]\!] - \left[\emptyset_P + \frac{1}{4}(\emptyset_W - \emptyset_E)\right][\![-F_w, 0]\!] \quad (5.38b)$$

Substituting these values in Eq. (5.16) and rearranging we get

$$-a_w\emptyset_W + a_P\emptyset_P - a_e\emptyset_E = \bar{b}_P$$

$$a_P = a_w + a_e + (F_e - F_w) - S_P$$

$$a_w = [\![F_w, 0]\!] + D_w; a_e = [\![-F_e, 0]\!] + D_e \quad (5.39a)$$

$$a_P = a_w + a_e + (F_e - F_w) - S_P$$

$$\bar{b}_P = b_P + b_{FR}$$

$$\begin{aligned} b_{FR} = &-\left[\frac{1}{4}(\emptyset_E - \emptyset_W)\right][\![F_e, 0]\!] + \left[\frac{1}{4}(\emptyset_P - \emptyset_{EE})\right][\![-F_e, 0]\!] \\ &+\left[\frac{1}{4}(\emptyset_P - \emptyset_{WW})\right][\![F_w, 0]\!] - \left[\frac{1}{4}(\emptyset_W - \emptyset_E)\right][\![-F_w, 0]\!] \end{aligned} \quad (5.39b)$$

So, comparing all the higher-order schemes, we can observe that the basic equation and the coefficients remain the same; only the deferred correction value changes with the convective scheme adapted.

5.3.3 Extension to Two Dimension

The general conservation equation for two-dimension may be written as

$$(flux_e - flux_w) + (flux_n - flux_s) = source$$

The discretized equation in two dimensions can be written as

$$\begin{aligned} &[F_e\emptyset_e - D_e(\emptyset_E - \emptyset_P)] - [F_w\emptyset_w - D_w(\emptyset_P - \emptyset_W)] + [F_n\emptyset_n - D_n(\emptyset_N - \emptyset_P)] \\ &-[F_s\emptyset_s - D_s(\emptyset_P - \emptyset_S)] = b_P + S_P\emptyset_P \end{aligned} \quad (5.40)$$

Like one-dimension problem, we need to approximate the face values for the convective part in terms of nodal values.

5.3.3.1 UDS

For UDS, the face values can be written as

$$\emptyset_e = \emptyset_P \quad \text{if } F_e > 0;\ \emptyset_e = \emptyset_E \quad \text{if } F_e < 0$$

$$\emptyset_w = \emptyset_W \quad \text{if } F_w > 0;\ \emptyset_w = \emptyset_P \quad \text{if } F_w < 0$$

$$\emptyset_n = \emptyset_P \quad \text{if } F_n > 0;\ \emptyset_n = \emptyset_N \quad \text{if } F_n < 0$$

$$\emptyset_s = \emptyset_S \quad \text{if } F_s > 0;\ \emptyset_s = \emptyset_P \quad \text{if } F_s < 0$$

Combining the flow directions, we get

$$F_w\emptyset_w = \emptyset_W[\![F_w, 0]\!] - \emptyset_P[\![-F_w, 0]\!]$$

$$F_e\emptyset_e = \emptyset_P[\![F_e, 0]\!] - \emptyset_E[\![-F_e, 0]\!]$$

$$F_s\emptyset_s = \emptyset_S[\![F_s, 0]\!] - \emptyset_P[\![-F_s, 0]\!]$$

$$F_n\emptyset_n = \emptyset_P[\![F_n, 0]\!] - \emptyset_N[\![-F_n, 0]\!]$$

Putting the values in the discretized equation Eq. (5.40) and combining, we get

$$-a_s\emptyset_S - a_w\emptyset_W + a_P\emptyset_P - a_e\emptyset_E - a_n\emptyset_N = \bar{b}_P$$

or

$$a_P\emptyset_P - \sum_{nb} a_{nb}\emptyset_{nb} = \bar{b}_P$$

$$a_w = [\![F_w, 0]\!] + D_w;\ a_e = [\![-F_e, 0]\!] + D_e$$

$$a_s = [\![F_s, 0]\!] + D_s;\ a_n = [\![-F_n, 0]\!] + D_n \tag{5.41}$$

$$a_P = a_w + a_e + a_n + a_s - S_P + [(F_e - F_w) + (F_n - F_s)]$$

$$\bar{b}_P = b_P + b_{DEF}$$

Where subscript *nb* denotes neighboring points and b_{DEF} is the term arising due to deferred correction.

For UDS, $b_{DEF} = 0$

5.3.3.2 CDS

As explained earlier, the CDS in deferred correction form can be written as

$$\emptyset_e = \emptyset_P + \frac{1}{2}\left(\emptyset_E - \emptyset_P\right) \quad \text{if } F_e > 0; \; \emptyset_e = \emptyset_E + \frac{1}{2}\left(\emptyset_P - \emptyset_E\right) \quad \text{if } F_e < 0$$

$$\emptyset_w = \emptyset_W + \frac{1}{2}\left(\emptyset_P - \emptyset_W\right) \quad \text{if } F_w > 0; \; \emptyset_w = \emptyset_P + \frac{1}{2}\left(\emptyset_W - \emptyset_P\right) \quad \text{if } F_w < 0$$

$$\emptyset_n = \emptyset_P + \frac{1}{2}\left(\emptyset_N - \emptyset_P\right) \quad \text{if } F_n > 0; \; \emptyset_n = \emptyset_N + \frac{1}{2}\left(\emptyset_P - \emptyset_N\right) \quad \text{if } F_n < 0$$

$$\emptyset_s = \emptyset_S + \frac{1}{2}\left(\emptyset_P - \emptyset_S\right) \quad \text{if } F_s > 0; \; \emptyset_s = \emptyset_P + \frac{1}{2}\left(\emptyset_S - \emptyset_P\right) \quad \text{if } F_s < 0$$

Combining the flow directions, we get

$$F_e\emptyset_e = \left[\emptyset_P + \frac{1}{2}\left(\emptyset_E - \emptyset_P\right)\right][\![F_e, 0]\!] - \left[\emptyset_E + \frac{1}{2}\left(\emptyset_P - \emptyset_E\right)\right][\![-F_e, 0]\!]$$

$$F_w\emptyset_w = \left[\emptyset_W + \frac{1}{2}\left(\emptyset_P - \emptyset_W\right)\right][\![F_w, 0]\!] - \left[\emptyset_P + \frac{1}{2}\left(\emptyset_W - \emptyset_P\right)\right][\![-F_w, 0]\!]$$

$$F_n\emptyset_n = \left[\emptyset_P + \frac{1}{2}\left(\emptyset_N - \emptyset_P\right)\right][\![F_n, 0]\!] - \left[\emptyset_N + \frac{1}{2}\left(\emptyset_P - \emptyset_N\right)\right][\![-F_n, 0]\!]$$

$$F_s\emptyset_s = \left[\emptyset_S + \frac{1}{2}\left(\emptyset_P - \emptyset_S\right)\right][\![F_s, 0]\!] - \left[\emptyset_P + \frac{1}{2}\left(\emptyset_S - \emptyset_P\right)\right][\![-F_s, 0]\!]$$

Substituting in the discretized equation, Eq. (5.40) and combining we get the Eq. (5.41), where the coefficients remain same except for deferred correction term b_{DEF}. For CDS

$$b_{DEF} = b_{CD}$$

$$\begin{aligned} b_{CD} = &- \left[\frac{1}{2}(\emptyset_E - \emptyset_P)\right][\![F_e, 0]\!] + \left[\frac{1}{2}(\emptyset_P - \emptyset_E)\right][\![-F_e, 0]\!] + \left[\frac{1}{2}(\emptyset_P - \emptyset_W)\right][\![F_w, 0]\!] \\ &- \left[\frac{1}{2}(\emptyset_W - \emptyset_P)\right][\![-F_w, 0]\!] - \left[\frac{1}{2}(\emptyset_N - \emptyset_P)\right][\![F_n, 0]\!] \\ &+ \left[\frac{1}{2}(\emptyset_P - \emptyset_N)\right][\![-F_n, 0]\!] + \left[\frac{1}{2}(\emptyset_P - \emptyset_S)\right][\![F_s, 0]\!] \\ &- \left[\frac{1}{2}(\emptyset_S - \emptyset_P)\right][\![-F_s, 0]\!] \end{aligned} \tag{5.42}$$

5.3.3.3 SOU

The convective flux at cell faces in deferred correction form can be written as

$$\emptyset_e = \emptyset_P + \frac{1}{2}(\emptyset_P - \emptyset_W) \text{ for } F_e > 0;\ \emptyset_e = \emptyset_E + \frac{1}{2}(\emptyset_E - \emptyset_{EE}) \text{ for } F_e < 0$$

$$\emptyset_w = \emptyset_W + \frac{1}{2}(\emptyset_W - \emptyset_{WW}) \text{ for } F_w > 0;\ \emptyset_w = \emptyset_P + \frac{1}{2}(\emptyset_P - \emptyset_E) \text{ for } F_w < 0$$

$$\emptyset_n = \emptyset_P + \frac{1}{2}(\emptyset_P - \emptyset_S) \text{ for } F_n > 0;\ \emptyset_n = \emptyset_N + \frac{1}{2}(\emptyset_N - \emptyset_{NN}) \text{ for } F_n < 0$$

$$\emptyset_s = \emptyset_S + \frac{1}{2}(\emptyset_S - \emptyset_{SS}) \text{ for } F_s > 0;\ \emptyset_s = \emptyset_P + \frac{1}{2}(\emptyset_P - \emptyset_N) \text{ for } F_s < 0$$

Combining flow directions, we get

$$F_e\emptyset_e = \left[\emptyset_P + \frac{1}{2}(\emptyset_P - \emptyset_W)\right][\![F_e, 0]\!] - \left[\emptyset_E + \frac{1}{2}(\emptyset_E - \emptyset_{EE})\right][\![-F_e, 0]\!]$$

$$F_w\emptyset_w = \left[\emptyset_W + \frac{1}{2}(\emptyset_W - \emptyset_{WW})\right][\![F_w, 0]\!] - \left[\emptyset_P + \frac{1}{2}(\emptyset_P - \emptyset_E)\right][\![-F_w, 0]\!]$$

$$F_n\emptyset_n = \left[\emptyset_P + \frac{1}{2}(\emptyset_P - \emptyset_S)\right][\![F_n, 0]\!] - \left[\emptyset_N + \frac{1}{2}(\emptyset_N - \emptyset_{NN})\right][\![-F_n, 0]\!]$$

$$F_s\emptyset_s = \left[\emptyset_S + \frac{1}{2}(\emptyset_S - \emptyset_{SS})\right][\![F_s, 0]\!] - \left[\emptyset_P + \frac{1}{2}(\emptyset_P - \emptyset_N)\right][\![-F_s, 0]\!]$$

Hence, substituting these values in Eq. (5.40) and rearranging we get Eq. (5.41), where the coefficients remain same, like UDS, except for deferred correction term b_{DEF}. For SOU

$$b_{DEF} = b_{SOU}$$

$$\begin{aligned} b_{SOU} = &- \left[\frac{1}{2}(\emptyset_P - \emptyset_W)\right][\![F_e, 0]\!] + \left[\frac{1}{2}(\emptyset_E - \emptyset_{EE})\right][\![-F_e, 0]\!] + \left[\frac{1}{2}(\emptyset_W - \emptyset_{WW})\right][\![F_w, 0]\!] \\ &- \left[\frac{1}{2}(\emptyset_P - \emptyset_E)\right][\![-F_w, 0]\!] - \left[\frac{1}{2}(\emptyset_P - \emptyset_S)\right][\![F_n, 0]\!] \\ &+ \left[\frac{1}{2}(\emptyset_N - \emptyset_{NN})\right][\![-F_n, 0]\!] + \left[\frac{1}{2}(\emptyset_S - \emptyset_{SS})\right][\![F_s, 0]\!] \\ &- \left[\frac{1}{2}(\emptyset_P - \emptyset_N)\right][\![-F_s, 0]\!] \end{aligned} \tag{5.43}$$

5.3.3.4 QUICK

The convective flux at cell faces in deferred correction form can be written as

$$\emptyset_e = \emptyset_P + \frac{1}{8}(3\emptyset_E - 2\emptyset_P - \emptyset_W)\ for\ F_e > 0;$$

$$\emptyset_e = \emptyset_E + \frac{1}{8}(3\emptyset_P - 2\emptyset_E - \emptyset_{EE})\ for\ F_e < 0$$

$$\emptyset_w = \emptyset_W + \frac{1}{8}(3\emptyset_P - 2\emptyset_W - \emptyset_{WW})\ for\ F_w > 0;$$

$$\emptyset_w = \emptyset_P + \frac{1}{8}(3\emptyset_W - 2\emptyset_P - \emptyset_E)\ for\ F_w < 0.$$

$$\emptyset_n = \emptyset_P + \frac{1}{8}(3\emptyset_N - 2\emptyset_P - \emptyset_S)\ for\ F_n > 0;$$

$$\emptyset_n = \emptyset_N + \frac{1}{8}(3\emptyset_P - 2\emptyset_N - \emptyset_{NN})\ for\ F_n < 0$$

$$\emptyset_s = \emptyset_S + \frac{1}{8}(3\emptyset_P - 2\emptyset_S - \emptyset_{SS})\ for\ F_s > 0;$$

$$\emptyset_s = \emptyset_P + \frac{1}{8}(3\emptyset_S - 2\emptyset_P - \emptyset_N)\ for\ F_s < 0$$

Combining the flow directions, we get

$$F_e\emptyset_e = \left[\emptyset_P + \frac{1}{8}(3\emptyset_E - 2\emptyset_P - \emptyset_W)\right][\![F_e, 0]\!] - \left[\emptyset_E + \frac{1}{8}(3\emptyset_P - 2\emptyset_E - \emptyset_{EE})\right][\![-F_e, 0]\!]$$

$$F_w\emptyset_w = \left[\emptyset_W + \frac{1}{8}(3\emptyset_P - 2\emptyset_W - \emptyset_{WW})\right][\![F_w, 0]\!] - \left[\emptyset_P + \frac{1}{8}(3\emptyset_W - 2\emptyset_P - \emptyset_E)\right][\![-F_w, 0]\!]$$

$$F_n\emptyset_n = \left[\emptyset_P + \frac{1}{8}(3\emptyset_N - 2\emptyset_P - \emptyset_S)\right][\![F_n, 0]\!] - \left[\emptyset_N + \frac{1}{8}(3\emptyset_P - 2\emptyset_N - \emptyset_{NN})\right][\![-F_n, 0]\!]$$

$$F_s\emptyset_s = \left[\emptyset_S + \frac{1}{8}(3\emptyset_P - 2\emptyset_S - \emptyset_{SS})\right][\![F_s, 0]\!] - \left[\emptyset_P + \frac{1}{8}(3\emptyset_S - 2\emptyset_P - \emptyset_N)\right][\![-F_s, 0]\!]$$

Hence, substituting these values in Eq. (5.40) and rearranging we get Eq. (5.41), where the coefficients remain same, like UDS, except for deferred correction term b_{DEF}. For QUICK

$$b_{DEF} = b_{QU}$$

$$b_{QU} = -\left[\frac{1}{8}(3\emptyset_E - 2\emptyset_P - \emptyset_W)\right][\![F_e, 0]\!] + \left[\frac{1}{8}(3\emptyset_P - 2\emptyset_E - \emptyset_{EE})\right][\![-F_e, 0]\!]$$
$$+\left[\frac{1}{8}(3\emptyset_P - 2\emptyset_W - \emptyset_{WW})\right][\![F_w, 0]\!] - \left[\frac{1}{8}(3\emptyset_W - 2\emptyset_P - \emptyset_E)\right][\![-F_w, 0]\!]$$
$$-\left[\frac{1}{8}(3\emptyset_N - 2\emptyset_P - \emptyset_S)\right][\![F_n, 0]\!] + \left[\frac{1}{8}(3\emptyset_P - 2\emptyset_N - \emptyset_{NN})\right][\![-F_n, 0]\!]$$
$$+\left[\frac{1}{8}(3\emptyset_P - 2\emptyset_S - \emptyset_{SS})\right][\![F_s, 0]\!] - \left[\frac{1}{8}(3\emptyset_S - 2\emptyset_P - \emptyset_N)\right][\![-F_s, 0]\!] \quad (5.44)$$

5.3.3.5 FROMM

The convective flux at cell faces in deferred correction form can be written as

$$\emptyset_e = \emptyset_P + \frac{1}{4}(\emptyset_E - \emptyset_W)\ for\ F_e > 0;\ \emptyset_e = \emptyset_E + \frac{1}{4}(\emptyset_P - \emptyset_{EE})\ for\ F_e < 0$$

$$\emptyset_w = \emptyset_W + \frac{1}{4}(\emptyset_P - \emptyset_{WW})\ for\ F_w > 0;\ \emptyset_w = \emptyset_P + \frac{1}{4}(\emptyset_W - \emptyset_E)\ for\ F_w < 0$$

$$\emptyset_n = \emptyset_P + \frac{1}{4}(\emptyset_N - \emptyset_S)\ for\ F_n > 0;\ \emptyset_n = \emptyset_N + \frac{1}{4}(\emptyset_P - \emptyset_{NN})\ for\ F_n < 0$$

$$\emptyset_s = \emptyset_S + \frac{1}{4}(\emptyset_P - \emptyset_{SS})\ for\ F_s > 0;\ \emptyset_s = \emptyset_P + \frac{1}{4}(\emptyset_S - \emptyset_N)\ for\ F_s < 0$$

Combining for the flow direction, we get

$$F_e\emptyset_e = \left[\emptyset_P + \frac{1}{4}(\emptyset_E - \emptyset_W)\right][\![F_e, 0]\!] - \left[\emptyset_E + \frac{1}{4}(\emptyset_P - \emptyset_{EE})\right][\![-F_e, 0]\!]$$

$$F_w\emptyset_w = \left[\emptyset_W + \frac{1}{4}(\emptyset_P - \emptyset_{WW})\right][\![F_w, 0]\!] - \left[\emptyset_P + \frac{1}{4}(\emptyset_W - \emptyset_E)\right][\![-F_w, 0]\!]$$

$$F_n\emptyset_n = \left[\emptyset_P + \frac{1}{4}(\emptyset_N - \emptyset_S)\right][\![F_n, 0]\!] - \left[\emptyset_N + \frac{1}{4}(\emptyset_P - \emptyset_{NN})\right][\![-F_n, 0]\!]$$

$$F_s\emptyset_s = \left[\emptyset_S + \frac{1}{4}(\emptyset_P - \emptyset_{SS})\right][\![F_s, 0]\!] - \left[\emptyset_P + \frac{1}{4}(\emptyset_S - \emptyset_N)\right][\![-F_s, 0]\!]$$

Substituting in discretized equation we get Eq. (5.41) where the coefficients remain same, like UDS, only the deferred correction term b_{DEF} changes which for FROMM scheme is given by

$$b_{DEF} = b_{FR}$$

$$\begin{aligned} b_{FR} = &-\left[\frac{1}{4}(\emptyset_E - \emptyset_W)\right][\![F_e,0]\!] + \left[\frac{1}{4}(\emptyset_P - \emptyset_{EE})\right][\![-F_e,0]\!] + \left[\frac{1}{4}(\emptyset_P - \emptyset_{WW})\right][\![F_w,0]\!] \\ &-\left[\frac{1}{4}(\emptyset_W - \emptyset_E)\right][\![-F_w,0]\!] - \left[\frac{1}{4}(\emptyset_N - \emptyset_S)\right][\![F_n,0]\!] + \left[\frac{1}{4}(\emptyset_P - \emptyset_{NN})\right][\![-F_n,0]\!] \\ &+\left[\frac{1}{4}(\emptyset_P - \emptyset_{SS})\right][\![F_s,0]\!] - \left[\frac{1}{4}(\emptyset_S - \emptyset_N)\right][\![-F_s,0]\!] \end{aligned} \tag{5.45}$$

5.3.4 Extension to Three Dimension

The general conservation equation for three-dimension may be written as

$$(flux_e - flux_w) + (flux_n - flux_s) + (flux_t - flux_b) = source$$

The discretized equation in three-dimension can be written as

$$\begin{aligned} &[F_e\emptyset_e - D_e(\emptyset_E - \emptyset_P)] - [F_w\emptyset_w - D_w(\emptyset_P - \emptyset_W)] + [F_n\emptyset_n - D_n(\emptyset_N - \emptyset_P)] \\ &-[F_s\emptyset_s - D_s(\emptyset_P - \emptyset_S)] + [F_t\emptyset_t - D_t(\emptyset_T - \emptyset_P)] - [F_b\emptyset_b - D_b(\emptyset_P - \emptyset_B)] \\ &= b_P + S_P\emptyset_P \end{aligned} \tag{5.46}$$

We have seen that the extension from one-dimension to two-dimension is straight forward. Similar way it can be extended to three-dimension. Here we shall discuss only about the coefficients.

Combining the flow directions of cell face velocities and substituting them in Eq. (5.46), the discretized equation and the coefficients for various convective schemes will remain same, as in the case for two-dimension, only the deferred correction term will change. The discretized equation and the coefficients for various convective scheme are

$$a_P\emptyset_P - \sum_{nb} a_{nb}\emptyset_{nb} = \bar{b}_P$$

$$a_w = [\![F_w,0]\!] + D_w;\, a_e = [\![-F_e,0]\!] + D_e;\, a_s = [\![F_s,0]\!] + D_s$$

$$a_n = [\![-F_n,0]\!] + D_n;\;\; a_b = [\![F_b,0]\!] + D_b;\, a_t = [\![-F_t,0]\!] + D_t \tag{5.47}$$

$$a_P = a_w + a_e + a_n + a_s + a_t + a_b + [(F_e - F_w) + (F_n - F_s) + (F_t - F_b)] - S_P$$

$$\bar{b}_P = b_P + b_{DEF}$$

5.3.4.1 UDS

$$b_{DEF} = 0$$

5.3.4.2 CDS

$$\begin{aligned} b_{DEF} = b_{CD} = & - \left[\frac{1}{2}(\emptyset_E - \emptyset_P)\right] \llbracket F_e, 0 \rrbracket + \left[\frac{1}{2}(\emptyset_P - \emptyset_E)\right] \llbracket -F_e, 0 \rrbracket \\ & + \left[\frac{1}{2}(\emptyset_P - \emptyset_W)\right] \llbracket F_w, 0 \rrbracket - \left[\frac{1}{2}(\emptyset_W - \emptyset_P)\right] \llbracket -F_w, 0 \rrbracket \\ & - \left[\frac{1}{2}(\emptyset_N - \emptyset_P)\right] \llbracket F_n, 0 \rrbracket + \left[\frac{1}{2}(\emptyset_P - \emptyset_N)\right] \llbracket -F_n, 0 \rrbracket \\ & + \left[\frac{1}{2}(\emptyset_P - \emptyset_S)\right] \llbracket F_s, 0 \rrbracket - \left[\frac{1}{2}(\emptyset_S - \emptyset_P)\right] \llbracket -F_s, 0 \rrbracket \\ & - \left[\frac{1}{2}(\emptyset_T - \emptyset_P)\right] \llbracket F_t, 0 \rrbracket + \left[\frac{1}{2}(\emptyset_P - \emptyset_T)\right] \llbracket -F_t, 0 \rrbracket \\ & + \left[\frac{1}{2}(\emptyset_P - \emptyset_B)\right] \llbracket F_b, 0 \rrbracket - \left[\frac{1}{2}(\emptyset_B - \emptyset_P)\right] \llbracket -F_b, 0 \rrbracket \end{aligned} \tag{5.48}$$

5.3.4.3 SOU

$$\begin{aligned} b_{DEF} = b_{SOU} = & - \left[\frac{1}{2}(\emptyset_P - \emptyset_W)\right] \llbracket F_e, 0 \rrbracket + \left[\frac{1}{2}(\emptyset_E - \emptyset_{EE})\right] \llbracket -F_e, 0 \rrbracket \\ & + \left[\frac{1}{2}(\emptyset_W - \emptyset_{WW})\right] \llbracket F_w, 0 \rrbracket - \left[\frac{1}{2}(\emptyset_P - \emptyset_E)\right] \llbracket -F_w, 0 \rrbracket \\ & - \left[\frac{1}{2}(\emptyset_P - \emptyset_S)\right] \llbracket F_n, 0 \rrbracket + \left[\frac{1}{2}(\emptyset_N - \emptyset_{NN})\right] \llbracket -F_n, 0 \rrbracket \\ & + \left[\frac{1}{2}(\emptyset_S - \emptyset_{SS})\right] \llbracket F_s, 0 \rrbracket - \left[\frac{1}{2}(\emptyset_P - \emptyset_N)\right] \llbracket -F_s, 0 \rrbracket \\ & - \left[\frac{1}{2}(\emptyset_P - \emptyset_B)\right] \llbracket F_t, 0 \rrbracket - \left[\frac{1}{2}(\emptyset_T - \emptyset_{TT})\right] \llbracket -F_t, 0 \rrbracket \\ & + \left[\frac{1}{2}(\emptyset_B - \emptyset_{BB})\right] \llbracket F_b, 0 \rrbracket - \left[\frac{1}{2}(\emptyset_P - \emptyset_T)\right] \llbracket -F_b, 0 \rrbracket \end{aligned} \tag{5.49}$$

5.3.4.4 Quick

$$\begin{aligned} b_{DEF} = b_{QU} = &- \left[\frac{1}{8}(3\emptyset_E - 2\emptyset_P - \emptyset_W)\right][\![F_e, 0]\!] + \left[\frac{1}{8}(3\emptyset_P - 2\emptyset_E - \emptyset_{EE})\right][\![-F_e, 0]\!] \\ &+ \left[\frac{1}{8}(3\emptyset_P - 2\emptyset_W - \emptyset_{WW})\right][\![F_w, 0]\!] - \left[\frac{1}{8}(3\emptyset_W - 2\emptyset_P - \emptyset_E)\right][\![-F_w, 0]\!] \\ &- \left[\frac{1}{8}(3\emptyset_N - 2\emptyset_P - \emptyset_S)\right][\![F_n, 0]\!] + \left[\frac{1}{8}(3\emptyset_P - 2\emptyset_N - \emptyset_{NN})\right][\![-F_n, 0]\!] \\ &+ \left[\frac{1}{8}(3\emptyset_P - 2\emptyset_S - \emptyset_{SS})\right][\![F_s, 0]\!] - \left[\frac{1}{8}(3\emptyset_S - 2\emptyset_P - \emptyset_N)\right][\![-F_s, 0]\!] \\ &- \left[\frac{1}{8}(3\emptyset_T - 2\emptyset_P - \emptyset_B)\right][\![F_t, 0]\!] + \left[\frac{1}{8}(3\emptyset_P - 2\emptyset_T - \emptyset_{TT})\right][\![-F_t, 0]\!] \\ &+ \left[\frac{1}{8}(3\emptyset_P - 2\emptyset_B - \emptyset_{BB})\right][\![F_b, 0]\!] - \left[\frac{1}{8}(3\emptyset_B - 2\emptyset_P - \emptyset_T)\right][\![-F_b, 0]\!] \end{aligned} \tag{5.50}$$

5.3.4.5 FROMM

$$\begin{aligned} b_{DEF} = b_{FR} = &- \left[\frac{1}{4}(\emptyset_E - \emptyset_W)\right][\![F_e, 0]\!] + \left[\frac{1}{4}(\emptyset_P - \emptyset_{EE})\right][\![-F_e, 0]\!] \\ &+ \left[\frac{1}{4}(\emptyset_P - \emptyset_{WW})\right][\![F_w, 0]\!] - \left[\frac{1}{4}(\emptyset_W - \emptyset_E)\right][\![-F_w, 0]\!] \\ &- \left[\frac{1}{4}(\emptyset_N - \emptyset_S)\right][\![F_n, 0]\!] + \left[\frac{1}{4}(\emptyset_P - \emptyset_{NN})\right][\![-F_n, 0]\!] \\ &+ \left[\frac{1}{4}(\emptyset_P - \emptyset_{SS})\right][\![F_s, 0]\!] - \left[\frac{1}{4}(\emptyset_S - \emptyset_N)\right][\![-F_s, 0]\!] \\ &- \left[\frac{1}{4}(\emptyset_T - \emptyset_B)\right][\![F_t, 0]\!] + \left[\frac{1}{4}(\emptyset_P - \emptyset_{TT})\right][\![-F_t, 0]\!] \\ &+ \left[\frac{1}{4}(\emptyset_P - \emptyset_{BB})\right][\![F_b, 0]\!] - \left[\frac{1}{4}(\emptyset_B - \emptyset_T)\right][\![-F_b, 0]\!] \end{aligned} \tag{5.51}$$

5.3.5 High Resolution and Bounded Convective Schemes

So far, we have discussed various convective schemes and have seen that lower-order schemes like UDS, hybrid and power law are unconditionally stable. However, these schemes are highly diffusive when the flow is skewed with respect to grid lines. Higher-order schemes like SOU, QUICK and FROMM were developed to overcome this deficiency and improve accuracy. However, these schemes suffer from boundedness problems. Their solutions lead to oscillation/wiggles in the region of sharp gradient and possibly numerical instability. Hence, higher-order bounded convective schemes were developed to allow for solutions free of oscillations. We shall now discuss higher-order bounded convective schemes. As background, we shall discuss normalized variable formulation (NVF) and convective boundedness criteria (CBC) before discussing high-resolution (HR) schemes.

5.3.5.1 Normalized Variable Formulation (NVF)

NVF was first introduced by Leonard [1987, 1988] and further simplified by Gaskell and Lau [1988] with CBC. NVF is a face formulation in which face value $\emptyset_f$ depends on upwind ($\emptyset_U$), downwind ($\emptyset_D$) and far upstream ($\emptyset_{UU}$) value of nodes as shown in Figure 5.18. It may be noted that many authors used the notations $\emptyset_C$ and $\emptyset_U$ where C is denoted as centre node and U as far upstream node. This is shown in Figure 5.19. Comparing the figures, we can observe that nodes C and U used in Figure 5.19 and U and UU used in Figure 5.18 are the same. Hence, we shall be following the notations used in Figure 5.18 for continuity with earlier sections.

The normalized value can be expressed as

$$\widehat{\emptyset} = \frac{\emptyset - \emptyset_{UU}}{\emptyset_D - \emptyset_{UU}} \quad and \quad \widehat{x} = \frac{x - x_{UU}}{x_D - x_{UU}} \tag{5.52}$$

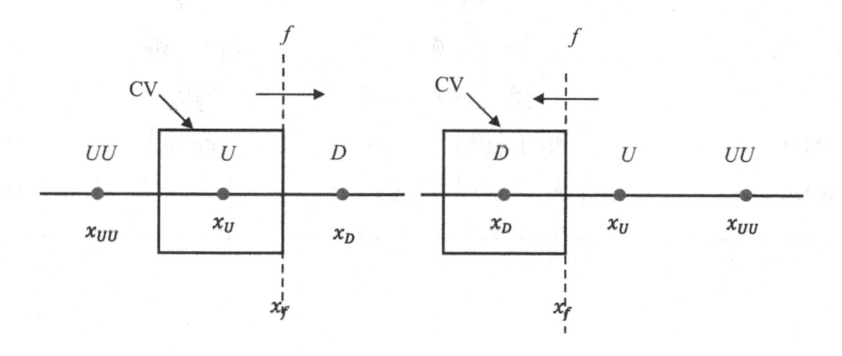

FIGURE 5.18 Notations Used for Normalized Variable Formulation.

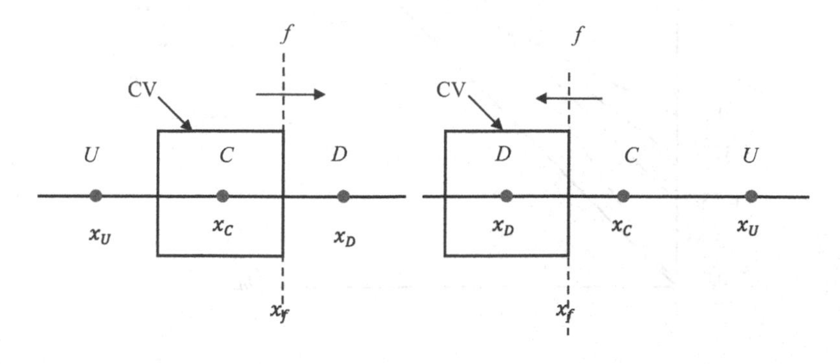

FIGURE 5.19 Notations Used for NVF by other authors.

Now, $\emptyset_f$ is a function of $\emptyset_U$, $\emptyset_D$ and $\emptyset_{UU}$. From Eq. (5.52) we can see that $\widehat{\emptyset}_{UU}=0$ and $\widehat{\emptyset}_D=1$. Similarly, $\widehat{x}_{UU}=0$ and $\widehat{x}_D=1$. Hence, $\widehat{\emptyset}_f=f\left(\widehat{\emptyset}_U,\hat{x}_U,\hat{x}_f\right)$.

Normalized value of $\emptyset_U$ $\left(\widehat{\emptyset}_U\right)$ is an indicator of smoothness of $\emptyset$ field. $0<\widehat{\emptyset}_U<1$ indicated a monotonic profile whereas $\widehat{\emptyset}_U\approx 0$ or $\widehat{\emptyset}_U\approx 1$ indicate a gradient jump. Normalisation is also useful in transforming relations of Higher Order (HO) schemes into linear relations between $\widehat{\emptyset}_f$ and $\widehat{\emptyset}_U$. For example, the approximations of face value with nodal values for various convective schemes for uniform grid are given in Table 5.2.

Thus, we can observe that for HO schemes, $\widehat{\emptyset}_f$ is a linear function of $\widehat{\emptyset}_U$. If we plot the functional relationship between $\widehat{\emptyset}_f$ and $\widehat{\emptyset}_U$, the resultant diagram is known as a normalized variable diagram (NVD) and is shown in Figure 5.20.

TABLE 5.2
Relationships between Normalized Face Value and Nodal Value for a Uniform Grid

Convective Scheme	Value of $\emptyset_f$	Value of $\widehat{\emptyset}_f$	
UDS	$\emptyset_U$	$\widehat{\emptyset}_U$	(5.53)
CDS	$\frac{1}{2}(\emptyset_U+\emptyset_D)$	$\frac{1}{2}\left(1+\widehat{\emptyset}_U\right)$	(5.54)
SOU	$\frac{3}{2}\emptyset_U-\frac{1}{2}\emptyset_{UU}$	$\frac{3}{2}\widehat{\emptyset}_U$	(5.55)
FROMM	$\emptyset_U+\frac{1}{4}(\emptyset_D-\emptyset_{UU})$	$\widehat{\emptyset}_U+\frac{1}{4}$	(5.56)
QUICK	$\frac{3}{4}\,\emptyset_U+\frac{3}{8}\,\emptyset_D-\frac{1}{8}\,\emptyset_{UU}$	$\frac{3}{4}\,\widehat{\emptyset}_U+\frac{3}{8}$	(5.57)

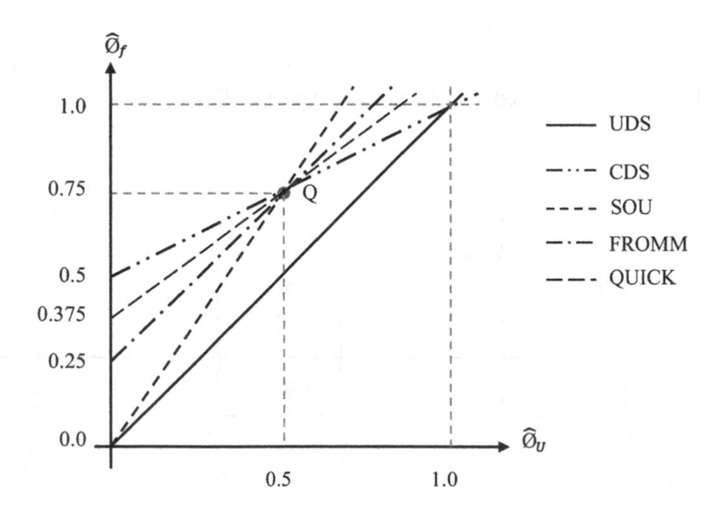

FIGURE 5.20 NVD of Some HO Convective Schemes.

For a uniform grid, all HO schemes pass through the point Q (0.5, 0.75). Leonard [1987] laid down a condition that for a scheme to be third-order accurate, its slope at Q should be 0.75.

For a non-uniform grid, the normalized face values for various HO schemes are given in Table 5.3.

As discussed earlier, HO schemes increase accuracy while retaining stability. However, they suffer from the problem of unboundedness giving rise to oscillations near a sudden jump. This may lead to unsatisfactory results, especially in turbulent flow calculations.

5.3.5.2 Convective Boundedness Criteria (CBC)

Any numerical scheme must reflect the physical phenomena it is trying to approximate. Convective schemes should satisfy the tranportiveness property as convection transports fluid from upstream to downstream. So, while approximating the convective schemes, in addition to nodal values $\emptyset_U$ and $\emptyset_D$, far upstream value $\emptyset_{UU}$ is considered.

The CBC, developed by Gaskell and Lau [1988] for implicit steady-state flow calculation, states that for a convective scheme to have the boundedness property, the functional relationship should be bounded by $\widehat{\emptyset}_U$ and 1.0, and should pass through points (0,0) and (1,1) in the monotonic range $0<\widehat{\emptyset}_U\leq 1$. For $\widehat{\emptyset}_U>1$ or $\widehat{\emptyset}_U<0$, functional relationship $f\left(\widehat{\emptyset}_U\right)$ should be equal to $\widehat{\emptyset}_U$. This is shown as a shaded area in Figure 5.21.

5.3.5.3 High-Resolution (HR) Schemes

Bounded Higher-Order (HR) schemes can be easily constructed by using the NVD. HR schemes were developed to improve convergence behavior. The HR schemes in the monotonic range $0<\widehat{\emptyset}_U<1$ should pass through points (0,0)

TABLE 5.3

Relationships between Normalized Face Value and Nodal Value for a Non-uniform Grid

Convective Scheme	Value of $\widehat{\emptyset}_f$	
UDS	$\widehat{\emptyset}_f=\widehat{\emptyset}_U$	(5.58)
CDS	$\widehat{\emptyset}_f=\dfrac{\widehat{x}_f-\widehat{x}_U}{1-\widehat{x}_U}+\dfrac{\widehat{x}_f-1}{\widehat{x}_U-1}\widehat{\emptyset}_U$	(5.59)
SOU	$\widehat{\emptyset}_f=\dfrac{\widehat{x}_f}{\widehat{x}_U}\widehat{\emptyset}_U$	(5.60)
FROMM	$\widehat{\emptyset}_f=\widehat{\emptyset}_U+\left(\widehat{x}_f-\widehat{x}_U\right)$	(5.61)
QUICK	$\widehat{\emptyset}_f=\widehat{x}_f+\dfrac{\widehat{x}_f(\widehat{x}_f-1)}{\widehat{x}_U(\widehat{x}_U-1)}\left(\widehat{\emptyset}_U-\widehat{x}_U\right)$	(5.62)

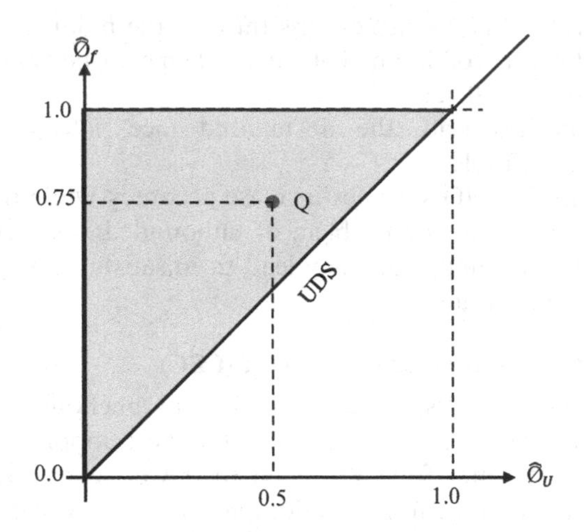

FIGURE 5.21 CBC on NVD.

and (1,1) and remain in the shaded region of the NVD in Figure 5.20. For $\widehat{\emptyset}_U > 1$ or $\widehat{\emptyset}_U < 0$, the profile should follow UDS. Some HR schemes are given in Table 5.4; the NVD of some HR schemes are shown in Figure 5.22.

TABLE 5.4

Some HR Schemes

Convective Schemes	Normalized Value			
MinMod	$\widehat{\emptyset}_f =$	$\frac{3}{2}\widehat{\emptyset}_U$	$0 \leq \widehat{\emptyset}_U \leq \frac{1}{2}$	(5.63)
Roe [1986,] Hirsch [1990]		$\frac{1}{2}\widehat{\emptyset}_U + \frac{1}{2}$	$\frac{1}{2} \leq \widehat{\emptyset}_U \leq 1$	
		$\widehat{\emptyset}_U$	*elsewhere*	
Bounded CD	$\widehat{\emptyset}_f =$	$\frac{1}{2}\widehat{\emptyset}_U + \frac{1}{2}$	$\frac{1}{2} \leq \widehat{\emptyset}_U \leq 1$	(5.64)
		$\widehat{\emptyset}_U$	*elsewhere*	
SMART	$\widehat{\emptyset}_f =$	$\frac{3}{4}\widehat{\emptyset}_U + \frac{3}{8}$	$0 \leq \widehat{\emptyset}_U \leq \frac{5}{6}$	(5.65)
Gaskell and Lau [1988]		1	$\frac{5}{6} \leq \widehat{\emptyset}_U \leq 1$	
		$\widehat{\emptyset}_U$	*elsewhere*	
Modified SMART	$\widehat{\emptyset}_f =$	$3\widehat{\emptyset}_U$	$0 \leq \widehat{\emptyset}_U \leq \frac{1}{6}$	(5.66)
		$\frac{3}{4}\widehat{\emptyset}_U + \frac{3}{8}$	$\frac{1}{6} \leq \widehat{\emptyset}_U \leq \frac{7}{10}$	
		$\frac{1}{3}\widehat{\emptyset}_U + \frac{2}{3}$	$\frac{7}{10} \leq \widehat{\emptyset}_U \leq 1$	
		$\widehat{\emptyset}_U$	*elsewhere*	

(Continued)

TABLE 5.4 (Cont.)

Convective Schemes	Normalized Value			
MUSCL	$\widehat{\emptyset}_f =$	$2\widehat{\emptyset}_U$	$0 \le \widehat{\emptyset}_U \le 1/4$	(5.67)
Van Leer [1979]		$\widehat{\emptyset}_U + \frac{1}{4}$	$\frac{1}{4} \le \widehat{\emptyset}_U \le \frac{3}{4}$	
		1	$\frac{3}{4} \le \widehat{\emptyset}_U \le 1$	
		$\widehat{\emptyset}_U$	*elsewhere*	

5.3.5.4 The TVD Framework

HR convective schemes can be developed using the total variation diminishing (TVD) framework. The total variation (TV) for the convective discretized solutions can be expressed as

$$TV = \sum_i |\emptyset_{i+1} - \emptyset_i| \tag{5.68}$$

where i represents the index of a node in the spatial solution domain. A numerical scheme is said to be TVD if the TV in the solution diminishes with time. For a solution to be stable and non-oscillatory, the desirable property of HO schemes must be monotonicity preserving. A monotonicity-preserving scheme does not create local extrema so that the value of local minimum must be non-decreasing and local maximum must be non-increasing.

As we have seen, UDS is very stable but highly diffusive, whereas second-order CDS is highly dispersive. Hence, the task is to find a convective scheme that will have the stability of UDS and the accuracy of CDS. Let us take the example of CDS for constructing such a scheme. We know for CDS that the cell face value $\emptyset_f$ is given by

$$\emptyset_f = \frac{1}{2}(\emptyset_U + \emptyset_D) = \emptyset_U + \frac{1}{2}(\emptyset_D - \emptyset_U) \tag{5.69}$$

The first term of the right-hand side of Eq. (5.69) is the value of the upwind node, and the last term can be thought of as an anti-diffusive term. Hence, the CDS can be written as the sum of the upwind scheme and an anti-diffusive flux, which makes the scheme second-order accurate. However, this flux leads to undesirable oscillations in the solution due to a decrease in numerical diffusion. One way of avoiding the unphysical oscillations is to use a limiter function or flux limiter, $\Psi(r)$, multiplied with the anti-diffusive flux. With this, the cell face value $\emptyset_f$ can be calculated as

$$\emptyset_f = \emptyset_U + \frac{1}{2}\Psi(r)(\emptyset_D - \emptyset_U) \ with \ r = \frac{\emptyset_U - \emptyset_{UU}}{\emptyset_D - \emptyset_U} \tag{5.70}$$

For preserving the sign of anti-diffusive flux, $\Psi(r)$ is to be positive. Now for UDS we know that $\emptyset_f = \emptyset_U$. Hence, $\Psi(r) = 0$ for UDS.

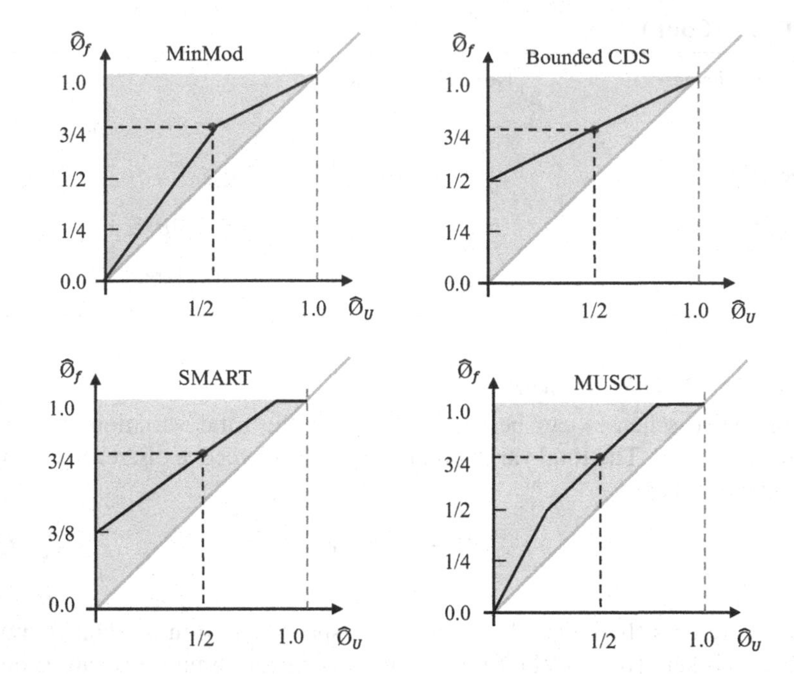

FIGURE 5.22 NVD of Some HR Schemes.

For CDS, $\emptyset_f = \emptyset_U + {}^1\!/_2(\emptyset_U - \emptyset_U)$ which leads to $\Psi(r) = 1$ for CDS.
For SOU, the face value $\emptyset_f$ is given by $\emptyset_f = \emptyset_U + {}^1\!/_2(\emptyset_U - \emptyset_{UU})$
Hence, $${}^1\!/_2(\emptyset_U - \emptyset_{UU}) = {}^1\!/_2\Psi(r)(\emptyset_D - \emptyset_U)$$
or

$$\Psi(r) = \frac{(\emptyset_U - \emptyset_{UU})}{(\emptyset_D - \emptyset_U)} = r$$

Similarly, we can show that for FROMM $\Psi(r) = {}^{(1+r)}\!/_2$ and for QUICK $\Psi(r) = {}^{(3+r)}\!/_4$. The $\Psi(r)$ values for HO schemes are summarized in Table 5.5.

As per Sweby [1984], the necessary and sufficient conditions for a scheme to be TVD are (as shown in Figure 5.23)

- For $0 < r < 1$, the upper limit is $\Psi(r) = 2r$, hence for TVD schemes $\Psi(r) \le 2r$.
- For $r \ge 1$, the upper limit is $\Psi(r) = 2$, so that for TVD schemes $\Psi(r) \le 2$.

Figure 5.24 shows the $r - \Psi$ relationship of various linear HO convective schemes in the TVD region.

TABLE 5.5

The Values of $\Psi(r)$ for Various HO Convective Schemes

Convective Scheme	Values of $\Psi(r)$
a) Linear HO Schemes	
UDS	0
CDS	1
SOU	r
FROMM	$\frac{1+r}{2}$
QUICK	$\frac{3+r}{4}$
b) Non-Linear HO Schemes	
Min Mod	$max(0, min(1, r))$
Van Leer	$\frac{r+\vert r\vert}{1+\vert r\vert}$
MUSCL	$max(0, min(2r, 0.5r + 0.5, 2))$
SMART	$max(0, min(2r, 0.75r + 0.25, 4))$

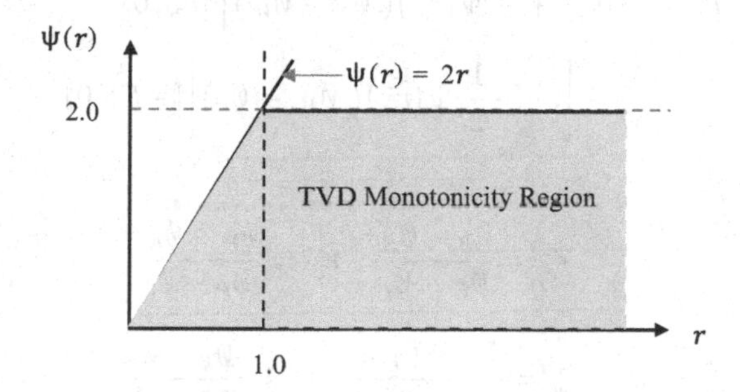

FIGURE 5.23 TVD Monotonicity region in a $r - \Psi$ diagram.

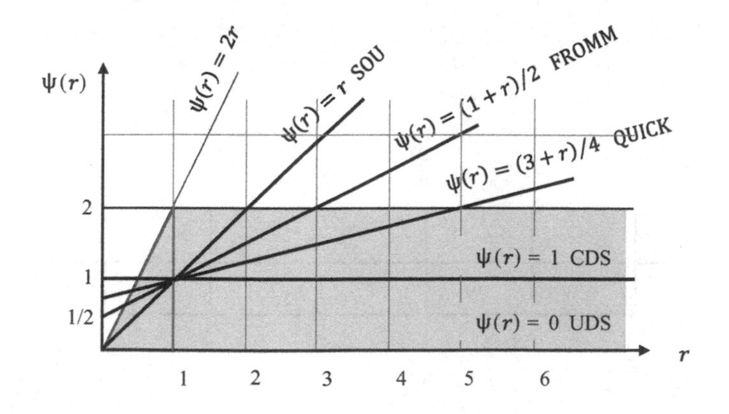

FIGURE 5.24 The $r - \Psi$ Relationship of Various Convective Schemes in the TVD Region.

The figure shows that

- UDS is TVD
- CDS is not TVD for $r<0.5$.
- SOU is not TVD for $r>2$.
- FROMM is not TVD for $r<1/3$ and $r>3$.
- QUICK is not TVD for $r<3/7$ and $r>5$.

5.3.5.5 Implementation of Various Convective Schemes in Code

Let us consider the flow domain in one dimension as shown in Figure 5.25.

The convective fluxes at the faces of control volume after taking the flow directions is given by

$$\begin{aligned} F_e\emptyset_e = &\left[\emptyset_P + \frac{1}{2}\,\Psi\left(r_e^+\right)(\,\emptyset_E - \emptyset_P)\right][\![F_e, 0]\!] \\ &- \left[\emptyset_E + \frac{1}{2}\,\Psi\left(r_e^-\right)(\,\emptyset_P - \emptyset_E)\right][\![-F_e, 0]\!] \end{aligned} \tag{5.71a}$$

$$\begin{aligned} F_w\emptyset_w = &\left[\emptyset_W + \frac{1}{2}\,\Psi\left(r_w^+\right)(\,\emptyset_P - \emptyset_W)\right][\![F_w, 0]\!] \\ &- \left[\emptyset_P + \frac{1}{2}\,\Psi\left(r_w^-\right)(\,\emptyset_W - \emptyset_P)\right][\![-F_w, 0]\!] \end{aligned} \tag{5.71b}$$

where

$$r_e^+ = \frac{\emptyset_P - \emptyset_W}{\emptyset_E - \emptyset_P};\; r_e^- = \frac{\emptyset_E - \emptyset_{EE}}{\emptyset_P - \emptyset_E} \tag{5.72a}$$

$$r_w^+ = \frac{\emptyset_W - \emptyset_{WW}}{\emptyset_P - \emptyset_W};\; r_w^- = \frac{\emptyset_P - \emptyset_E}{\emptyset_W - \emptyset_P} \tag{5.72b}$$

Substituting these values in Eq. (5.16) and rearranging we get

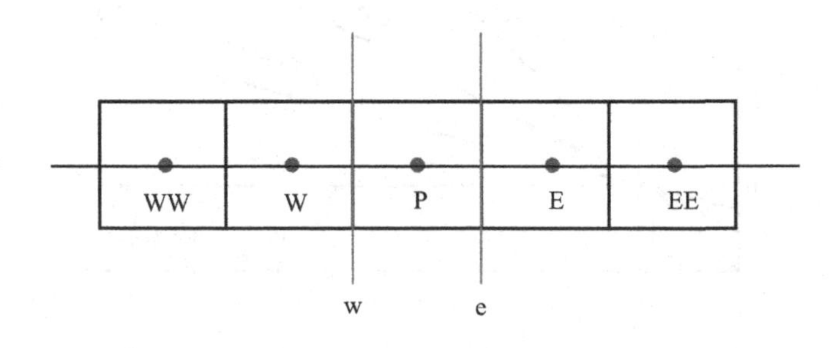

FIGURE 5.25 Convective Fluxes in One Dimension.

$$-a_w\emptyset_W + a_P\emptyset_P - a_e\emptyset_E = \bar{b}_P$$
$$a_P = a_w + a_e + (F_e - F_w) - S_P$$
$$a_w = [\![F_w, 0]\!] + D_w; \; a_e = [\![-F_e, 0]\!] + D_e \tag{5.72c}$$
$$a_P = a_w + a_e + (F_e - F_w) - S_P$$
$$\bar{b}_P = b_P + b_{DEF}$$

$$b_{DEF} = -\left[\frac{1}{2}\,\Psi(r_e^+)(\emptyset_E - \emptyset_P)\right][\![F_e, 0]\!] + \left[\frac{1}{2}\,\Psi(r_e^-)(\emptyset_P - \emptyset_E)\right][\![-F_e, 0]\!]$$
$$+\left[\frac{1}{2}\,\Psi(r_w^+)(\emptyset_P - \emptyset_W)\right][\![F_w, 0]\!] - \left[\frac{1}{2}\,\Psi(r_w^-)(\emptyset_W - \emptyset_P)\right][\![-F_w, 0]\!] \tag{5.72d}$$

Here the term b_{DEF} is treated explicitly.

Similarly, this concept can be easily extended to two and three dimensions.

5.4 TIME-DEPENDENT METHODS

So far, we have discussed the discretization of the steady-state transport equation. The time-dependent scalar-transport equation for a control volume can be given by

$$\frac{d}{dt}(\rho V\emptyset) + \sum_{faces}\left(F\emptyset - \Gamma\frac{\partial\emptyset}{\partial x_j}\right)A = S$$

Earlier, we discussed the method of discretization of the convective, diffusive fluxes and source term. Now, we shall discuss about the discretization of time derivative term.

Let us consider a first-order differential equation

$$\frac{d\emptyset}{dt} = F(t, \emptyset) \tag{5.73}$$

with initial condition $\emptyset(0) = \emptyset_0$, where F is an arbitrary function of t and $\emptyset$.

This is an *initial-value problem* and can be solved with the time marching method. There are two ways we can solve Eq. (5.73) numerically:

- One-step method, in which only the values from the previous time are used
- Multi-step method, in which we use values from several previous times

In this chapter, we shall discuss only one-step methods, as most commercial codes follow these.

5.4.1 One-Step Methods

Let us consider the first-order differential equation

$$\frac{d\emptyset}{dt} = F$$

In the one-step method, the values of $\emptyset$ at new time level, $t^{(n+1)}$ can be found out from the values of $\emptyset$ at old time level $t^{(n)}$ according to the relation

$$\emptyset^{new} = \emptyset^{old} + F\Delta t$$

where

$\emptyset^{new}$ = the value of $\emptyset$ at the new time level $t^{(n+1)}$ which we want to calculate, and

$\emptyset^{old}$ = the value of $\emptyset$ at the previous time level $t^{(n)}$ which we already know.

This time-stepping method is widely used in general-purpose CFD codes. There are three methods by which the values of derivative can be obtained.

5.4.1.1 Forward Differencing (Euler Method)

$$\emptyset^{new} = \emptyset^{old} + F^{old}\Delta t \tag{5.74}$$

This is an explicit method and can be implemented very easily. The values of the right-hand side of Eq. (5.74) are calculated from the old (i.e. previous) time-step level, which is already known. Hence, it is known as the *explicit method*. However, the method is only first-order accurate in time, and time-step restrictions are imposed due to stability considerations, as already discussed in the chapter 4 involving finite difference discretization methods.

5.4.1.2 Backward Differencing (Backward Euler)

$$\emptyset^{new} = \emptyset^{old} + F^{new}\Delta t \tag{5.75}$$

This is an implicit method and there is no time-step restrictions. However, the method is only first order accurate in time and requires solution a system of algebraic equations.

5.4.1.3 Central Differencing (Crank-Nicolson)

$$\emptyset^{new} = \emptyset^{old} + \frac{1}{2}\left(F^{new} + F^{old}\right)\Delta t \tag{5.76}$$

This is second-order accurate in time. However, computational effort is higher than backward Euler.

5.5 TIME DISCRETIZATION METHODS APPLIED TO THE GENERAL SCALAR TRANSPORT EQUATION

Let us now consider how time discretization methods are applied to the general scalar transport equation. The general scalar-transport equation is given by

$$\frac{d}{dt}(\rho V \emptyset_P) + netflux - source = 0$$

In one-step methods, the time derivative is discretized as

$$\frac{d}{dt}(\rho V \emptyset_P) \rightarrow \frac{(\rho V \emptyset_P)^{new} - (\rho V \emptyset_P)^{old}}{\Delta t}$$

Flux and source terms can be discretized as:

$$netflux - source = a_P \emptyset_P - \sum_{nb} a_{nb} \emptyset_{nb} - b_P$$

However, the time level of flux and source term depends on whether the scheme is explicit or implicit.

5.5.1 Forward Differencing – Explicit Scheme

$$\frac{(\rho V \emptyset_P)^{new} - (\rho V \emptyset_P)^{old}}{\Delta t} + \left[a_P \emptyset_P - \sum_{nb} a_{nb} \emptyset_{nb}\right]^{old} = b_P$$

or

$$\frac{\rho V}{\Delta t} \emptyset_P^{new} = \left[\left(\frac{\rho V}{\Delta t} - a_P\right)\emptyset_P + \sum_{nb} a_{nb} \emptyset_{nb}\right]^{old} + b_P \tag{5.77}$$

This method is explicit. Since the values on the right-hand side is known, we do not require simultaneous solution of systems of equation. However, for stability consideration, we require positive coefficient of $\emptyset_p^{old}$ which can be achieved by

$$\frac{\rho V}{\Delta t} - a_P \geq 0$$

5.5.2 Backward Differencing- Implicit Scheme

$$\frac{(\rho V \emptyset_P)^{new} - (\rho V \emptyset_P)^{old}}{\Delta t} + \left[a_P \emptyset_P - \sum_{nb} a_{nb} \emptyset_{nb}\right]^{new} = b_P$$

or

$$\left[\left(\frac{\rho V}{\Delta t}+a_P\right)\emptyset_P-\sum_{nb}a_{nb}\emptyset_{nb}\right]^{new}=b_P+\left(\frac{\rho V}{\Delta t}\emptyset_P\right)^{old} \tag{5.78}$$

As this method is implicit scheme, we require solving the system of equations simultaneously. There is no time step restriction from stability consideration. However, the time step is restricted by the transient phenomena.

The method is straight forward to implement by changing the coefficients as:

$$a_P\rightarrow a_P+\frac{\rho V}{\Delta t}\ and\ b_P\rightarrow b_P+\left(\frac{\rho V}{\Delta t}\emptyset_P\right)^{old}$$

5.5.3 Crank-Nicolson – Central Difference Scheme:

$$\frac{(\rho V\emptyset_P)^{new}-(\rho V\emptyset_P)^{old}}{\Delta t}+\frac{1}{2}\left[a_P\emptyset_P-\sum_{nb}a_{nb}\emptyset_{nb}\right]^{new}+\frac{1}{2}\left[a_P\emptyset_P-\sum_{nb}a_{nb}\emptyset_{nb}\right]^{old}=b_P$$

or

$$\left[\left(\frac{\rho V}{\Delta t}+\frac{1}{2}a_P\right)\emptyset_P-\frac{1}{2}\sum_{nb}a_{nb}\emptyset_{nb}\right]^{new}=b_P+\left[\left(\frac{\rho V}{\Delta t}-\frac{1}{2}a_P\right)\emptyset_P+\frac{1}{2}\sum_{nb}a_{nb}\emptyset_{nb}\right]^{old}$$

Multiplying the above equation by two we get

$$\left[\left(2\frac{\rho V}{\Delta t}+a_P\right)\emptyset_P-\sum_{nb}a_{nb}\emptyset_{nb}\right]^{new}=2b_P+\left[\left(2\frac{\rho V}{\Delta t}-a_P\right)\emptyset_P+\sum_{nb}a_{nb}\emptyset_{nb}\right]^{old} \tag{5.79}$$

This method can be implemented by simply changing the coefficients as

$$a_P\rightarrow a_P+2\frac{\rho V}{\Delta t}\ and\ b_P\rightarrow 2b_P+\left[\left(2\frac{\rho V}{\Delta t}-a_P\right)\emptyset_P+\sum_{nb}a_{nb}\emptyset_{nb}\right]^{old}$$

The method is second order accurate in time and unconditionally stable.

In general, we can use method of weighted averages, as discussed in the finite difference chapter 4 as:

$$\frac{d}{dt}(\rho V\emptyset_P)=\ \beta F^{new}+(1-\beta)F^{old}$$

where $0 \leq \beta \leq 1$.

For, the values of β for various difference schemes are:

Forward Differencing	$\beta = 0$,
Backward Differencing	$\beta = 1$ and
Crank-Nicolson	$\beta = ½$.

This method can be implemented by a simple change in matrix coefficients.

For $\beta \neq 1$, time step restriction imposed by stability condition as

$$\Delta t < \frac{1}{1-\beta}\left(\frac{\rho V}{a_P}\right)^{old}$$

5.6 COURANT NUMBER

The *Courant number c is defined by:*

$$c = \frac{u\Delta t}{\Delta x}$$

This is the ratio of distance of travel at by a fluid particle with speed *u in one-time step ($u\,\Delta t$)* to the mesh spacing Δx.

For the fully explicit method, the Courant-number is restricted by the condition $c < 1$. This means that for one time-step the information advected should not exceed the mesh spacing. For Crank-Nicolson scheme, Courant-number restriction is slightly milder.

5.7 USES OF TIME-MARCHING IN CFD

Time-dependent schemes are used when we are solving the unsteady phenomena like initial value problems. In this case, stability criteria are imposed for finding the time step requirement. In some cases, even though the time step requirements are met from the stability consideration, ultimate time step requirement is obtained from the transient phenomena. Sometimes we may require using C-N method to improve the accuracy of the scheme.

There are some cases like solving parabolic or hyperbolic equations where we need to use time marching procedures. For example, we need to use time marching procedure for transonic calculations involving compressible flows.

Sometimes, even for incompressible steady flows, we use time marching procedure to accelerate the convergence rate. In this case, we can have higher time step as accuracy at each time step is not important.

5.8 IMPLEMENTATION OF BOUNDARY CONDITIONS IN CODE

5.8.1 Generalized Boundary Conditions

The generalized boundary conditions can be specified in generalized form as

$$C_1\emptyset - C_2\frac{\partial\emptyset}{\partial n} = C_3$$

Where, C_1, C_2 and C_3 are constant whose values depend on the type of boundary conditions specified.

i) For Dirichlet boundary condition:

$$C_1 = 1,\ C_2 = 0 \text{ and } C_3 = \textit{specified value}.$$

ii) For Neumann boundary condition

$$C_1 = 0,\ C_2 = 1 \text{ and } C_3 = 0.$$

iii) For Mixed boundary condition (for heat transfer):

$$C_1 = 1,\ C_2 = k \text{ and } C_3 = h\Delta t.$$

5.8.2 Convective Boundary Conditions

For border control volumes, the use of a two-point scheme like first order upwind poses no problem as the interfaces lie on the boundary where the values of $\emptyset$ are specified by the given boundary conditions. The same is true for QUICK scheme. However, at the interface between the border control volume and adjacent interior control volume, same treatment as internal control volumes cannot be applied, if the direction of convective flux is inward from the boundary. For instance, QUICK uses three-point stencil and its implementation at the boundary would require grid points outside the physical domain. See Figure 5.26 for a treatment that can be carried out.

For border control volume (Figure 5.26a) on the west boundary, the west boundary, $\emptyset_w$ is prescribed and hence known. However, for east face $\emptyset_e$ is approximated as

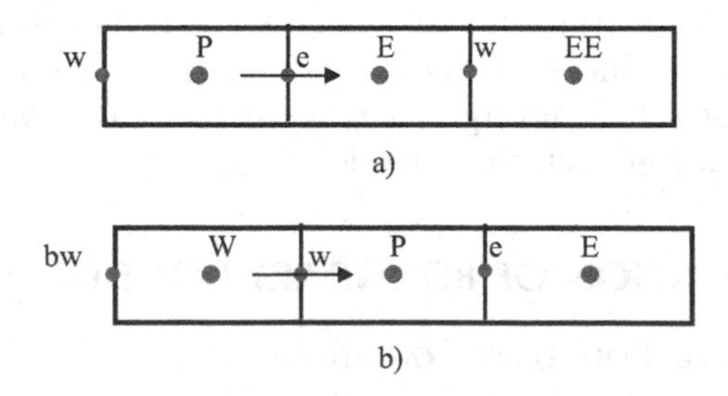

FIGURE 5.26 Boundary Control Volumes: a) Border Control Volume; b) Adjacent Control Volume.

$$\emptyset_e = \emptyset_P + \frac{1}{8}(3\emptyset_E - 2\emptyset_P - \emptyset_w) \qquad for\ F_e > 0$$

$$\emptyset_e = \emptyset_E + \frac{1}{8}(3\emptyset_P - 2\emptyset_E - \emptyset_{EE}) \qquad for\ F_e < 0$$

The corresponding flux is given as

$$F_e\emptyset_e = \left[\emptyset_P + \frac{1}{8}(3\emptyset_E - 2\emptyset_P - \emptyset_w)\right][\![F_e, 0]\!] - \left[\emptyset_E + \frac{1}{8}(3\emptyset_P - 2\emptyset_E - \emptyset_{EE})\right][\![-F_e, 0]\!]$$

For adjacent control volume (Figure 5.26b), the approximation of $\emptyset_w$ for $F_w > 0$ requires the value of $\emptyset_{WW}$ which is outside the boundary. In this case, $\emptyset_{WW}$ can be approximated as $\emptyset_{bw}$, resulting in

$$\emptyset_w = \emptyset_W + \frac{1}{8}(3\emptyset_P - 2\emptyset_W - \emptyset_{bw}) \qquad for\ F_w > 0$$

$$\emptyset_w = \emptyset_P + \frac{1}{8}(3\emptyset_W - 2\emptyset_P - \emptyset_E) \qquad for\ F_w < 0$$

Combining the flow directions, we get

$$F_w\emptyset_w = \left[\emptyset_W + \frac{1}{8}(3\emptyset_P - 2\emptyset_W - \emptyset_{bw})\right][\![F_w, 0]\!] - \left[\emptyset_P + \frac{1}{8}(3\emptyset_W - 2\emptyset_P - \emptyset_E)\right][\![-F_w, 0]\!]$$

5.9 EXAMPLES

Example 5.1 Heat is generated uniformly in a stainless-steel plate (k = 20 W/mK), having a thickness of 1cm. The heat generation rate is 500MW/m^3. The left side of the plate is kept at 200^{0}C; the right side is subjected to convection by a fluid at 100^{0}C with heat transfer coefficient h = 4000W/m^{2}K. Find the temperature distribution inside the plate. Take 4 equal divisions.

Answer: Refer to Figure 5.27.

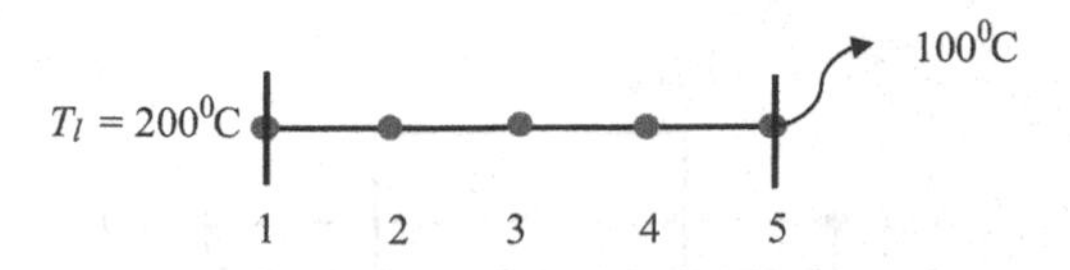

FIGURE 5.27 Problem Definition for Example 5.1.

Given

$$L = 0.01m; \dot{q} = 500 \times 10^6\, W/m^3; k = W/mK; h = 4000\, W/m^2K;$$
$$T_l = 200^0 C; T_\infty = 100^0 C; ndiv = 4$$

Hence,

$$\Delta x = \frac{0.01}{4} = 0.0025;\ Total\ number\ of\ nodes\ nx = 4 + 1 = 5.$$

The governing equation for interior nodes (see Figure 5.28) is

$$\frac{d}{dx}\left(k\frac{dT}{dx}\right) + \dot{q} = 0$$

Integrating over the control volume shown in Figure 5.28 we get

$$\int\limits_{\Delta V} \frac{d}{dx}\left(\Gamma\frac{d\emptyset}{dx}\right) dV + \int\limits_{\Delta V} \dot{q} dV = 0$$

$$\int\limits_{\Delta V} \left(\Gamma\frac{d\emptyset}{dx}\right) dA + \int\limits_{\Delta V} \dot{q} dV = 0$$

$$\left(\Gamma A\frac{d\emptyset}{dx}\right)_e - \left(\Gamma A\frac{d\emptyset}{dx}\right)_w + \dot{q}A\Delta x = 0$$

$$Flux_e - Flux_w + S = 0$$

$$\left(\Gamma A\frac{d\emptyset}{dx}\right)_e = \left(\frac{\Gamma A}{\Delta x}\right)_e (\emptyset_E - \emptyset_P) = D_e(\emptyset_E - \emptyset_P)$$

$$\left(\Gamma A\frac{d\emptyset}{dx}\right)_w = \left(\frac{\Gamma A}{\Delta x}\right)_w (\emptyset_P - \emptyset_W) = D_w(\emptyset_P - \emptyset_W)$$

where

$$D_e = \left(\frac{\Gamma A}{\Delta x}\right)_e \ and\ D_w = \left(\frac{\Gamma A}{\Delta x}\right)_w$$

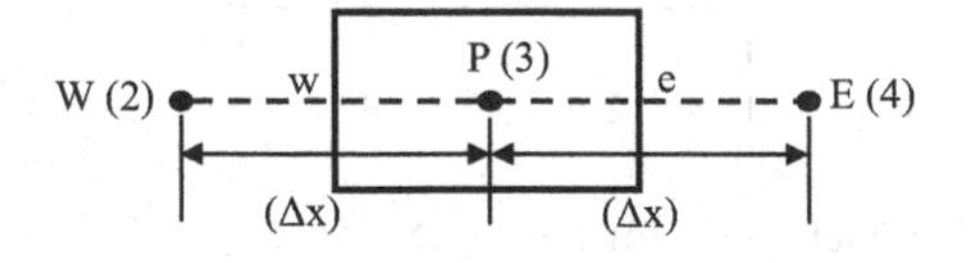

FIGURE 5.28 Interior Control Volume for Example 5.1.

Hence, we get

$$D_w\emptyset_W - (D_w + D_e - S_P)\emptyset_P + D_e\emptyset_E + b_P = 0$$

or

$$-a_w\emptyset_W + a_P\emptyset_P - a_e\emptyset_E = b_P$$

where $a_w = D_w;\ a_e = D_e;\ a_P = D_w + D_e - S_P$ *and* $b_P = \dot{q}A\Delta x$

Since the temperature for node 1 is given, we need discretized equations for nodes 2 to 5.

Node 2:

$$-a_w(2)T_1 + a_P(2)T_2 - a_e(2)T_3 = b_2$$

Since, the temperature of node 1 is known, the above equation becomes

$$a_P(2)T_2 - a_e(2)T_3 = b_2 + a_w(2)T_1$$

Node 3:

$$-a_w(3)T_2 + a_P(3)T_3 - a_e(3)T_4 = b_3$$

Node 4:

$$-a_w(4)T_3 + a_P(4)T_4 - a_e(4)T_5 = b_4$$

Node 5 is the right boundary (Figure 5.29).

Governing equation:

$$-\left(kA\frac{dT}{dx}\right)_w + \dot{q}A\frac{\Delta x}{2} + hA(T_P - T_\infty) = 0$$

Or

$$-a_w\emptyset_W + a_P\emptyset_P - a_e\emptyset_E = b_P$$
$$-a_w(5)T_4 + a_P(5)T_5 = b_5$$

where,

$$a_w = D_w = \frac{kA}{\Delta x};\ a_e = 0;\ a_P = a_w + a_e - s_P;\ s_P = -hA;\ b_P = \dot{q}A\frac{\Delta x}{2} + hAT_\infty$$

Let us assume, $A = 1m^2$
Then we get

$$\frac{kA}{\Delta x} = 8000;\ \dot{q}A\Delta x = 1.25 \times 10^6;\ hA = 4000;\ \dot{q}A\frac{\Delta x}{2} + hAT_\infty = 1.025 \times 10^6$$

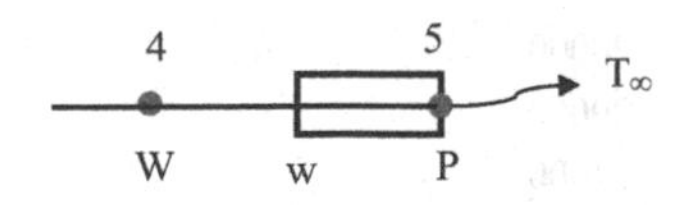

FIGURE 5.29 Right Boundary for Example 5.1.

Table 5.6 gives the calculated values of the coefficients at various nodes.

Solving we get

$$T_2 = 521.9; T_3 = 687.54; T_4 = 696.94; T_5 = 550$$

We can increase the number of divisions to 10. In that case $\Delta x = 0.001\ m$ *and* $nx = 11$.

Then we get

$$\frac{kA}{\Delta x} = 20000; \dot{q}A\Delta x = 0.5 \times 10^6; hA = 4000; \dot{q}A\frac{\Delta x}{2} + hAT_\infty = 0.65 \times 10^6$$

Table 5.7 gives the calculated values of the coefficients at various nodes.

Solving, we get

$$T_2 = 347.49; T_3 = 469.99; T_4 = 567.48; T_5 = 639.94; T_6 = 687.40$$

$$T_7 = 709.88; T_8 = 707.40; T_9 = 679.91; T_{10} = 627.44 \textit{ and } T_{11} = 549.96$$

TABLE 5.6

Values of Coefficients for Various Nodes for Example 5.1 with $\Delta x = 0.0025\ m$

Node	a_w	a_e	S_P	a_P	b_P
2	0	8000	0	16000	2.85×10^6
3	8000	8000	0	16000	1.25×10^6
4	8000	8000	0	16000	1.25×10^6
5	8000	0	- 4000	12000	1.025×10^6

TABLE 5.7

Values of Coefficients for Various Nodes for Example 5.1 with $\Delta x = 0.001\ m$

Node	a_w	a_e	S_P	a_P	b_P
2	0	20000	0	40000	4.5×10^6
3	20000	20000	0	40000	0.5×10^6
4	20000	20000	0	40000	0.5×10^6
5	20000	20000	0	40000	0.5×10^6
6	20000	20000	0	40000	0.5×10^6
7	20000	20000	0	40000	0.5×10^6
8	20000	20000	0	40000	0.5×10^6
9	20000	20000	0	40000	0.5×10^6
10	20000	20000	0	40000	0.5×10^6
11	20000	0	- 4000	24000	0.65×10^6

A comparison of results of with different numbers of nodes is shown in Table 5.8.

Example 5.2 Solve the problem given in Example 5.1 when the left side of the plate is insulated. Find the temperature distribution inside the plate. Take 4 equal divisions.

Answer:
In this problem, the discretized equation for node 2 to 5 will be same as in Example 5.1.

Node 2: $$-a_w(2)T_1 + a_P(2)T_2 - a_e(2)T_3 = b_2$$

Node 3: $$-a_w(3)T_2 + a_P(3)T_3 - a_e(3)T_4 = b_3$$

Node 4: $$-a_w(4)T_3 + a_P(4)T_4 - a_e(4)T_5 = b_4$$

Node 5: $$-a_w(5)T_4 + a_P(5)T_5 = b_5$$

The insulated left boundary is shown in Figure 5.30.

Governing equation:

$$Flux_e - Flux_w + S = 0$$

TABLE 5.8
Comparison of Results of 5 Nodes and 11 Nodes

5 Nodes		11 Nodes	
Distance	**Value**	**Distance**	**Value**
0.0000	0.00	0.000	0.0
0.0025	521.90	0.001	347.49
0.0050	687.54	0.002	469.99
0.0075	696.94	0.003	567.48
0.01	550.00	0.004	639.94
		0.005	687.40
		0.006	709.88
		0.007	707.40
		0.008	679.91
		0.009	627.44
		0.01	549.96

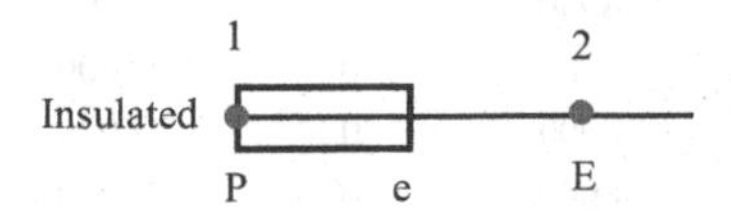

FIGURE 5.30 Left Boundary for Example 5.2.

Now, $Flux_w = 0$; *since the boundary is insulated.*

$$\left(kA\frac{dT}{dx}\right)_e + \dot{q}A\frac{\Delta x}{2} = 0$$

Or

$$-a_w\emptyset_W + a_P\emptyset_P - a_e\emptyset_E = b_P$$

$$a_P(1)T_1 - a_e(1)T_2 = b_1$$

Now

$$\frac{kA}{\Delta x} = 8000; \dot{q}A\Delta x = 1.25 \times 10^6; hA = 4000; \dot{q}A\frac{\Delta x}{2} + hAT_\infty = 1.025 \times 10^6$$

Table 5.9 gives the values of coefficients at different nodes.
Solving, we get

$$T_1 = 2600; T_2 = 2521.9; T_3 = 2287.5; T_4 = 1896.9; T_5 = 1350.$$

Example 5.3 A two-dimensional plate is of length 4cm and height 3cm. It is maintained at temperatures 100^{0}C at the north, 100^{0}C at the east, 0^{0}C at the south and 40^{0}C at the west face as shown in Figure 5.31. The thermal conductivity of the plate material is 20W/mK. Take $\Delta x = \Delta y = 1cm$. Find the temperature of the interior points.

Answer:
Given

$$L = 0.04m; h = 0.03\ cm;\ k = 20\ W/mK;\ \Delta x = \Delta y = 0.01m;$$

$$T_n = 100^0C; T_e = 100^0C;\ T_s = 0^0C;\ T_w = 40^0C$$

Hence,

$$x = \frac{0.04}{0.01} + 1 = 5\ and\ ny = \frac{0.03}{0.01} + 1 = 4.\ Total\ number\ of\ nodes = 5 \times 4 = 20$$

TABLE 5.9

Values of Coefficients for Various Nodes for Example 5.2

Node	a_W	a_e	S_P	a_P	b_P
1	0	8000	0	8000	0.625×10^6
2	8000	8000	0	16000	1.25×10^6
3	8000	8000	0	16000	1.25×10^6
4	8000	8000	0	16000	1.25×10^6
5	8000	0	− 4000	12000	1.025×10^6

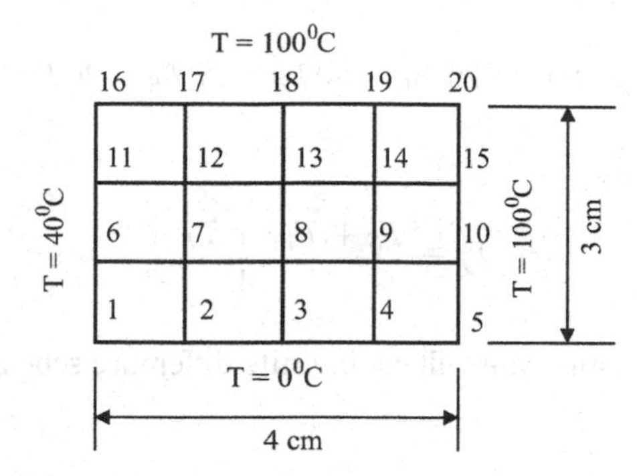

FIGURE 5.31 Problem Definition for Example 5.3.

Since the temperature at boundary nodes are specified, we must find the temperature at nodes 7, 8, 9, 12, 13 and 14 which are interior nodes.

Governing equation:

$$\frac{\partial}{\partial x}\left(k\frac{\partial T}{\partial x}\right)+\frac{\partial}{\partial y}\left(k\frac{\partial T}{\partial y}\right)=0$$

Discretizing the above equation, we get

$$-a_sT_S-a_wT_W+a_PT_P-a_eT_E-a_nT_N=b_P$$

where

$$D_w=\left(\frac{kA}{\Delta x}\right)_w;\ D_e=\left(\frac{kA}{\Delta x}\right)_e;\ D_s=\left(\frac{kA}{\Delta y}\right)_s;D_n=\left(\frac{kA}{\Delta y}\right)_n$$

$$a_w=D_w;\ a_e=D_e;\ a_s=D_s;\ a_n=D_n\ \textit{and}\ a_P=\ D_w+D_e+D_s+D_n-S_P$$

Now, assuming unit width

$$A_e=A_w=\Delta y\times 1=0.01\ m^2\ \textit{and}\ A_n=A_s=\Delta x\times 1=0.01\ m^2;\ b_P=0$$

Hence,

$$a_w=D_w=20;\ a_e=D_e=20;\ a_s=D_s=20;\ a_n=D_n=20\ \textit{and}$$

$$a_P=\ D_w+D_e+D_s+D_n-S_P=80;\ \textit{since}\ S_P=0$$

Hence, we get

$$-20T_S - 20T_W + 80T_P - 20T_E - 20T_N = 0$$

or

$$T_P = \frac{T_S + T_W + T_E + T_N}{4}$$

Note that similar equation you will get in finite difference scheme with equal grids.

So, we get

Node7:

$$T_7 = \frac{T_2 + T_6 + T_8 + T_{12}}{4}$$

Node8:

$$T_8 = \frac{T_3 + T_7 + T_9 + T_{13}}{4}$$

Node9:

$$T_9 = \frac{T_4 + T_8 + T_{10} + T_{14}}{4}$$

Node12:

$$T_{12} = \frac{T_7 + T_{11} + T_{13} + T_{17}}{4}$$

Node13:

$$T_{13} = \frac{T_8 + T_{12} + T_{14} + T_{18}}{4}$$

Node 14:

$$T_{14} = \frac{T_9 + T_{13} + T_{15} + T_{19}}{4}$$

The above equations can be solved iteratively by point Gauss-Seidel method till convergence.

Let us assume the temperatures of these nodes are initially at 20^0C.

The temperatures at the interior nodes against iterations are given in Table 5.10.

We can see from the table that after the 13th iteration, the maximum error in predicting the temperatures is 0.17%.

TABLE 5.10

Temperature at Interior Nodes vs. Iteration

Iteration	1	2	3	4	5	6	7	8	9	10	11	12	13
T7	20.00	20.00	25.00	28.44	31.48	32.89	34.04	34.56	34.99	35.18	35.33	35.40	35.46
T8	20.00	15.00	23.75	30.94	34.30	37.01	38.24	39.23	39.67	40.03	40.20	40.33	40.39
T9	20.00	35.00	43.75	48.13	51.41	52.87	54.04	54.56	54.99	55.18	55.33	55.40	55.46
T12	20.00	45.00	50.00	55.00	57.27	59.16	60.02	60.72	61.03	61.29	61.40	61.50	61.54
T13	20.00	40.00	55.00	60.63	65.16	67.19	68.83	69.57	70.17	70.43	70.65	70.75	70.83
T14	20.00	60.00	68.75	74.69	77.19	79.14	80.01	80.72	81.03	81.29	81.40	81.50	81.54

If we want to solve the problem by the line Gauss-Seidel method, the discretized equation for each node point becomes

Node 7: $$4T_7 - T_8 = T_2 + T_6 + T_{12}$$

Node 8: $$-T_7 + 4T_8 - T_9 = T_3 + T_{13}$$

Node 9: $$-T_8 + 4T_9 = T_4 + T_{10} + T_{14}$$

Node 12: $$4T_{12} - T_{13} = T_7 + T_{11} + T_{17}$$

Node 13: $$-T_{12} + 4T_{13} - T_{14} = T_8 + T_{18}$$

Node 14: $$-T_{13} + 4T_{14} = T_9 + T_{15} + T_{19}$$

Let us assume the temperatures of these nodes are initially 20^{0}C. Then the equations can be solved iteratively by the Thomas algorithm until convergence.

Example 5.4 A property $\emptyset$ is transported by means of convection and diffusion through the one-dimensional domain.

i) Write down the transport equation for steady state
ii) Using 5 equal spaced cells, write down the finite volume discretized equations for each cell using central difference scheme for both convection and diffusion terms and find the value of $\emptyset$ at each node.

Take $u = 0.1$ m/s, Length, $L = 1.0$ m, $\rho = 1.0$ kg/m^3, $\Gamma = 0.1$ kg/m/s, $\emptyset|_{x=0} = 1$ and $\emptyset|_{x=L} = 0$

Answer:
Governing equation:

$$\frac{d}{dx}(\rho u \emptyset) = \frac{d}{dx}\left(\Gamma \frac{d\emptyset}{dx}\right) + S$$

The total flux consists of convective flux ($\rho u \emptyset$) and diffusive flux ($\Gamma\ (d\emptyset)/dx$).
Integrating the fluxes over the control volume

$$\left(\rho Au\emptyset - \Gamma A\frac{\partial\emptyset}{\partial x}\right)_e - \left(\rho Au\emptyset - \Gamma A\frac{\partial\emptyset}{\partial x}\right)_w = S\Delta x$$

Let $F = \rho Au$

Then the above equation becomes

$$[F_e\emptyset_e - D_e(\emptyset_E - \emptyset_P)] - [F_w\emptyset_w - D_w(\emptyset_P - \emptyset_W)] = b_P + S_P\emptyset_P$$

In *central differencing scheme,* we approximate the cell-face value by the average of values at the adjacent nodes, i.e.

$$\emptyset_e = \frac{1}{2}(\emptyset_E + \emptyset_P)\ etc.$$

Substituting we get

$$\frac{1}{2}F_e(\emptyset_E + \emptyset_P) - D_e(\emptyset_E - \emptyset_P) - \frac{1}{2}F_w(\emptyset_W + \emptyset_P) + D_w(\emptyset_P - \emptyset_W) + b_P + S_P\emptyset_P = 0$$

$$-a_w\emptyset_W + a_P\emptyset_P - a_e\emptyset_E = b_P$$

where

$$a_w = \frac{1}{2}F_w + D_w; \quad a_e = -\frac{1}{2}F_e + D_e$$

$$a_P = a_w + a_e - S_P + (F_e - F_w)$$

Given: u = 0.1 m/s, Length, L = 1.0 m, Number of cells = 5, ρ = 1.0 kg/m^3, Γ = 0.1 kg/ms, $\emptyset|_{x=0} = 1$ and $\emptyset|_{x=L} = 0$. Refer to Figure 5.32.

Now the values of node 1 and 6 is known. So, the discretized equation becomes

$$-0.55\emptyset_W + \emptyset_P - 0.45\emptyset_E = 0$$

Node 2:

$$-a_w(2)\emptyset_1 + a_P(2)\emptyset_2 - a_e(2)\emptyset_3 = 0$$

Since $\emptyset_1$ is known, the above equation becomes

$$a_P(2)\emptyset_2 - a_e(2)\emptyset_3 = a_w(2)\emptyset_1$$

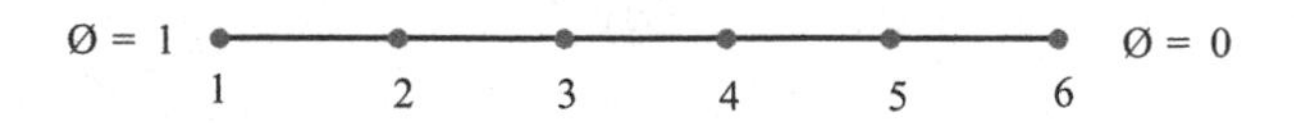

FIGURE 5.32 Problem Definition for Example 5.4.

or $$\emptyset_2 - 0.45\emptyset_3 = 0.55$$

Node 3: $$-0.55\emptyset_2 + \emptyset_3 - 0.45\emptyset_4 = 0$$

Node 4: $$-0.55\emptyset_3 + \emptyset_4 - 0.45\emptyset_5 = 0$$

Node 5: $$-0.55\emptyset_4 + \emptyset_5 - 0.45\emptyset_6 = 0$$

Now $\emptyset_6$ is known, the above equation becomes

$$-0.55T_4 + \emptyset_5 = 0$$

Solving the above equations, we get

$$\emptyset_2 = 0.871; \emptyset_3 = 0.714; \emptyset_4 = 0.522; \emptyset_5 = 0.287$$

The analytical solution is given by

$$\frac{\emptyset - \emptyset_0}{\emptyset_L - \emptyset_0} = \frac{exp\left(\rho ux/_{\Gamma}\right) - 1}{exp\left(\rho uL/_{\Gamma}\right) - 1} \; or \; \emptyset = 1 - 0.582(e^x - 1)$$

A comparison of analytical and computed results is shown in Table 5.11.

Program

Comments: Specify dimensions of aw, ae, SP, ap, bP, b, d, a, c and T. Dimension should be at least equal to number of nodes. Arbitrarily let us assume 100.

Dimension aw(100), ae(100), SP(100), aP(100), bP(100), b(100), d(100), a(100), c(100), phi(100)

Comments: Read the specified values.

Read L, ncell, rho, gamma, phio, phil, u

Comments: Calculate distance between nodes (delx), total number of nodes (nnode), F and D

TABLE 5.11

Comparison of Analytical and Computed Results

Distance	Analytical	Computed
0.0	1.0	1.0
0.2	0.871	0.871
0.4	0.714	0.7138
0.6	0.522	0.5215
0.8	0.287	0.2876
1.0	0.0	0.0

```
delx = L/ncell
nnode = ncell+1
F = rho*u
D = gamma/delx
```

Comments: IMIN is subscript of first equation (2 in this example)
Comments: IMAX is the subscript of last equation (nnode in this case)

```
ip = IMIN+1
IMAX = nnode
phi(1)= phi0
phi(nnode) = phil
```

Comments: Calculate the values of aw, ae, SP, aP, bP, b, d, a, c

```
i = IMIN
do while (i ≤ IMAX-1)
{
  aw(i) = F/2+D
  ae(i) = -F/2+D
  SP(i) = 0.0
  aP(i) = aw(i)+ae(i)-SP(i)
  bP(i)=0.0
  i++
}
```

Comments: Calculate the values of b, d, a, c

```
b(IMIN) = 0.0
d(IMIN) = aP(IMIN)
a(IMIN) = -ae(IMIN)
c(IMIN) – bP(IMIN)+aw(IMIN)*phi(IMIN-1)
i =IMIN+1
do while (i ≤ IMAX-2)
{
  b(i) = -aw(i)
  d(i) = ap(i)
  a(i) = -ae(i)
  c(i) = bP(i)
  i++
}
i = IMAX-1
b(i) = -aw(i)
d(i) = aP(i)
c(i) = bP(i) + ae(i)*phi(i+1)
```

Comments: Solution by Thomas algorithm

```
i =IMIN+1
do while (i ≤ IMAX-1)
```

```
{
   r = b(i)/d(i-1)
   d(i) = d(i) – r*a(i-1)
   c(i) = c(i) – r*c(i-1)
   i++
}
```

Comments: Back substitution

```
iu=IMAX-1
phi(iu) = c(iu)/d(iu)
i =IMIN+1
do while (i ≤ IMAX-1)
{
   j = iu -i +IMIN
   phi(j) = (c(j)- a(j)*phi(j+1))/d(j)
   i++
}
end
```

Example 5.5 A thin plate of thermal conductivity 10W/mK is initially at a uniform temperature of 200^0C. At time t = 0, the right side of the plate is suddenly reduced to 0^0C while the other surface is insulated. Taking 5 equal control volumes, find the temperature of the nodes after 40 seconds using explicit FVM. Take L = 2 cm, ρc = 10×10^6 J/m^3K.

Answer:

Governing equation:

$$\rho c\frac{\partial T}{\partial t}=\frac{\partial}{\partial x}\left(k\frac{\partial T}{\partial x}\right)$$

Initial condition: $T=200^0C\ at\ t=0$

Boundary conditions:

$$\frac{\partial T}{\partial x}=0\ at\ x=0\ and\ t>0$$

$$T=0^0C\ at\ x=L\ and\ t>0$$

Refer to Figure 5.33.

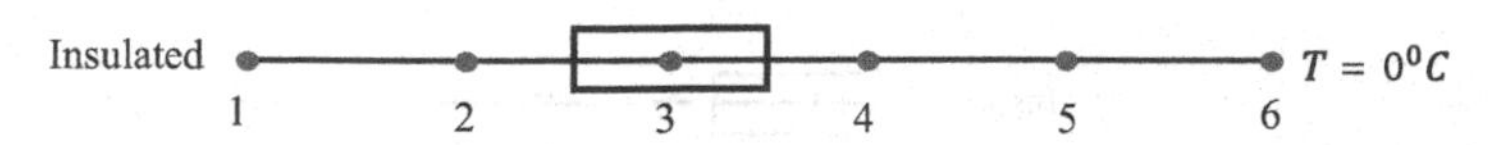

FIGURE 5.33 Problem Definition for Example 5.5.

Integrating over control volume for interior nodes, we get

$$\rho c\frac{(T_P^n - T_P^0)}{\Delta t}\Delta V = \left(kA\frac{\partial T}{\partial x}\right)_e - \left(kA\frac{\partial \emptyset}{\partial x}\right)_w$$

Now, $A = 1$ *and* $\Delta V = \Delta x$.

Hence,

$$\rho c\frac{(T_P^n - T_P^0)}{\Delta t}\Delta x = \left(\frac{k}{\Delta x}\right)_e (T_E^0 - T_P^0) - \left(\frac{k}{\Delta x}\right)_w (T_P^0 - T_W^0)$$

or $$a_P T_P^n = a_w T_W^0 + a_e T_E^0 + \left[a_P^0 - (a_w + a_e)\right] T_P^0 + b_P$$

where

$$a_P = a_P^0 = \frac{\rho C \Delta x}{\Delta t};\ a_w = a_e = \frac{k}{\Delta x}$$

Node 1 is insulated (Figure 5.34).
Node 1 is insulated (Figure 5.34). Hence, $flux_w = 0$
So

$$\rho c\frac{(T_P^n - T_P^0)}{2\Delta t}\Delta x = \left(\frac{k}{\Delta x}\right)_e (T_E^0 - T_P^0)$$

or

$$a_P T_P^n = a_e T_E^0 + \left[a_P^0 - a_e\right] T_P^0 + b_P$$

where

$$a_P = a_P^0 = \frac{\rho C \Delta x}{2\Delta t};\ a_e = \frac{k}{\Delta x}$$

$\Delta x = L/5 = 0.02/5 = 0.004$ and $b_P = 0$.

Now the time step is subject to the condition

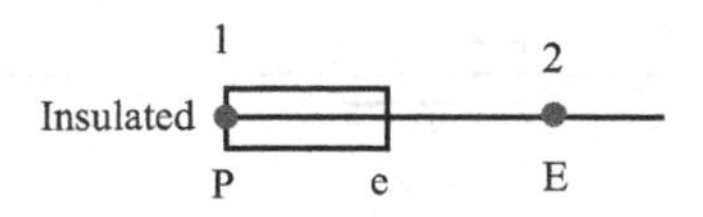

FIGURE 5.34 Left Boundary for Example 5.5.

$$\Delta t < \frac{\rho c \Delta x^2}{2k} < \frac{10 \times 10^6 \times (0.004)^2}{2 \times 10} < 8s$$

Let us assume, $\Delta t = 4\ s$.

Then

$$\frac{\rho C \Delta x}{\Delta t} = \frac{10 \times 10^6 \times 0.004}{4} = 10000 \textit{ and } \frac{k}{\Delta x} = \frac{10}{0.004} = 2500$$

So, we get

Node 1: $5000T_1^n = 2500T_2^0 + 2500T_1^0$

Node 2: $10000T_2^n = 2500T_1^0 + 2500T_3^0 + 5000T_2^0$

Node 3: $10000T_3^n = 2500T_2^0 + 2500T_4^0 + 5000T_3^0$

Node 4: $10000T_4^n = 2500T_3^0 + 2500T_5^0 + 5000T_4^0$

Node 5: $10000T_5^n = 2500T_4^0 + 2500T_6^0 + 5000T_5^0$

Since the temperature of node 6 is given, we get

$$10000T_5^n = 2500T_4^0 + 5000T_5^0$$

Solving the above equations, the temperatures at the nodes after 40s are

$$T_1 = 189.36^0C;\ T_2 = 182.89^0C; T_3 = 161.85^0C; T_4 = 123.29^0C; T_5 = 67.27^0C$$

$$T_6 = 0^0C$$

5.10 SUMMARY

- The generic scalar-transport equation for a control volume has the form

$$\frac{\partial}{\partial t}(\rho\emptyset) + \frac{\partial}{\partial x_j}(\rho u_j\emptyset) = \frac{\partial}{\partial x_j}\left(\Gamma\frac{\partial\emptyset}{\partial x_j}\right) + s$$

- Discretization of the scalar-transport equation yields an equation of form

$$a_P\emptyset_P - \sum_{nb} a_{nb}\emptyset_{nb} = b_p$$

- Source terms are linearized as:

$$\textit{Source},\ S = b_P + S_P\emptyset_P;\ \textit{where } S_P \le 0$$

- Central differencing is used for discretizing the diffusive fluxes.
- For a stable a numerical scheme, the discretization scheme should be: *Consistent, conservative, bounded, stable, transportive.*
- The central differencing of convective term lacks transportiveness and leads to instability at higher cell Peclet number.
- Upwind, hybrid and power-law differencing scheme guarantees conservativeness, boundedness and transportiveness and are highly stable. However, they are of first order accurate and suffer from false diffusion.
- SOU, FROMM, QUICK schemes are higher order schemes and can minimize false diffusion errors. Although they are stable, they suffer from boundedness property leading to solutions with unphysical oscillations in the region of the steep gradient.
- Higher order bounded convective schemes also called High Resolution schemes have been developed for obtaining solutions free of oscillations.
- For a solution to be stable and non-oscillatory, the desirable property of HO schemes must be monotonicity preserving.
- A flux limiter, $\psi(r)$, is used for avoiding the unphysical oscillations. With this, the cell face value $\emptyset_f$ can be calculated as

$$\emptyset_f = \emptyset_U + \frac{1}{2}\psi(r)(\emptyset_D - \emptyset_U) \; with \; r = \frac{\emptyset_U - \emptyset_{UU}}{\emptyset_D - \emptyset_U}$$

- All TVD discretization schemes based on the flux limiter give second order accuracy without any undesirable oscillations, which is essential for a general purpose CFD computation.
- The transient/unsteady equations are solved by time-marching procedure.
- Time-marching schemes may be *explicit* or *implicit.*
- For explicit scheme (Euler forward differencing), stability criteria determine time step requirement and is normally not used due to increase in computational effort.
- Implicit schemes like Euler Backward or Crank-Nicolson differencing is widely used for time dependent fluid flow problems. They are unconditionally stable.
- While Euler forward and Backward differencing are first order accurate in time, Crank-Nicolson differencing is second order accurate. This means for C-N differencing, we need less time steps to achieve same accuracy in time.
- We can easily implement implicit schemes by simply changing the matrix coefficients. For the Backward-Differencing scheme, this can be achieved by:

$$a_P \rightarrow a_P + \frac{\rho V}{\Delta t} \; and \; b_P \rightarrow b_P + \left(\frac{\rho V}{\Delta t}\emptyset_P\right)^{old}$$

QUESTIONS

1) State the desirable properties of a discretization scheme.
2) Propose a general expression for the transport of the entity ∅.
3) Define cell Peclet number and state its significance.
4) What is an upwind scheme? Which type of problem uses upwind scheme?
5) What are the various convective schemes? State the advantages and disadvantages of various methods.
6) Calculate the value of $\Psi(r)$ for FROMM and QUICK scheme.
7) An insulated rod of length 1.0 m and 1.0 cm X 1.0 cm square cross section has its temperature fixed at 100^0C at one end and 650^0C at the other end. Assuming steady state heat conduction in one dimension, find the temperature at each node by finite volume method and compare with the analytical solution. Divide the rod in five equal parts and take conductivity of the material k = 1000 W/mK.
8) Consider a cylindrical fin of length L and with uniform cross-sectional area A. The base is at a temperature of 400^0C and the end is insulated. The fin is exposed to an ambient temperature of 25^0C. Assuming one-dimensional steady state heat transfer

    a. Write the governing equation
    b. Using finite volume method, write the discretized equations for five equally spaced cells.
    c. Take L= 1.0 m, and hP/kA = 25/m^2 where kA is constant.

9) Consider the two-dimensional steady state heat conduction equation

$$\frac{\partial^2 T}{\partial x^2} + \frac{\partial^2 T}{\partial y^2} = 0$$

   Discretize this equation by FVM. How many boundary conditions are needed for solving this equation? Can this be solved by only specifying the Neumann boundary conditions? Explain.

10) A two-dimensional square plate of length 3cm and thickness 1cm is insulated at the west and south faces. The plate is uniformly generating heat at the rate of 90MW/m^3. The east face is maintained at 100^0C, whereas the north face is exposed to air at 20^0C with a heat transfer coefficient of 100W/m^{2}K as shown in Figure 5.35. The thermal conductivity of the plate material is 20W/mK. Take $\Delta x = \Delta y = 1 cm$. Find the temperature of the interior points.
11) Consider one-dimensional steady state equation

$$\frac{d}{dx}(\rho u \emptyset) = \frac{d}{dx}\left(\Gamma \frac{d\emptyset}{dx}\right)$$

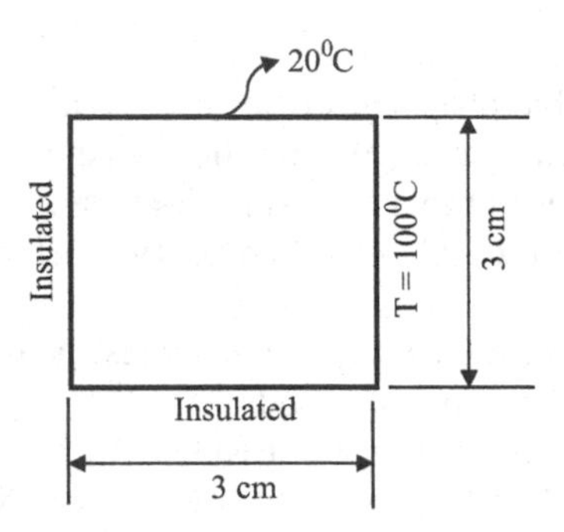

FIGURE 5.35 Problem Definition for Exercise 10.

with $\emptyset|_{x=0} = 1$ and $\emptyset|_{x=L} = 0$. Using 5 equally spaced cells, UDS for convection term and CDS for diffusion term, write down the finite volume discretized equations for each cell and find the values in each node. Compare with the analytical results. If you increase the number of cells to 10, what happens? Take $u = 2$m/s; L = 1.0m; $\rho = 1.0\text{kg/m}^3$; $\Gamma = 0.1$kg/m/s.

12) For the above-mentioned problem 11, can you use central differencing both for advection and diffusion terms? If not, explain why it is so and suggest a method for solving this.
13) Solve the problem given in Example 5.5 by implicit method. Other data remain the same.
14) A large uranium plate of thickness L= 8cm with thermal conductivity k = 20W/mK and thermal diffusivity $\alpha = 12.5 \times 10^{-6}$ m²/s is initially at temperature 100°C. The plate is uniformly generating heat at a constant rate of 1000kW/m³. At time t = 0, the left side of the plate is insulated, and the right side is subjected to convection with a fluid of 30°C with h = 35W/m²K. Take 5 equal cells. Find the temperature distribution inside the plate after 5 minutes using the i) explicit method and ii) implicit method. Compare the results.

6 Solution of Incompressible Navier-Stokes Equations

6.1 INTRODUCTION

In the previous chapter, we discussed the application of the finite volume method for diffusion and advection-diffusion problems. In the advection-diffusion problem, it has been assumed that the velocity field can somehow be determined. However, the user most commonly does not know the velocity from the start and must solve for it as well. Solving the velocity field requires some extra attention, which we will examine now. The Navier-Stokes (N-S) equations are as follows:

$$\frac{\partial}{\partial x}(\rho u) + \frac{\partial}{\partial y}(\rho v) + \frac{\partial}{\partial z}(\rho w) = 0 \tag{6.1}$$

The X-momentum equation:

$$\frac{\partial}{\partial t}(\rho u) + \frac{\partial}{\partial x}(\rho uu) + \frac{\partial}{\partial y}(\rho uv) + \frac{\partial}{\partial z}(\rho uw) = -\frac{\partial p}{\partial x} + \frac{\partial}{\partial x}\left(\mu \frac{\partial u}{\partial x}\right) + \frac{\partial}{\partial y}\left(\mu \frac{\partial u}{\partial y}\right) + \frac{\partial}{\partial z}\left(\mu \frac{\partial u}{\partial z}\right) + f_x \tag{6.2a}$$

Similarly, the Y-momentum equation:

$$\frac{\partial}{\partial t}(\rho v) + \frac{\partial}{\partial x}(\rho uv) + \frac{\partial}{\partial y}(\rho vv) + \frac{\partial}{\partial z}(\rho vw) = -\frac{\partial p}{\partial y} + \frac{\partial}{\partial x}\left(\mu \frac{\partial v}{\partial x}\right) + \frac{\partial}{\partial y}\left(\mu \frac{\partial v}{\partial y}\right) + \frac{\partial}{\partial z}\left(\mu \frac{\partial v}{\partial z}\right) + f_y \tag{6.2b}$$

and Z-momentum equation:

$$\frac{\partial}{\partial t}(\rho w) + \frac{\partial}{\partial x}(\rho uw) + \frac{\partial}{\partial y}(\rho vw) + \frac{\partial}{\partial z}(\rho ww) = -\frac{\partial p}{\partial z} + \frac{\partial}{\partial x}\left(\mu \frac{\partial w}{\partial x}\right) + \frac{\partial}{\partial y}\left(\mu \frac{\partial w}{\partial y}\right) + \frac{\partial}{\partial z}\left(\mu \frac{\partial w}{\partial z}\right) + f_z \tag{6.2c}$$

Each momentum component satisfies its own scalar-transport equation. For one cell, the scalar-transport equations for momentum are given by:

$$\frac{d}{dt}(\rho V\emptyset) + \sum_{faces}\left(F\emptyset - \Gamma\frac{\partial\emptyset}{\partial n}\right)A = s \tag{6.3}$$

where, $\emptyset$ represents velocity components, *u, v or w*; Γ represents diffusivity, μ viscosity, and source, s represents non-viscous forces like pressure forces and body forces. All of these equations are written in 'primitive variable' form.

However, the scalar-transport equations for momentum equations are non-linear, coupled and required to satisfy mass conservation. Thus, the momentum equations are different from other scalar-transport equations.

For example, the *x-momentum flux through an x-directed face is* $FA = (\rho uA)u$. Now, the mass flux F changes with velocity u. Hence, the momentum equation is non-linear and must be solved iteratively. Also, solution of the v equation depends on the solution of the u equation and vice versa. Hence, the momentum equations are coupled, and they must be solved together. In each momentum equation, pressure also appears. This further couples the equations, and we need some means of determining pressure.

For a compressible flow, density is computed from a continuity equation and temperature from the energy (enthalpy) equation; pressure is then computed from the equation of the state $p = p(\rho, T)$.

For incompressible flow, density is constant, and there is no equation for solving pressure. As pressure appears in all three momentum equations, the lack of an independent equation for pressure complicates the solution of the N-S equation. For a given pressure field, the velocities are calculated from the momentum equations. However, the pressure field is not known beforehand. The velocity field must satisfy the continuity equation, and for incompressible flow, the continuity equation becomes a compatibility condition. Even though there is no explicit equation for pressure, we have four equations (i.e. continuity and three momentum equations for four variables *u, v, w and p*). The set of equations can be closed by deriving a pressure equation from the continuity equation. In other words, satisfying mass conservation will lead to a pressure equation. Now we shall find out how velocity and pressure are related and how to obtain the pressure equation.

6.2 PRESSURE-VELOCITY COUPLING

We have already seen that the discretized equation for the convection-diffusion problem is given by

$$a_P u_P - \sum_{nb} a_{nb} u_{nb} = b_P$$

In the momentum equation, pressure forces appear in the source term. Considering control volume as shown in Figure 6.1,

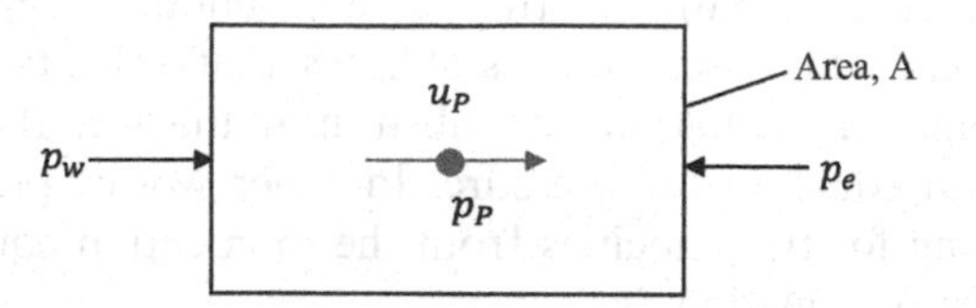

FIGURE 6.1 Notations for Pressure-Velocity Coupling.

Net pressure force = $(p_w - p_e)\,A$.
Hence, the discretized momentum equation takes the form

$$a_P u_P - \sum_{nb} a_{nb} u_{nb} = A(p_w - p_e) + other\,forces \tag{6.4}$$

or,

$$u_P = {A}/{a_P}(p_w - p_e) + \;other\;terms = \;d_P(p_w - p_e) + other\;terms \tag{6.5}$$

where, $d_p = {A}/{a_p}$.
Eq. (6.5) shows how the velocity and pressure are related in the momentum equation. It can be observed that velocity depends on the difference between pressure values at the cell faces (i.e. $p_w - p_e$).
Let us consider Figure 6.2.
The continuity equation is given by $0 = (\rho u A)_e - \; (\rho u A)_w$
From Eq. (6.3) we get

$$u_e = d_e(p_P - p_E)\;and\;u_w = d_w(p_W - p_P).$$

Substituting for velocity in the continuity equation, we get

$$0 = (\rho u A)_e - \;(\rho u A)_w = (\rho A d)_e(p_P - p_E) - \;(\rho A d)_w(p_W - p_P)$$

or

$$-a_w p_W + a_P p_P - a_e p_E = 0 \tag{6.6}$$

where, $a_w = (\rho A d)_w; a_e = (\rho A d)_e \;and\; a_P = \;a_w + a_e$

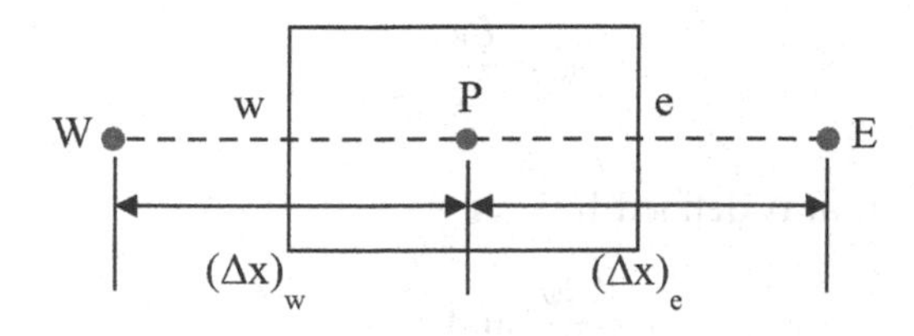

FIGURE 6.2 Typical Control Volume.

We can see that Eq. (6.6) has the same algebraic form as other scalar-transport equations. This discussion establishes that velocity and pressure are related in the momentum equation. By substituting this into the continuity equation, we obtain an equation for pressure. In other words, pressure equation is obtained by solving for the velocities from the momentum equation that satisfy the mass conservation equation.

6.3 THE VORTICITY-STREAM FUNCTION METHOD

Earlier we discussed the problem associated with the solution of the N-S equation in a primitive variable form. The difficulty arises due to determination of the pressure for which no obvious equation exists. This has led to development of the vorticity-stream function method, in which the pressure terms in the momentum equations are eliminated by cross differentiation. This method is applicable only for two dimensions. The main advantage of this method is that we need to solve only two equations to obtain stream function and vorticity. The disadvantages of this methods are

i) the values of vorticity at the wall are difficult to specify;
ii) the determination of the pressure field is very desirable, so we need an equation of pressure in terms of vorticity and stream function
iii) the extension of this method to three dimensions is not possible.

We shall now discuss the method in detail.

In two-dimensional incompressible flow, in the absence of body forces, N-S equations become

$$\frac{\partial u}{\partial x} + \frac{\partial v}{\partial y} = 0 \tag{6.7}$$

$$\frac{\partial u}{\partial t} + u\frac{\partial u}{\partial x} + v\frac{\partial u}{\partial y} = -\frac{1}{\rho}\frac{\partial p}{\partial x} + \nu\left[\frac{\partial^2 u}{\partial x^2} + \frac{\partial^2 u}{\partial y^2}\right] \tag{6.8}$$

$$\frac{\partial v}{\partial t} + u\frac{\partial v}{\partial x} + v\frac{\partial v}{\partial y} = -\frac{1}{\rho}\frac{\partial p}{\partial y} + \nu\left[\frac{\partial^2 v}{\partial x^2} + \frac{\partial^2 v}{\partial y^2}\right] \tag{6.9}$$

For two dimensions, vorticity is given by

$$\xi = \frac{\partial v}{\partial x} - \frac{\partial u}{\partial y} \tag{6.10}$$

Now stream function is defined by

$$u = \frac{\partial \psi}{\partial y} \text{ and } v = -\frac{\partial \psi}{\partial x} \tag{6.11}$$

The pressure term in momentum equations can be eliminated by cross-differentiating momentum equations. Differentiating Eq. (6.8) with respect to y and Eq. (6.9) with respect to x, we get

$$\frac{\partial}{\partial y}\left[\frac{\partial u}{\partial t}+u\frac{\partial u}{\partial x}+v\frac{\partial u}{\partial y}=-\frac{1}{\rho}\frac{\partial p}{\partial x}+\nu\left(\frac{\partial^2 u}{\partial x^2}+\frac{\partial^2 u}{\partial y^2}\right)\right] \tag{6.12a}$$

$$\frac{\partial}{\partial x}\left[\frac{\partial v}{\partial t}+u\frac{\partial v}{\partial x}+v\frac{\partial v}{\partial y}=-\frac{1}{\rho}\frac{\partial p}{\partial y}+\nu\left(\frac{\partial^2 v}{\partial x^2}+\frac{\partial^2 v}{\partial y^2}\right)\right] \tag{6.12b}$$

Subtracting Eqs. (6.12a) from Eq. (6.12b) we get

$$\frac{\partial}{\partial t}\left(\frac{\partial v}{\partial x}-\frac{\partial u}{\partial y}\right)+\frac{\partial u}{\partial x}\left(\frac{\partial v}{\partial x}-\frac{\partial u}{\partial y}\right)+u\frac{\partial}{\partial x}\left(\frac{\partial v}{\partial x}-\frac{\partial u}{\partial y}\right)+\frac{\partial v}{\partial y}\left(\frac{\partial v}{\partial x}-\frac{\partial u}{\partial y}\right)+v\frac{\partial}{\partial y}\left(\frac{\partial v}{\partial x}-\frac{\partial u}{\partial y}\right)$$
$$=\nu\left[\frac{\partial^2}{\partial x^2}\left(\frac{\partial v}{\partial x}-\frac{\partial u}{\partial y}\right)+\frac{\partial^2}{\partial y^2}\left(\frac{\partial v}{\partial x}-\frac{\partial u}{\partial y}\right)\right]$$

Using the definition of vorticity, we get the vorticity transport equation which is parabolic.

$$\frac{\partial \xi}{\partial t}+u\frac{\partial \xi}{\partial x}+v\frac{\partial \xi}{\partial y}=\nu\left(\frac{\partial^2 \xi}{\partial x^2}+\frac{\partial^2 \xi}{\partial y^2}\right) \tag{6.13}$$

Again, substituting the values of u and v in terms of stream function in Eq. (6.10) we get following elliptic (Poisson) equation

$$\frac{\partial^2 \psi}{\partial x^2}+\frac{\partial^2 \psi}{\partial y^2}=-\xi \tag{6.14}$$

The two equations are coupled because of the appearance of u and v in the vorticity transport equation. These are related to the stream function, and vorticity acts as source term of the Poisson equation. Velocity components can be obtained from Eq. (6.11). It is desirable to know the pressure field. This is obtained by solving the pressure Poisson equation (PPE). The PPE can be obtained by differentiating Eq. (6.8) with respect to x and Eq. (6.9) with respect to y and combining them. Hence, we get

$$\frac{\partial}{\partial t}\left(\frac{\partial u}{\partial x}+\frac{\partial v}{\partial y}\right)+\left(\frac{\partial u}{\partial x}\right)^2+\left(\frac{\partial v}{\partial y}\right)^2+2\frac{\partial v}{\partial x}\frac{\partial u}{\partial y}+u\left(\frac{\partial^2 u}{\partial x^2}+\frac{\partial^2 v}{\partial x\partial y}\right)+v\left(\frac{\partial^2 u}{\partial x\partial y}+\frac{\partial^2 v}{\partial y^2}\right)$$
$$=-\frac{1}{\rho}\left(\frac{\partial^2 p}{\partial x^2}+\frac{\partial^2 p}{\partial y^2}\right)+v\left[\frac{\partial}{\partial x}\left(\frac{\partial^2 u}{\partial x^2}+\frac{\partial^2 u}{\partial y^2}\right)+\frac{\partial}{\partial y}\left(\frac{\partial^2 v}{\partial x^2}+\frac{\partial^2 v}{\partial y^2}\right)\right]$$

Simplifying, one obtains

$$\left(\frac{\partial u}{\partial x}\right)^2+\left(\frac{\partial v}{\partial y}\right)^2+2\frac{\partial v}{\partial x}\frac{\partial u}{\partial y}=-\frac{1}{\rho}\left(\frac{\partial^2 p}{\partial x^2}+\frac{\partial^2 p}{\partial y^2}\right)$$

or

$$\nabla^2 p=2\rho\left(\frac{\partial u}{\partial x}\frac{\partial v}{\partial y}-\frac{\partial v}{\partial x}\frac{\partial u}{\partial y}\right)$$

In terms of stream function, the above equation can be written as

$$\nabla^2 p=2\rho\left[\frac{\partial^2 \psi}{\partial x^2}+\frac{\partial^2 \psi}{\partial y^2}-\left(\frac{\partial^2 \psi}{\partial x \partial y}\right)^2\right] \tag{6.15}$$

PPE is an elliptic equation. A steady-state problem needs to be solved once after solving for ξ and ψ . Hence, the solution procedure is

1) Specify initial values of the velocity field, ξ and ψ.
2) Solve vorticity transport equation (6.13) for $t+\Delta t$.
3) Solve elliptic equation (6.14) to obtain ψ.
4) Update velocity field using relations given in Eq. (6.11).
5) Return to step 2 if results are not converged.

Now solution of Eqs. (6.13) and (6.14) require proper boundary conditions for ξ and ψ as they affect the stability and accuracy of the solution.

6.3.1 Boundary Conditions

For wall boundaries, $u=0$ *and* $v=0$. Hence, ψ is constant along the walls. Since the flow is parallel to the solid wall, wall can be treated as streamlines and the constant can be set to zero.

The vorticity values at the walls is found out from stream function. As stream function at the wall is constant, all the derivatives of stream function along the wall is zero. Hence, from Eq. (6.14) we get

$$\xi_{wall}=-\frac{\partial^2 \psi}{\partial x^2} \text{ along left and right boundaries.}$$

$$\xi_{wall}=-\frac{\partial^2 \psi}{\partial y^2} \text{ along top and bottom boundaries.}$$

For solving Pressure Poisson Equation, we need to know the pressure at the wall. This can be obtained from the tangential momentum equation applied to the fluid element adjacent to the wall. This will lead to

$$\left(\frac{\partial p}{\partial s}\right)_{wall} = \mu\left(\frac{\partial^2 u_n}{\partial n^2}\right)_{wall} = -\mu\left(\frac{\partial \xi}{\partial n}\right)_{wall}$$

where n is the normal direction to the surface and s is tangential to the surface.

The vorticity-stream function method was widely used earlier. However, as discussed, this method is applicable only for two-dimensional cases, and most flow is three-dimensional in nature. Extension of this approach to three dimension is not straightforward. Hence, the primitive variable approach is widely used today.

6.4 PRIMITIVE VARIABLE METHODS

N-S equations are described in Eqs. (6.1) and (6.2). In these equations, the dependable variables are velocities u, v, w and pressure p. These values can be directly measured from laboratory experiments and determined by solving those equations. Hence, they are called "*primitive*" variables. As discussed, we can obtain the velocities from momentum equations. Although there is no direct equation for pressure, the pressure equation can be indirectly obtained by enforcing mass conservation. Today, the primitive variable form of the N-S equations is widely used, even in commercial software.

The problem of pressure-velocity coupling is specific to the incompressible N-S equations while for the compressible flow situation this problem is non-existent. The following sub-sections shall explore the problems associated with grid structure and the various methods available for solving the primitive form of N-S equations.

6.4.1 Co-located Storage of Variables

We shall now find the merits and demerits of specifying velocity and pressure at the same node (co-located) or at different nodes (staggered). Let us assume

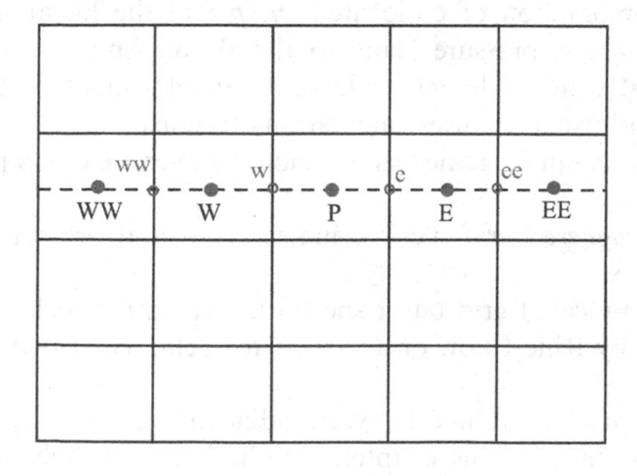

FIGURE 6.3 Co-located Storage of Variables.

that pressure and velocity are *co-located* (i.e. stored at the same positions P), as shown in Figure 6.3.

Now we have $u_P = d_P(p_w - p_e) + \textit{other terms}$

To find the velocity u_P, we need the values of p_w and p_e. The pressure values at cell faces e and w can only be obtained by interpolation from the nodal values, as the face values are not known. If we use linear interpolation, we get

$$(p_w - p_e) = \frac{1}{2}(p_W + p_P) - \frac{1}{2}(p_P + p_E) = \frac{1}{2}(p_W - p_E)$$

Hence,

$$u_P = \frac{1}{2}d_P(p_W - p_E) + \textit{other terms}$$

In the continuity equation, the net outward mass flux depends on u_w and u_e. Again, if we carry out linear interpolation for obtaining the values of u_w and u_e, we get

$$(u_e - u_w) = \frac{1}{2}(u_P + u_E) - \frac{1}{2}(u_W + u_P) = \frac{1}{2}(u_E - u_W)$$

Now,

$$u_E = \frac{1}{2}d_E(p_P - p_{EE}); \textit{and } u_W = \frac{1}{2}d_W(p_{WW} - p_P)$$

Hence,

$$(u_e - u_w) = \frac{1}{2}(u_E - u_W) = \frac{1}{4}[d_E(p_P - p_{EE}) - d_W(p_{WW} - p_P)]$$

Thus, we can see that both mass and momentum equations produce coupling between pressures and velocities at alternate nodes, leading to odd-even decoupling. This may lead to the existence of a wavy pressure/velocity field, which is unrealistic and will make the solution diverge.

Thus, the combination of co-located u, p *and* the linear interpolation of cell face velocity and pressure leads to the decoupling of odd nodal values from even nodal values. This effect leads to indeterminate oscillations in the pressure field and usually causes calculations to crash.

There are two common remedies available to overcome this problem:

1) Using a *staggered grid* (i.e. storing velocity and pressure at different locations) or
2) Using a co-located grid but a special interpolation technique like that developed by Rhie-Chow or a variant for cell-face velocities.

Both methods provide a link between adjacent pressure nodes, preventing odd-even decoupling. In this chapter, we shall discuss only the first option. The co-located grid will be discussed in Chapter 7.

6.4.2 Staggered Grid (Harlow and Welch [1965])

In the staggered-velocity-grid arrangement, while the pressures are stored at the centre of the cell, the velocity components are stored at the cell as shown in Figure 6.4.

This arrangement leads to different sets of control volumes for pressure and velocities. Other scalars are stored at the same position as pressure.

Now let us consider Figure 6.5 in which the velocities are stored at points w, e, etc. (i.e. at the cell faces) whereas the pressures are stored at W, P, E, etc.

Hence, from the momentum equation, we can write $u_w = d_w(p_W - p_P)$.

That is, the velocity at point w (west cell face) is driven by the pressures at nodes W and P. As we saw earlier, the continuity equation can be expressed as

$$u_e - u_w = d_e(p_P - p_E) - d_w(p_W - p_P)$$

Hence, no interpolation is required for cell-face values in either case; there is a strong linkage between pressure and velocity nodes, thus avoiding odd-even decoupling. The advantage of using the staggered grid is that we do not require any interpolation, as variables are stored where they are needed. This avoids the problem of odd-even pressure decoupling.

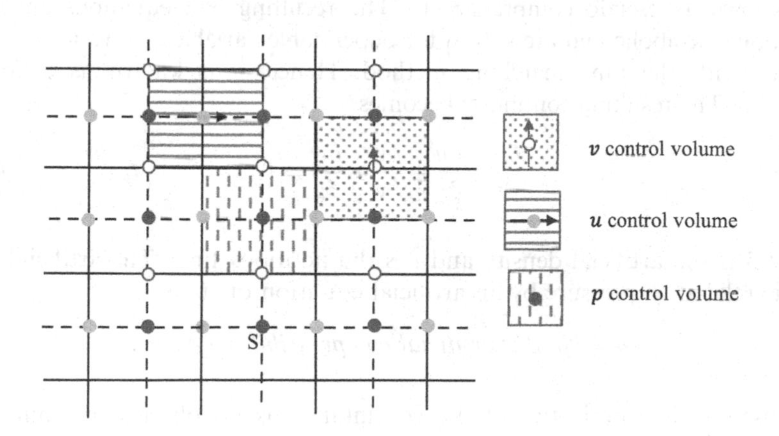

FIGURE 6.4 The Staggered Grid Arrangement.

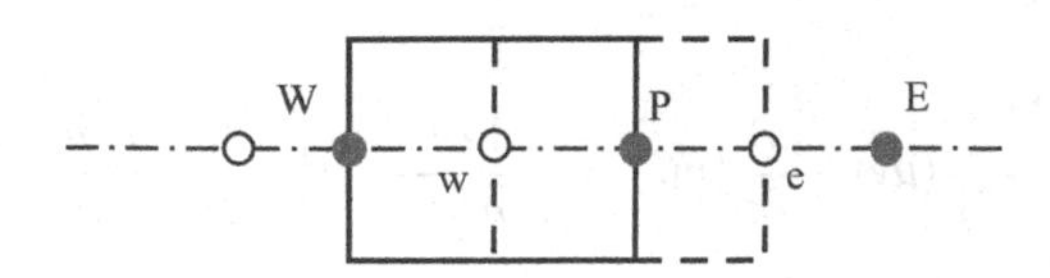

FIGURE 6.5 Pressure and Velocity Locations.

The staggered-grid system has the following disadvantages:

- We get four sets of control volumes for u, v, w and p, thereby requiring additional computational complexity for calculating the geometrical details.
- For meshes other than Cartesian, the velocity nodes may not lie between the pressure nodes that drive them.

6.5 SOLUTION METHODS FOR THE PRIMITIVE VARIABLE FORM OF N-S EQUATIONS

Methods of solution for N-S equations in primitive variable form can be grouped in two categories:

1) The artificial compressibility method (Chorin [1967])
2) The pressure correction approach

6.5.1 The Artificial Compressibility Method

In incompressible N-S equations, there is no obvious equation for pressure. Chorin [1967] modified the continuity by introducing an artificial compressibility, also known as pseudo compressibility. The resulting N-S equations are mixed hyperbolic-parabolic and are solved for dependable variables (u, v, w, p) simultaneous with the time marching method. Hence, it is known as a coupled approach. The resulting continuity becomes

$$\frac{\partial \tilde{\rho}}{\partial \tilde{t}} + \frac{\partial u}{\partial x} + \frac{\partial v}{\partial y} + \frac{\partial w}{\partial z} = 0 \tag{6.16}$$

where $\tilde{\rho}$ is the artificial density and $\tilde{t}$ is the fictitious time. The artificial density $\tilde{\rho}$ is related to pressure by an artificial equation of state

$$p = \beta\tilde{\rho};\ \beta \textit{ is artificial compressibility } (AC). \tag{6.17}$$

The pseudo-time derivative of the continuity has no physical meaning until we obtain the steady-state solution. Hence, this method is applicable only for steady flow. The optimum value of β is somewhat problem dependent.

For time-accurate solutions of N-S equations, the pseudo-time derivative is also introduced in the momentum equations. Thus, the momentum equations become

$$\frac{\partial u}{\partial \tilde{t}} + \frac{\partial}{\partial x}(uu) + \frac{\partial}{\partial y}(uv) + \frac{\partial}{\partial z}(uw) = -\frac{1}{\rho}\frac{\partial p}{\partial x} + \nu\left(\frac{\partial^2 u}{\partial x^2} + \frac{\partial^2 u}{\partial y^2} + \frac{\partial^2 u}{\partial z^2}\right) \tag{6.18a}$$

$$\frac{\partial v}{\partial \tilde{t}} + \frac{\partial}{\partial x}(uv) + \frac{\partial}{\partial y}(vv) + \frac{\partial}{\partial z}(vw) = -\frac{1}{\rho}\frac{\partial p}{\partial y} + \nu\left(\frac{\partial^2 v}{\partial x^2} + \frac{\partial^2 v}{\partial y^2} + \frac{\partial^2 v}{\partial z^2}\right) \quad (6.18b)$$

$$\frac{\partial w}{\partial \tilde{t}} + \frac{\partial}{\partial x}(uw) + \frac{\partial}{\partial y}(vw) + \frac{\partial}{\partial z}(ww) = -\frac{1}{\rho}\frac{\partial p}{\partial z} + \nu\left(\frac{\partial^2 w}{\partial x^2} + \frac{\partial^2 w}{\partial y^2} + \frac{\partial^2 w}{\partial z^2}\right) \quad (6.18c)$$

In each physical time step, the governing equations are solved iteratively in pseudo-time until convergence is achieved. The AC method provides good results for stationary flows. However, for unsteady and/or low Reynolds number flows, the choice of the AC parameter can become too restrictive, leading to slow convergence rates.

6.5.2 The Pressure Correction Approach

In this approach, the derived equation is used to solve the pressure. The velocities are obtained sequentially from the momentum equations. These velocity components do not satisfy mass conservation. Hence, a Poisson equation is derived for pressure (or changes in pressure); it enforces velocities to satisfy mass conservation. These are also known as segregated methods.

We saw that a pressure equation can be derived from the continuity equation. However, it is better to use the pressure correction equation rather than the pressure equation. Let us now consider how change in pressure can enforce mass conservation. Whenever there is a net mass flux in the control volume, there will be increase cell pressure. This increase in cell pressure will drive mass out. Similarly, whenever there is a net mass flux out in the control volume (CV), the mass will be sucked in the CV due to the decrease in cell pressure.

Pressure-correction methods are pressure-based iterative methods in which velocity and pressure fields are updated satisfying both mass and momentum equations. The steps are as follows:

- For current pressure field, compute velocity by solving the momentum equation.
- This velocity field may not satisfy the mass conservation requirement.
- Hence, transform the continuity equation into pressure-correction equation using the relationship between velocity and pressure in the momentum equation.
- Compute pressure correction p' by solving the pressure-correction equation, which pushes the velocity field to satisfy mass conservation.

There are various methods developed based on the pressure correction method. A few will be discussed here.

6.5.2.1 The MAC Method

The MAC (Marker And Cell) Method was originally developed by Harlow and Welch [1965] for solving the incompressible N-S equations for free surface flows. The MAC method uses a staggered grid and a Poisson equation of pressure. It is assumed that the initial velocity is known and divergence free. The pressure field is calculated so that the rate of change of the divergence of velocity is 0. This requires the solution of Poisson equation.

The steps are as follows:

- Assume at t = n^th time, we have converged solution.
- At next time step

$$\frac{u^{n+1}-u^n}{\Delta t} = -\left[\frac{\partial}{\partial x}(uu)+\frac{\partial}{\partial y}(uv)+\frac{\partial}{\partial z}(uw)\right]^n - \frac{1}{\rho}\left(\frac{\partial p}{\partial x}\right)^n + \nu\left(\frac{\partial^2 u}{\partial x^2}+\frac{\partial^2 u}{\partial y^2}+\frac{\partial^2 u}{\partial z^2}\right)^n \quad (6.19)$$

Similarly, we can find the values of v^{n+1} *and* w^{n+1}.

The new velocity values may not satisfy the continuity equation; this indicates an improper distribution of pressure field. The pressure is corrected such the velocity field is divergence free. This is done by solving the PPE by taking the derivatives of momentum equations, as shown in Section 6.3. The viscous term in the PPE can be neglected with strict convergence criteria.

6.5.2.2 The Fractional Step Pressure Projection Method (Chorin [1968]; Temam [1969])

The fractional step (FS) pressure projection (PP) method was developed originally for a co-located grid for solving time-dependent incompressible N-S equations. In this method, the velocity vector field $\boldsymbol{u}$ can be decomposed into divergence free (solenoidal) part $\boldsymbol{u}_{sol}$ and an irrotational part $\boldsymbol{u}_{irr}$.

Thus,

$$\boldsymbol{u} = \boldsymbol{u}_{sol} + \boldsymbol{u}_{irr} = \boldsymbol{u}_{sol} + \nabla\Phi; \ \textit{where } \Phi \textit{ is velocity potential.}$$

Now

$$\nabla \times \nabla\Phi = 0; \ \text{so} \nabla.\boldsymbol{u} = \ \nabla^2\Phi \ \textit{since } \nabla.\boldsymbol{u}_{sol} = 0$$

Thus

$$\boldsymbol{u}_{sol} = \boldsymbol{u} - \nabla\Phi$$

Chorin constructed a PPE considering pressure a potential function of the irrotational velocity field. Now the X-momentum equation is given by

$$\frac{\partial u}{\partial t}+\frac{\partial}{\partial x}(uu)+\frac{\partial}{\partial y}(uv)+\frac{\partial}{\partial z}(uw)=-\frac{1}{\rho}\frac{\partial p}{\partial x}+\nu\left(\frac{\partial^2 u}{\partial x^2}+\frac{\partial^2 u}{\partial y^2}+\frac{\partial^2 u}{\partial z^2}\right) \tag{6.20}$$

In original version, first compute intermediate velocity u^*explicitly to obtain the final solution of u^{n+1}.

i.e.

$$\frac{u^*-u^n}{\Delta t}=-\left[\frac{\partial}{\partial x}(uu)+\frac{\partial}{\partial y}(uv)+\frac{\partial}{\partial z}(uw)\right]^n+\nu\left(\frac{\partial^2 u}{\partial x^2}+\frac{\partial^2 u}{\partial y^2}+\frac{\partial^2 u}{\partial z^2}\right)^n \tag{6.21a}$$

Final velocity is then obtained from

$$u^{n+1}=u^*-\frac{\Delta t}{\rho}\left(\frac{\partial p}{\partial x}\right)^{n+1} \tag{6.21b}$$

Similar way we can obtain v^{n+1}andw^{n+1}.

This is an operator-splitting method that considers viscous forces in the first time step and pressure force in the second half step. Pressure can be obtained from the PPE

$$\nabla^2 p^{n+1}=\frac{\rho}{\Delta t}\nabla .u^* \tag{6.22}$$

Boundary condition for pressure is $\nabla p^{n+1}.n=0$. This pressure condition gives rise error which can be avoided by using staggered grid.

6.6 THE SIMPLE METHOD

The most commonly used pressure-correction methods are SIMPLE, its variants and PISO. We shall now discuss the SIMPLE method and its variants.

6.6.1 DERIVATION OF VELOCITY CORRECTION AND PRESSURE CORRECTION EQUATIONS

Spalding [1972] proposed the SIMPLE (Semi-Implicit Method for Pressure-Linked Equation) algorithm for correcting the pressure field using the continuity equation as a constraint. This method is discussed here. Consider Figure 6.6.

The momentum equation can be written as

$$a_e u_e=\sum a_{nb}u_{nb}+b_e+(p_P-p_E)A_e$$

similarly,

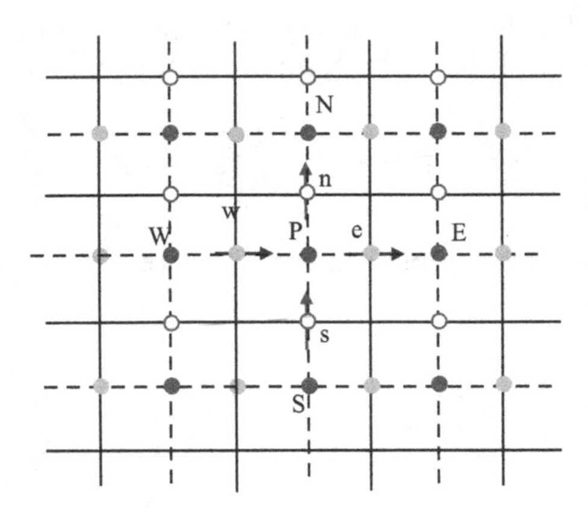

FIGURE 6.6 Pressure and Velocity Control Volumes.

$$a_n v_n = \sum a_{nb} v_{nb} + b_n + (p_P - p_N) A_n \tag{6.23}$$

and $a_t w_t = \sum a_{nb} w_{nb} + b_t + (p_P - p_T) A_t$

Now, momentum equations can be solved only when pressure field is known or somewhat estimated. Initially, we do not know the pressure field. Hence, based on guess pressure p^*, predicted velocities u^*, v^* and w^* are calculated from the equations:

$$\begin{aligned} a_e u_e^* &= \sum a_{nb} u_{nb}^* + b_e + (p_P^* - p_E^*) A_e \\ a_n v_n^* &= \sum a_{nb} v_{nb}^* + b_n + (p_P^* - p_N^*) A_n \\ a_t w_t^* &= \sum a_{nb} w_{nb}^* + b_t + (p_P^* - p_T^*) A_t \end{aligned} \tag{6.24}$$

Since the predicted velocities are calculated based on the guess pressure field, they may not satisfy the continuity equation. Hence, we must find a way to improve the guess pressure p^* such that the resulting starred velocity field comes closer to satisfying the continuity equation.

6.6.1.1 Pressure and Velocity Corrections

Let correct pressure p is obtained from, $p = p^* + p'$, where p' is called *pressure correction.*

Similarly, we can write corrected velocities as

$$u = u^* + u'; \; v = v^* + v' \text{ and } w = w^* + w'$$

Hence,

$$a_e u_e = \sum a_{nb} u_{nb} + b_e + (p_P - p_E) A_e \tag{6.25}$$

and

$$a_e u_e^* = \sum a_{nb} u_{nb}^* + b_e + (p_P^* - p_E^*) A_e \tag{6.26}$$

Subtracting Eq. (6.26) from Eq. (6.25) we get

$$a_e u'_e = \sum a_{nb} u'_{nb} + (p'_P - p'_E) A_e \tag{6.27}$$

In SIMPLE normally the term $\sum a_{nb} u'_{nb}$ is neglected. Then we get

$$a_e u'_e = (p'_P - p'_E) A_e$$

or

$$u'_e = (p'_P - p'_E) \frac{A_e}{a_e} = d_e (p'_P - p'_E); where\ d_e = \frac{A_e}{a_e}$$

The above equation is called velocity-correction formula. Hence, we can write

$$u_e = u_e^* + d_e (p'_P - p'_E)$$

Similarly,

$$v_n = v_n^* + d_n (p'_P - p'_N) \tag{6.28}$$

$$w_t = w_t^* + d_t (p'_P - p'_T)$$

6.6.2 Pressure Correction Equation

Now, the continuity equation is given by

$$\frac{\partial(\rho u)}{\partial x} + \frac{\partial(\rho v)}{\partial y} + \frac{\partial(\rho w)}{\partial z} = 0$$

Discretizing we get

$$(\rho u A)_e - (\rho u A)_w + (\rho v A)_n - (\rho v A)_s + (\rho w A)_t - (\rho w A)_b = 0$$

Putting the values of u, v and w from Eq. (6.28) we get

$$(\rho A)_e u_e^* + (\rho Ad)_e(p'_P - p'_E) - (\rho A)_w u_w^* - (\rho Ad)_w(p'_W - p'_P) + (\rho A)_n v_n^*$$
$$+(\rho Ad)_n(p'_P - p'_N) - (\rho A)_s v_s^* - (\rho Ad)_s(p'_S - p'_P) + (\rho A)_t w_t^*$$
$$+(\rho Ad)_t(p'_P - p'_T) - (\rho A)_b w_b^* - (\rho Ad)_b(p'_B - p'_P) = 0$$

or

$$\left[(\rho Ad)_e + (\rho Ad)_w + (\rho Ad)_n + (\rho Ad)_s + (\rho Ad)_t + (\rho Ad)_b\right]p'_P$$
$$= (\rho Ad)_e p'_E + (\rho Ad)_w p'_W + (\rho Ad)_n p'_N + (\rho Ad)_s p'_S + (\rho Ad)_t p'_T + (\rho Ad)_b p'_B$$
$$- \left[(\rho A)_e u_e^* - (\rho A)_w u_w^* + (\rho A)_n v_n^* - (\rho A)_s v_s^* + (\rho A)_t w_t^* - (\rho A)_b w_b^*\right]$$

or

$$\left[(\rho Ad)_e + (\rho Ad)_w + (\rho Ad)_n + (\rho Ad)_s + (\rho Ad)_t + (\rho Ad)_b\right]p'_P$$
$$= (\rho Ad)_e p'_E + (\rho Ad)_w p'_W + (\rho Ad)_n p'_N + (\rho Ad)_s p'_S + (\rho Ad)_t p'_T + (\rho Ad)_b p'_B$$
$$-\left[F_e^* - F_w^* + F_n^* - F_s^* + F_t^* - F_b^*\right]$$

That is

$$a_P p'_P = \sum a_{nb} p'_{nb} + b_P \tag{6.29}$$

where

$$\begin{aligned}
a_e &= (\rho Ad)_e;\ a_w = (\rho Ad)_w \\
a_n &= (\rho Ad)_n;\ a_s = (\rho Ad)_s \\
a_t &= (\rho Ad)_t;\ a_b = (\rho Ad)_b \\
a_P &= a_e + a_w + a_n + a_s + a_t + a_b \\
b_P &= -\left(F_e^* - F_w^* + F_n^* - F_s^* + F_t^* - F_b^*\right)
\end{aligned} \tag{6.30}$$

There is a reason behind the source term b_P being negative in the pressure correction equation. As discussed earlier, whenever there is a net mass flux in the control volume, there will be increased cell pressure. This increase in cell pressure will drive mass out of the cell and hence be a negative.

6.6.3 The SIMPLE Algorithm

Now the SIMPLE algorithm can be described as follows

1. Guess pressure field p^*
2. Based on this guess pressure, p^*, solve momentum equations (Eq. 6.24) to obtain u^*, v^*, w^*.

3. Calculate the coefficients of pressure correction and the source term b_P from Eq. (6.30).
4. Solve for pressure correction p' from pressure-correction equation, Eq. (6.29).
5. Calculate the corrected pressure $p = p^* + p'$
6. Calculate corrected velocities u, v, w from Eq. (6.25).
7. Calculate other ∅s such as temperature etc.

Take the corrected pressure p as guess pressure p^* and repeat steps 2 to 7 till convergence.

The flowchart of the SIMPLE algorithm is shown in Figure 6.7.

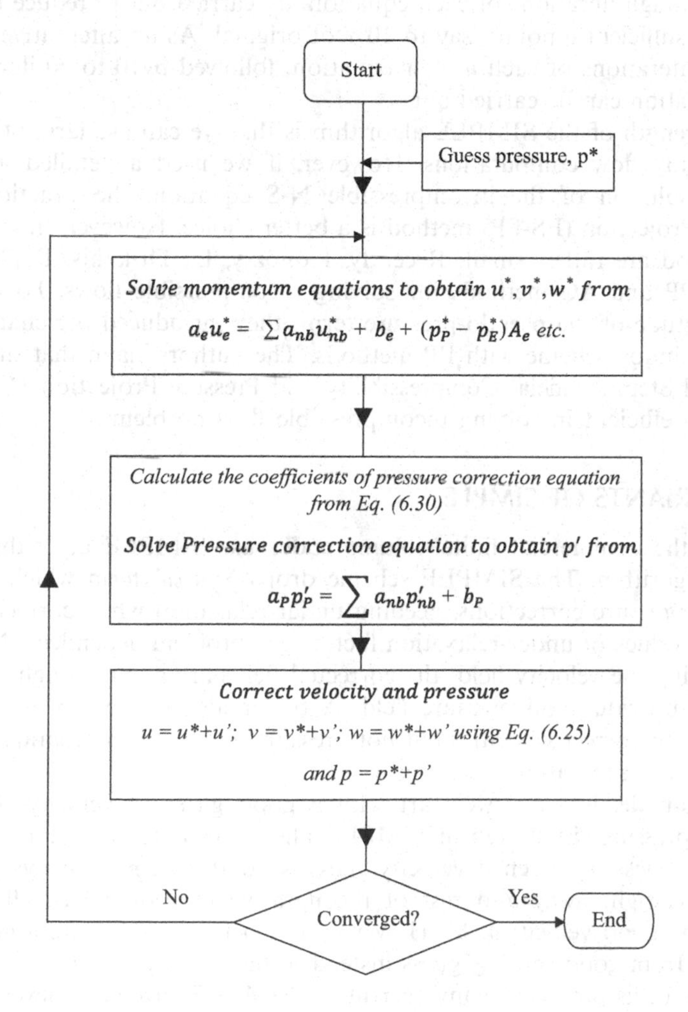

FIGURE 6.7 Flowchart of the SIMPLE Algorithm.

In step 5 of SIMPLE algorithm, the pressure updating is shown as $p = p^* + p'$. In practice, substantial under-relaxation is carried out while updating the pressure to prevent solution from diverging. In the correction step, i.e. step 5 the pressure update (but not the velocity) is done as follows: $p = p^* + \alpha_P p'$

where α_P is the relaxation parameter for pressure which lies *in the range 0.1 – 0.3.*

Velocity is also under-relaxed in the momentum equations, the values of the under-relaxation α_u generally lies in the range *0.6 – 0.8.*

SIMPLE algorithm is an iterative method. During each iteration, the equations are changed as the equations are non-linear and coupled. Hence, there will be little incentive in solving the equations exactly in each iteration. Instead enough iterations of each equation are carried out to reduce the residuals by a sufficient amount, say to 10% of original. As an alternative strategy, around 5 iterations of each *u, v, w* equation, followed by10 to 20 iterations of the p' equation can be carried out.

The strength of the SIMPLE algorithm is that we can use larger time steps for unsteady flow computations. However, if we need a detailed and time-accurate solution of the incompressible N-S equation, the Fractional Step Pressure Projection (FS-PP) method is a better choice. However, time-steps for this method are rather small. Recently, Könözsy, L., Drikakis, D [2014] unified FS, PP and AC methods for solving incompressible flows. For reducing the magnitude of the pseudo-pressure term, they introduced the characteristic based Gudunov scheme with PP methods. The authors claim that this unified Fractional Step, Artificial Compressibility and Pressure Projection (FSAC-PP) method is efficient in solving incompressible flow problems.

6.7 VARIANTS OF SIMPLE

Most of the commercial finite volume codes use SIMPLE algorithm as the default algorithm. The SIMPLE scheme drops $\sum a_{nb}u'_{nb}$term which produces too-large pressure corrections, needing under-relaxation when correcting pressure. The values of under-relaxation factors are problem dependent. Normally, for updating the velocity field, the corrected fields are good enough. However, it cannot provide good pressure field. A better idea will be to use pressure-correction to correct velocities but not pressure and create a separate equation for pressure computation.

For example, let's say we start with a good guess of velocity but a bad guess of pressure. First step in SIMPLE algorithm is to solve for u^*, v^*, w^* using bad guess p^*. Even if velocity guess is good, bad p^* destroys good velocity guess right away and rest of the iterative procedure basically tries to recover the good velocity field. However, it is better to compute a good pressure field from good velocity guess instead of having to guess it.

To avoid this problem, many variants of SIMPLE have been developed:

- SIMPLER (SIMPLE Revised – Patankar [1981]),
- SIMPLEC (Van Doormal and Raithby [1984]),
- PISO (Pressure Implicit with Splitting of Operators – Issa et al. [1986]).

6.7.1 The SIMPLER Algorithm

This algorithm assumes that the correction equation is good for updating the velocity field but not pressure and solves the pressure equation before solving the momentum and pressure-correction equations.

As we have seen, the discretized momentum equation Eq. (6.23) is given by

$$a_e u_e = \sum a_{nb} u_{nb} + b_e + (p_P - p_E)A_e$$
$$a_n v_n = \sum a_{nb} v_{nb} + b_n + (p_P - p_N)A_n$$
$$a_t w_t = \sum a_{nb} w_{nb} + b_t + (p_P - p_T)A_t$$

In terms of velocities

$$u_e = \frac{\sum a_{nb} u_{nb} + b_e}{a_e} + \frac{A_e}{a_e}(p_P - p_E) = \frac{\sum a_{nb} u_{nb} + b_e}{a_e} + d_e(p_P - p_E)$$
$$v_n = \frac{\sum a_{nb} v_{nb} + b_n}{a_n} + d_n(p_P - p_N)$$
$$w_t = \frac{\sum a_{nb} w_{nb} + b_t}{a_t} + d_t(p_P - p_T)$$

Let

$$\hat{u}_e = \frac{\sum a_{nb} u_{nb} + b_e}{a_e}; etc. \tag{6.31}$$

Then

$$u_e = \hat{u}_e + d_e(p_P - p_E)$$
$$v_n = \hat{v}_n + d_n(p_P - p_N) \tag{6.32}$$
$$w_t = \hat{w}_t + d_t(p_P - p_T)$$

The discretized continuity equation is given by

$$(\rho u A)_e - (\rho u A)_w + (\rho v A)_n - (\rho v A)_s + (\rho w A)_t - (\rho w A)_b = 0$$

Putting the velocities in the above equation we get

$$a_P p_P = \sum a_{nb} p_{nb} + b_P \tag{6.33}$$

where

$$
\begin{aligned}
a_e &= (\rho A d)_e;\ a_w = (\rho A d)_w \\
a_n &= (\rho A d)_n;\ a_s = (\rho A d)_s \\
a_t &= (\rho A d)_t;\ a_b = (\rho A d)_b \\
a_P &= a_e + a_w + a_n + a_s + a_t + a_b \\
b_P &= -\left(\hat{F}_e - \hat{F}_w + \hat{F}_n - \hat{F}_s + \hat{F}_t - \hat{F}_b\right)
\end{aligned}
\tag{6.34}
$$

Note:

- p coefficients are the same as p' coefficients
- b_P in pressure equation is in terms of $\hat{F}$, not F^* i.e. not mass imbalance
- There are no approximations – if velocity is exact, pressure is exact.

Overall algorithm:

1) Guess velocity field
2) Compute momentum coefficients and store. Compute $\hat{u}, \hat{v}, \hat{w}$ from Eq. (6.31).
3) Compute pressure coefficients from Eq. (6.34) and store. Solve pressure equation, Eq. (6.33) and obtain pressure.
4) Solve momentum equations using stored momentum coefficients and just-computed pressure. Find u^*, v^* *and* w^* from Eq. (6.24).
5) Find coefficients of pressure-correction equation from Eq. (6.30) using u^*, v^* *and* w^*.
6) Solve p' equation, Eq. (6.29) to find pressure correction field
7) Correct velocities from: $u = u^*+u'$; $v = v^*+v'$; $w = w^*+w'$. But do not correct pressure.
8) Compute other scalar fields if necessary
9) Check for convergence. If not converged, go to 2. Else stop.

Discussion

- SIMPLER is found to be faster than SIMPLE (about 30-50% fewer iterations).
- Does not need good pressure guess – as it finds pressure from velocity guess
- About 50% larger computational effort typically
  - → Two pressure-like equations
  - → Expensive because of lack of Dirichlet conditions
- Need extra storage for pressure and momentum coefficients

- P' equation is used only for velocity correction – pressure correction does not correct pressure
  - → No need to under-relax pressure correction
- May under-relax pressure equation directly, but usually not necessary
- May need to under-relax momentum equations to account for non-linearities, sequential solution procedure (as with SIMPLE)

6.7.2 The SIMPLEC Algorithm

SIMPLE-Corrected (SIMPLEC) algorithm seeks to diminish the effects of dropping velocity neighbor correction terms $\sum a_{nb}u'_{nb}$ in SIMPLE algorithm. In SIMPLE algorithm, the equation for corrected velocities are given by

$$a_e u'_e = \sum a_{nb}u'_{nb} + (p'_P - p'_E)A_e$$
$$a_n v'_n = \sum a_{nb}v'_{nb} + (p'_P - p'_N)A_n$$
$$a_t w'_t = \sum a_{nb}w'_{nb} + (p'_P - p'_T)A_t$$

SIMPLEC retains neighbor velocity correction terms, but makes an approximation:

$$\sum a_{nb}u'_{nb} \approx u'_e \sum a_{nb}$$
$$\sum a_{nb}v'_{nb} \approx v'_n \sum a_{nb}$$
$$\sum a_{nb}w'_{nb} \approx w'_t \sum a_{nb}$$

Thus

$$\left(a_e - \sum a_{nb}\right)u'_e = (p'_P - p'_E)A_e$$
$$\left(a_n - \sum a_{nb}\right)v'_n = (p'_P - p'_N)A_n \tag{6.35}$$
$$\left(a_t - \sum a_{nb}\right)w'_t = (p'_P - p'_T)A_t$$

Redefine d coefficients as

$$d_e = \frac{A_e}{(a_e - \sum a_{nb})}; etc. \tag{6.36}$$

Hence, we get $a_P p'_P = \sum a_{nb}p'_{nb} + b_P$
where

$$
\begin{aligned}
a_e &= (\rho A d)_e; a_w = (\rho A d)_w \\
a_n &= (\rho A d)_n; a_s = (\rho A d)_s \\
a_t &= (\rho A d)_t; a_b = (\rho A d)_b \\
a_P &= a_e + a_w + a_n + a_s + a_t + a_b \\
b_P &= -\left(F_e^* - F_w^* + F_n^* - F_s^* + F_t^* - F_b^*\right)
\end{aligned}
$$

Note that the form of equation is the same as for SIMPLE. b_p term is still the amount by which the starred velocities do not satisfy continuity. Only *"d"* coefficients are different from SIMPLE.

The overall algorithm is identical to that for SIMPLE. Only the d coefficients used to drive the p' equation are different. With SIMPLEC, because the terms $\sum a_{nb}u'_{nb}$ *etc.* are not dropped, there is no need to under-relax the p' correction. Thus: $p = p^* + p'$.

6.7.3 The PISO (Pressure Implicit with Split Operator) Algorithm

The PISO was proposed by Issa et al. [1986] for the non-iterative computation of unsteady compressible flow. The SIMPLE algorithm is a one-step predictor-corrector method. However, the PISO algorithm involves a two-stage correction method instead of a one step. Each time-step ($t^{old} \rightarrow t^{new}$) consists of a sequence of three stages:

1) **Predictor Step:**

With a guess pressure p^*, the values of velocities u^*, v^* and w^* are calculated from Eq. (6.24) as in SIMPLE algorithm.

2) **Corrector Step 1:**

Since, we have used the guess value of pressure, the velocity component obtained may not satisfy the continuity equation.

Let correct pressure p is obtained from, $p^{**} = p^*+p'$, where p' is called *pressure correction*. Similarly, we can write $u^{**}=u^*+u'$; $v^{**}=v^*+v'$ and $w^{**}=w^*+w'$.

So, we can write

$$
\begin{aligned}
u_e^{**} &= u_e^* + d_e(p'_P - p'_E) \\
v_n^{**} &= v_n^* + d_n(p'_P - p'_N) \\
w_t^{**} &= w_t^* + d_t(p'_P - p'_T)
\end{aligned}
\tag{6.37}
$$

Putting Eq. (6.37) in continuity equation, we get the pressure correction equation, Eq. (6.30) which is then solve to obtain p'. Then find out the corrected velocities u^{**}, v^{**} and w^{**} from Eq. (6.37).

3) **Corrector Step 2:**

A second corrector step is introduced by Issa to enhance the SIMPLE procedure. The discretized momentum equations for u^{**}, v^{**} and w^{**} are

$$\begin{aligned} a_e u_e^{**} &= \sum a_{nb} u_{nb}^{*} + b_e + (p_P^{**} - p_E^{**}) A_e \\ a_n v_n^{**} &= \sum a_{nb} v_{nb}^{*} + b_n + (p_P^{**} - p_N^{**}) A_n \\ a_t w_t^{**} &= \sum a_{nb} w_{nb}^{*} + b_t + (p_P^{**} - p_T^{**}) A_t \end{aligned} \tag{6.38}$$

Let us define second corrected values by

$u^{***} = u^{**} + u''$; $v^{***} = v^{**} + v''$; $w^{***} = w^{**} + w''$ *and* $p^{***} = p^{**} + p''$

Then the momentum equation becomes

$$\begin{aligned} a_e u_e^{***} &= \sum a_{nb} u_{nb}^{**} + b + (p_P^{***} - p_E^{***}) A_e \\ a_n v_n^{***} &= \sum a_{nb} v_{nb}^{**} + b + (p_P^{***} - p_N^{***}) A_n \\ a_t w_t^{***} &= \sum a_{nb} w_{nb}^{**} + b + (p_P^{***} - p_T^{***}) A_t \end{aligned} \tag{6.39}$$

Subtracting Eq. (6.38) from Eq. (6.39) we get

$$\begin{aligned} u_e^{***} &= u_e^{**} + \frac{\sum a_{nb}\left(u_{nb}^{**} - u_{nb}^{*}\right)}{a_e} + d_e(p''_P - p''_E) \\ v_n^{***} &= v_n^{**} + \frac{\sum a_{nb}\left(v_{nb}^{**} - v_{nb}^{*}\right)}{a_n} + d_n(p''_P - p''_N) \\ w_t^{***} &= w_t^{**} + \frac{\sum a_{nb}\left(w_{nb}^{**} - w_{nb}^{*}\right)}{a_t} + d_t(p''_P - p''_T) \end{aligned} \tag{6.40}$$

Putting the velocity values in continuity equation we get

$$a_P p''_P = \sum a_{nb} p''_{nb} + b_P \tag{6.41}$$

where

$$a_e = (\rho A d)_e;\ a_w = (\rho A d)_w$$

$$a_n = (\rho A d)_n;\ a_s = (\rho A d)_s$$

$$a_t = (\rho A d)_t;\ a_b = (\rho A d)_b$$

$$a_P = a_e + a_w + a_n + a_s + a_t + a_b$$

$$b_P = -\left[\left(\frac{\rho A}{d}\sum a_{nb}u'_{nb}\right)_e - \left(\frac{\rho A}{d}\sum a_{nb}u'_{nb}\right)_w + \left(\frac{\rho A}{d}\sum a_{nb}v'_{nb}\right)_n - \left(\frac{\rho A}{d}\sum a_{nb}v'_{nb}\right)_s + \left(\frac{\rho A}{d}\sum a_{nb}w'_{nb}\right)_t - \left(\frac{\rho A}{d}\sum a_{nb}w'_{nb}\right)_b\right]$$

In the source term $-\left(F_e^{**} - F_w^{**} + F_n^{**} - F_s^{**} + F_t^{**} - F_b^{**}\right)$ is zero as u^{**}, v^{**} and w^{**} satisfy the continuity equation.

After solving for p'' from Eq. (6.41), correct the velocities from Eq. (6.40). The corrected pressure $p^{***}=p^{**}+p''$ and final velocities and pressure *are* $u = u^{***}$; $v = v^{***}$; $w = w^{***}$ and $p = p^{***}$.

Steps 1 and 2 in the PISO algorithm are same as in SIMPLE. In SIMPLE, we require an iteration in each time step. Step 3 in the PISO algorithm eliminates the need for iteration at each time step. PISO is more efficient in time-dependent calculations; SIMPLE and its variants are better for steady-state calculations.

6.8 SUMMARY

- A single scalar-transport equation can be used to represent each component of a momentum equation, with the following parameters taken into account:
  - → Concentration, $\emptyset$ → velocity component (u, v *or* w)
  - → Diffusivity, Γ → viscosity, μ
  - → source, S → *non-viscous forces*
- Momentum equations are non-linear and coupled. The velocities obtained from a momentum equation must satisfy the continuity equation. Thus, they have to be solved iteratively with the continuity equation.
- For incompressible flow, the pressure equation is derived from the continuity equation.
- In the vorticity-stream function method, pressure terms are eliminated in the momentum equations. However, this method is applicable only for two dimensions.
- For a co-located grid, linear interpolation of pressure-gradient source terms can lead to odd- even decoupling, which can be avoided by using either a *staggered velocity grid* or a special interpolation like the *Rhie-Chow interpolation* for velocities for a co-located grid.
- Chorin introduced the artificial compressibility method by modifying continuity with pseudo compressibility.
  - → This method provides good results for stationary flows.
  - → However, for unsteady and/or low Reynolds-number flows, the AC parameter can be too restrictive.

- *Pressure-correction* methods make small corrections to pressure so that the velocity field becomes mass consistent whilst preserving a solution of the momentum equation.
- Various methods like MAC, FS-PP and SIMPLE and its variants have been developed based on the pressure correction method.
- In the SIMPLE algorithm, larger time steps can be used for unsteady flow computations. However, if we need a detailed and time-accurate solution of incompressible N-S equations, FS-PP method may be a better choice.
- *SIMPLE* (and its variants) and *PISO* are widely used pressure-correction algorithms. The first is an iterative scheme efficient for steady-state solutions; the second is a non-iterative, efficient time-dependent scheme.

QUESTIONS:

1) List the difficulties associated in solving incompressible N-S Equations.
2) What is a co-located grid? Can it be used in fluid-flow problems?
3) What is a staggered grid, and why is it needed?
4) In solving an incompressible fluid-flow problem, pressure and velocities are not specified in the same node. Why?
5) How does the "staggered-grid" arrangement of velocity and pressure overcome the "odd-even decoupling" problem connected with a co-located grid? Describe, with the help of diagrams.
6) In a finite-volume CFD calculation, flow variables are usually stored at the centre of computational cells. Explain briefly

    i) how *pressure-correction* methods may be used to generate a mass-consistent velocity field;
    ii) why simple *linear* interpolation may cause difficulties with the coupling of pressure and velocity if these variables are stored at the same locations.

7) In incompressible N-S equations, there is no equation for pressure. How will you solve for pressure? Explain in detail.
8) Describe in detail the method of deriving the pressure correction equation in the SIMPLE algorithm.
9) Explain how the "odd-even decoupling" problem in the discrete pressure field arises with co-located storage of velocity and pressure in a finite-volume mesh. Describe, with the help of diagrams, the "staggered-grid" arrangement of velocity and pressure to overcome this problem and state its advantages and disadvantages.
10) Use the vorticity-stream function method in solving a lid-driven square cavity problem for Re = 100. Assume the flow is steady and 21x21 grids. Compare your results with results of Ghia et al. [1982].
11) Use the artificial compressibility method and solve the above problem 10.

7 Finite Volume Method for Complex Geometries

7.1 INTRODUCTION

In earlier chapters, we discussed the application of the finite volume method to regular geometries. However, most of the practical problems involve complex geometries. Irregular boundary surfaces must be approximated in a stepwise manner. To gain a better description of a boundary, it is possible to let the wall boundary intersect the regular co-ordinate lines at irregular locations, resulting in irregular shapes and sizes of boundary cells. These cells need special discretization and can introduce considerable errors in the implementation of boundary conditions for such a scheme. There have been efforts to use body-fitted coordinate transformation for mapping complex geometries into regular shapes (Thompson et al. [1982]). In these methods, by suitable transformations, computational mesh is always made rectangular. However, the process of mapping from Cartesian coordinates to body-fitting curvilinear coordinates is very cumbersome; often there are difficulties in controlling the grid size and grid skewness, as necessitated by the variation of flow and temperature solutions.

The need for greater flexibility in the flow simulation for complex geometries has led to the development of solution algorithms based on non-orthogonal meshes. Quadrilaterals/triangles are used in two-dimensional and hexahedra/tetrahedra/arbitrary polyhedral are used in three-dimensional flows for non-orthogonal meshes. The flow solutions can be obtained by staggered or co-located grid arrangements. In this chapter, flow algorithms based on non-orthogonal structured grids will be initially discussed. The discretization procedure for unstructured grids will be discussed later.

7.2 STAGGERED GRID ALGORITHM

As discussed in Chapter 6, in the conventional staggered grid arrangement scalar quantities such as pressure, temperature and density are stored at the main grid point, whereas all Cartesian velocity components are stored at the cell faces of the control volume. However, this procedure breaks down when the grid is rotated by 90^0 (e.g. in the case of a curve duct), and the pressure-velocity staggered arrangement becomes incorrect. Refer to Figure 7.1.

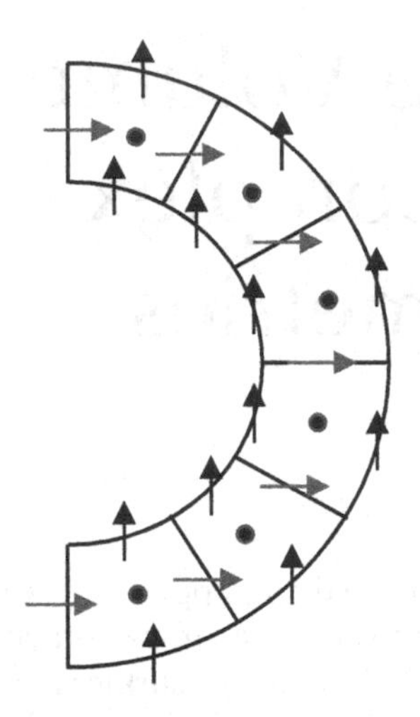

FIGURE 7.1 Pressure-Velocity Nodes for Staggered Arrangement.

To overcome this problem, covariant or contravariant velocity components are used to allow the discretization method to be independent of the grid orientation. However, covariant velocity components are not orthogonal to the faces; some averaging is necessary to calculate the mass fluxes from the boundaries of the control volume.

Although the staggered grid method was developed to prevent the appearance of the oscillating pressure field, this method has its disadvantages. Nodal staggering complicates the programming and implementation of the solution algorithm. In two dimensions, the algorithm must calculate four different sets of coefficients for each index location: one set for each horizontal and vertical velocity; one set for pressure or pressure correction equation and a final set for other scalar quantities. The use of a non-uniform grid system complicates the bookkeeping requirements since a different set of spatial dimensions must be defined for each of three control volumes. The use of multilevel, multigrid, grid-embedding methods further complicates the problem, as it requires multiple interpolations to transform information between the fine and course grids and to specify the internal boundaries of locally refined grids. The inconvenience associated with a staggered grid system is compounded in flow problems requiring three-dimensional generalized coordinate systems of time-dependent geometries.

7.3 THE CO-LOCATED GRID ALGORITHM

Co-located or non-staggered grid arrangement is popular because of its geometric simplicity and ease of programming and is adopted in all commercial codes. As discussed earlier, in the co-located arrangement, velocity components and all scalar quantities are stored at the same grid nodes as shown in Figure 7.2. Co-located grid arrangement is also the natural choice for multilevel, multigrid, grid embedding and adapted meshes, as it requires less interpolation to transfer the information.

Although co-located grid arrangement is very attractive due to its simplicity for location of variables, the main drawback of this arrangement is the occurrence of non-physical oscillations of pressure and/or velocity field. This is due to decoupling between pressure and the velocity field, if linear interpolation is used. Hence, a special interpolation technique as proposed by Rhie and Chow [1983] or other researchers is required for the success of the scheme.

Almost all the work done on the development of the flow algorithm based on co-located grid approach handles the non-orthogonality of the grid by body-fitted coordinate transformations. The choice of curvilinear contravariant or covariant velocity components as cell face velocities involves complicated tensor algebra to derive the momentum equations in the transformed domain. In addition, it is difficult to control or check physical parameters because the computation proceeds in a transformed domain. Also, the inherent benefit of the finite volume method is somewhat lost because the local flux balance cannot be checked during computation.

To overcome this problem, Roychowdhury et al. [1999] proposed a numerical scheme based on a non-orthogonal, co-located grid arrangement in which the governing equations are solved in the physical plane itself. Therefore, coefficients of flow variables in the discretized equations are directly linked to the corresponding areas or volumes, thereby retaining the physical grasp present in the finite volume method. It also preserves the conservation of the governing equations locally in each discretized cell and globally in the entire computational domain.

This chapter discusses the application of the finite volume method based on a non-orthogonal, co-located grid, beginning with the flow algorithm for two

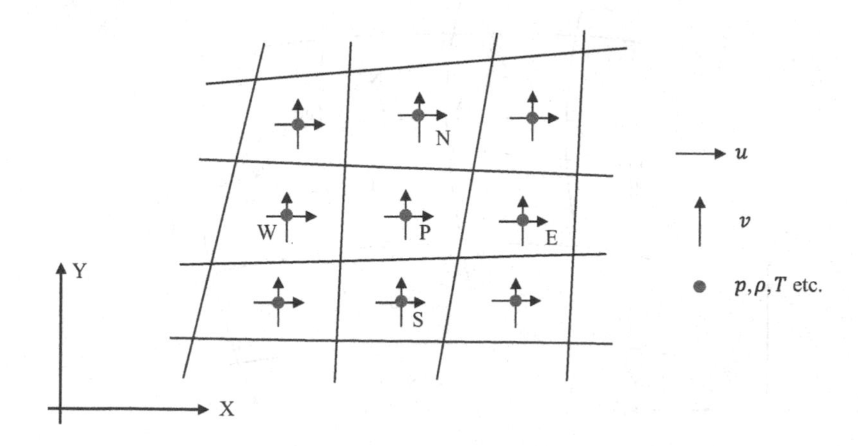

FIGURE 7.2 Co-located Grid Arrangement in Two Dimensions.

dimensions with arbitrary quadrilateral meshes to represent the complex geometry, because extension of this algorithm to three dimensions is straight forward. Flow algorithms for two dimension with unstructured grids are addressed later in the chapter.

7.4 DISCRETIZATION METHODS FOR NON-ORTHOGONAL STRUCTURED GRIDS

The governing equations in the integral form for flow simulation are given by

Continuity:

$$\int_A \rho u_j n_j dA = 0 \tag{7.1}$$

Momentum:

$$\frac{\partial}{\partial t}\int_V (\rho u_i) dV + \int_A (\rho u_i u_j) n_j dA = -\int_A p n_i dA + \int_A \tau_{ij} n_j dA + \int_V s dV \tag{7.2}$$

where

$$\tau_{ij} = \mu\left[\frac{\partial u_i}{\partial x_j} + \frac{\partial u_j}{\partial x_i}\right].$$

The integration of these equations over the control volume as shown in Figure 7.3 yields the following expressions:

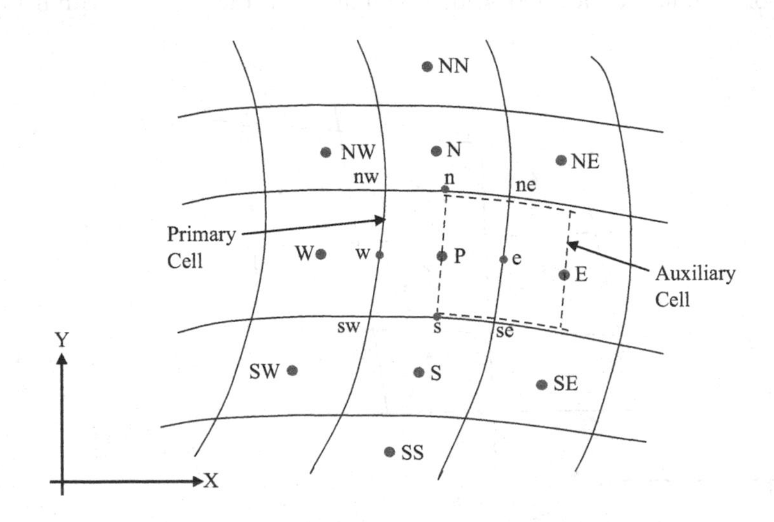

FIGURE 7.3 Typical Control Volume for the Non-Orthogonal Grid.

7.4.1 The Continuity Equation

$$\sum \langle \rho u \rangle_k A_{kx} + \langle \rho v \rangle_k A_{ky} = 0 \tag{7.3}$$

where k represents the faces e, w, n and s.

or

$$\left[\langle \rho u \rangle_e A^1_{ex} + \langle \rho v \rangle_e A^1_{ey}\right] - \left[\langle \rho u \rangle_w A^1_{wx} + \langle \rho v \rangle_w A^1_{wy}\right] + \left[\langle \rho u \rangle_n A^2_{nx} + \langle \rho v \rangle_n A^2_{ny}\right] - \left[\langle \rho u \rangle_s A^2_{sx} + \langle \rho v \rangle_s A^2_{sy}\right] = 0 \tag{7.4}$$

or

$$F_e - F_w + F_n - F_s = 0 \tag{7.5}$$

where $F_k = \sum \rho_k \emptyset_k A_{kj};\quad \emptyset \Rightarrow u, v;\quad k \Rightarrow e, w, n, s;\quad j \Rightarrow x, y$

and e, w, n, s indicates the control volume faces in east, west, north and south direction respectively. Here, F_k represents total mass flux entering or leaving the control volume normal to face k as shown in Figure 7.4.

In these expressions, A_{kj} represents the projected area in x or y direction for the kth face of the control volume as shown in Figure 7.5.

7.4.2 The Transport Equation

In general, the governing equation for the transport equation can be expressed as

$$\frac{\partial}{\partial t} \langle \rho \emptyset \rangle_P V_P + \sum \langle J \rangle_k = \langle S_\emptyset \rangle_P \tag{7.6}$$

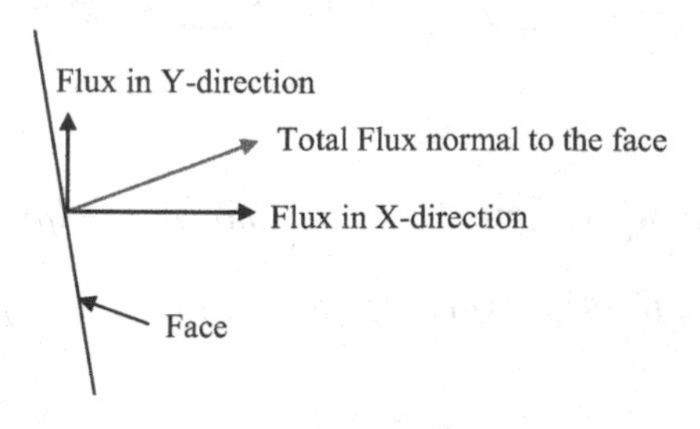

FIGURE 7.4 Fluxes Acting on the Face.

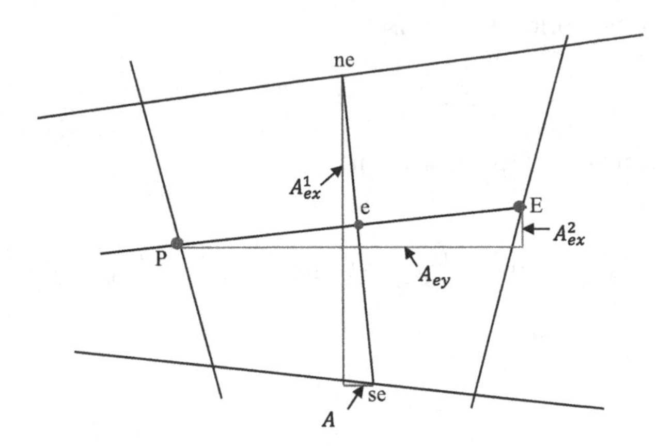

FIGURE 7.5 Area Vectors for Face e.

where

$\emptyset = u,\ v;\ \Gamma = \mu \, for\, momentum\, equation.$

$J_k = Total\, flux\, crossing\, the\, boundary = J_C + J_D;$

$J_C = Convective\, flux = \langle \rho u_j \emptyset \rangle_k A_{kj};$

$J_D = Diffusive\, flux = -\langle \Gamma \frac{\partial \emptyset}{\partial x_j} \rangle_k A_{kj};$

$S_\emptyset = Source\, term = body\, forces,\ s_i V_p + \delta p_i \ \ for\, momentum\, equation;$

V_p being the volume of the cell connected with node P.

7.4.2.1 Discretization of Convective Flux

The net convective flux through the control volume as shown in Figure 7.6 becomes

$$J_C = \sum \langle \rho u \rangle_k \emptyset_k A_{kx} + \sum \langle \rho v \rangle_k \emptyset_k A_{ky}$$

or

$$\begin{aligned} J_C = &\left[\langle \rho u \rangle_e A^1_{ex} + \langle \rho v \rangle_e A^1_{ey}\right] \emptyset_e - \left[\langle \rho u \rangle_w A^1_{wx} + \langle \rho v \rangle_w A^1_{wy}\right] \emptyset_w \\ &+ \left[\langle \rho u \rangle_n A^2_{nx} + \langle \rho v \rangle_n A^2_{ny}\right] \emptyset_n - \left[\langle \rho u \rangle_s A^2_{sx} + \langle \rho v \rangle_s A^2_{sy}\right] \emptyset_s \end{aligned} \tag{7.7}$$

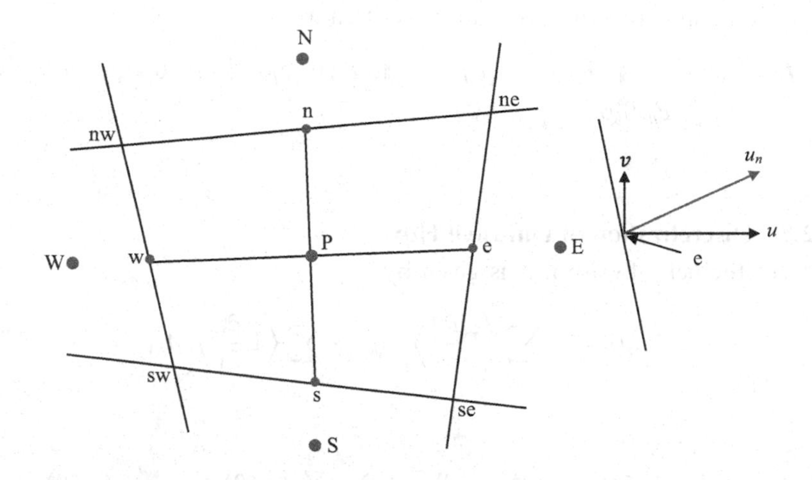

FIGURE 7.6 Control Volume Used for Discretization.

or

$$J_C = F_e\emptyset_e - F_w\emptyset_w + F_n\emptyset_n - F_s\emptyset_s \tag{7.8}$$

As earlier discussed, for the convective terms it is required to assume the value of $\emptyset_k$ at the interface. The value of the convective property $\emptyset_k$ at the interface can be assumed as the average of the nodal values (i.e. $\emptyset_e$ can be taken as the average of $\emptyset_E$ and $\emptyset_P$), which leads to central difference scheme. Already we have seen that use of central difference scheme for convective term is suitable for low Reynolds numbers only and for high Reynolds number, this scheme leads to physically inconsistent result. To avoid this, generally upwind schemes are used.

We already know that for UDS

$$\emptyset_e = \emptyset_P \quad \text{if } F_e < 0$$

and

$$\emptyset_e = \emptyset_E \quad \text{if } F_e > 0$$

The above statements can be written in compact form as

$$F_w\emptyset_w = \emptyset_W[\![F_w, 0]\!] - \emptyset_P[\![-F_w, 0]\!] \tag{7.9a}$$

$$F_e\emptyset_e = \emptyset_P[\![F_e, 0]\!] - \emptyset_E[\![-F_e, 0]\!] \tag{7.9b}$$

$$F_s\emptyset_s = \emptyset_S[\![F_s, 0]\!] - \emptyset_P[\![-F_s, 0]\!] \tag{7.9c}$$

$$F_n\emptyset_n = \emptyset_P[\![F_n, 0]\!] - \emptyset_N[\![-F_n, 0]\!] \tag{7.9d}$$

Hence, the total convective flux can be written as

$$J_C = [[\![F_e, 0]\!] + [\![-F_w, 0]\!] + [\![F_n, 0]\!] + [\![-F_s, 0]\!]]\emptyset_P - [\![-F_e, 0]\!]\emptyset_E - [\![F_w, 0]\!]\emptyset_W - [\![-F_n, 0]\!]\emptyset_N - [\![F_s, 0]\!]\emptyset_S \tag{7.10}$$

7.4.2.2 Discretization of Diffusive Flux

Similarly, the net diffusive flux is given by

$$J_D = -\sum \left\langle \Gamma \frac{\partial \emptyset}{\partial x} \right\rangle_k A_{kx} - \sum \left\langle \Gamma \frac{\partial \emptyset}{\partial y} \right\rangle_k A_{ky} \tag{7.11}$$

or

$$J_D = -\left[\left\{ \left\langle \Gamma \frac{\partial \emptyset}{\partial x} \right\rangle_e A^1_{ex} + \left\langle \Gamma \frac{\partial \emptyset}{\partial y} \right\rangle_e A^1_{ey} \right\} - \left\{ \left\langle \Gamma \frac{\partial \emptyset}{\partial x} \right\rangle_w A^1_{wx} + \left\langle \Gamma \frac{\partial \emptyset}{\partial y} \right\rangle_w A^1_{wy} \right\} + \left\{ \left\langle \Gamma \frac{\partial \emptyset}{\partial x} \right\rangle_n A^2_{nx} + \left\langle \Gamma \frac{\partial \emptyset}{\partial y} \right\rangle_n A^2_{ny} \right\} - \left\{ \left\langle \Gamma \frac{\partial \emptyset}{\partial x} \right\rangle_s A^2_{sx} + \left\langle \Gamma \frac{\partial \emptyset}{\partial y} \right\rangle_s A^2_{sy} \right\} \right] \tag{7.12}$$

Consider an auxiliary cell centered at face e as shown in Figure 7.3. Referring to Figure 7.5, the expression for the diffusive flux through east face can be obtained in terms of the projected areas and neighboring node values of the transport quantity $\langle \emptyset \rangle$ as:

$$J_{De} = \left\langle \Gamma \frac{\partial \emptyset}{\partial x} \right\rangle_e A^1_{ex} + \left\langle \Gamma \frac{\partial \emptyset}{\partial y} \right\rangle_e A^1_{ey}$$

$$J_{De} = \frac{\Gamma_e A^1_{ex}}{V_e} \left[A^1_{ex}(\emptyset_E - \emptyset_P) + A^2_{ex}(\emptyset_{ne} - \emptyset_{se}) \right] + \frac{\Gamma_e A^1_{ex}}{V_e} \left[A^1_{ey}(\emptyset_E - \emptyset_P) + A^2_{ey}(\emptyset_{ne} - \emptyset_{se}) \right]$$

$$= \frac{\Gamma_e}{V_e} \left[A^1_{ex}A^1_{ex} + A^1_{ey}A^1_{ey} \right] (\emptyset_E - \emptyset_P) + \frac{\Gamma_e}{V_e} \left[A^1_{ex}A^2_{ex} + A^1_{ey}A^2_{ey} \right] (\emptyset_{ne} - \emptyset_{se})$$

or

$$J_{De} = d^1_e(\emptyset_E - \emptyset_P) + d^2_e(\emptyset_{ne} - \emptyset_{se}) \tag{7.13a}$$

Similarly, we get

$$J_{Dw} = d^1_w(\emptyset_P - \emptyset_W) + d^2_w(\emptyset_{nw} - \emptyset_{sw}) \tag{7.13b}$$

$$J_{Dn} = d^1_n(\emptyset_N - \emptyset_P) + d^2_n(\emptyset_{ne} - \emptyset_{nw}) \tag{7.13c}$$

$$J_{Ds} = d^1_s(\emptyset_P - \emptyset_S) + d^2_s(\emptyset_{se} - \emptyset_{sw}) \tag{7.13d}$$

where

$$d_e^1 = \frac{\Gamma_e}{V_e}\left[A_{ex}^1 A_{ex}^1 + A_{ey}^1 A_{ey}^1\right] \tag{7.14a}$$

$$d_w^1 = \frac{\Gamma_w}{V_w}\left[A_{wx}^1 A_{wx}^1 + A_{wy}^1 A_{wy}^1\right] \tag{7.14b}$$

$$d_n^1 = \frac{\Gamma_n}{V_n}\left[A_{nx}^2 A_{nx}^2 + A_{ny}^2 A_{ny}^2\right] \tag{7.14c}$$

$$d_s^1 = \frac{\Gamma_s}{V_s}\left[A_{sx}^2 A_{sx}^2 + A_{sy}^2 A_{sy}^2\right] \tag{7.14d}$$

are the orthogonal parts of the diffusive fluxes, and

$$d_e^2 = \frac{\Gamma_e}{V_e}\left[A_{ex}^1 A_{ex}^2 + A_{ey}^1 A_{ey}^2\right] \tag{7.15a}$$

$$d_w^2 = \frac{\Gamma_w}{V_w}\left[A_{wx}^1 A_{wx}^2 + A_{wy}^1 A_{wy}^2\right] \tag{7.15b}$$

$$d_n^2 = \frac{\Gamma_n}{V_n}\left[A_{nx}^1 A_{nx}^2 + A_{ny}^1 A_{ny}^2\right] \tag{7.15c}$$

$$d_s^2 = \frac{\Gamma_s}{V_s}\left[A_{sx}^1 A_{sx}^2 + A_{sy}^1 A_{sy}^2\right] \tag{7.15d}$$

are the non-orthogonal parts of the diffusive flux (i.e. extra terms arising from the non-orthogonality of the grid). Areas $A^1$and A^2 are defined in Figure 7.5. The detailed calculation of area vectors and volumes is given in Appendix 1.

The net diffusive flux contribution for the control volume is

$$J_D = -[J_{De} - J_{Dw} + J_{Dn} - J_{Ds}]$$

or

$$\begin{aligned} J_D = & -\left[d_e^1 \emptyset_E + d_w^1 \emptyset_W + d_n^1 \emptyset_N + d_s^1 \emptyset_S - (d_e^1 + d_w^1 + d_n^1 + d_s^1)\emptyset_P \right. \\ & \left. + \{(d_e^2 + d_n^2)\emptyset_{ne} - (d_w^2 + d_n^2)\emptyset_{nw} - (d_e^2 + d_s^2)\emptyset_{se} + (d_w^2 + d_s^2)\emptyset_{sw}\}\right] \end{aligned}$$

or

$$J_D = -\left[d_e^1 \emptyset_E + d_w^1 \emptyset_W + d_n^1 \emptyset_N + d_s^1 \emptyset_S - (d_e^1 + d_w^1 + d_n^1 + d_s^1)\emptyset_P + (b_{no})\right] \tag{7.16}$$

where

$$b_{no} = (d_e^2 + d_n^2)\emptyset_{ne} - (d_w^2 + d_n^2)\emptyset_{nw} - (d_e^2 + d_s^2)\emptyset_{se} + (d_w^2 + d_s^2)\emptyset_{sw}.$$

The corner values ($\emptyset_{ne}, \emptyset_{nw}$ *etc.*) are unknown and can be approximated using the four surrounding node values. For example, the north-east corner value is obtained as

$$\emptyset_{ne} = \frac{1}{4}(\emptyset_N + \emptyset_E + \emptyset_P + \emptyset_{NE}) \tag{7.17}$$

The term b_{no} arises from the non-orthogonality of the grid and vanishes whenever the grid becomes orthogonal. The non-orthogonal diffusive flux terms can be treated either implicitly or explicitly. If treated implicitly, they give rise to a set of algebraic equations with nine non-zero coefficients in two dimensions, corresponding to the node P and its eight neighbors E, W, N, S, NE, NW, SE, SW. The resulting matrix may not be unconditionally diagonally dominant, since the coefficients of non-orthogonal terms may be of either sign. The treatment of non-orthogonal terms explicitly does not lower the accuracy of the scheme; it only affects the convergence rate. However, if non-orthogonal parts are smaller, the rate of convergence is not affected significantly. For achieving such a gain, non-orthogonal contributions can be lumped in the source term and handled explicitly.

7.4.2.3 Discretization of Pressure Terms

The pressure contribution for an X-momentum equation in discretized form is

$$\delta P_x = -\sum \langle p \rangle_k A_{kx} = -\left[(p_e - p_w)A_{Px}^1 + (p_n - p_s)A_{Px}^2\right]$$

or

$$\delta P_x = A_{Px}^1(p_w - p_e) + A_{Px}^2(p_s - p_n) \tag{7.18}$$

The pressure contribution for Y-momentum equation is similarly given by

$$\delta P_y = A_{Py}^1(p_w - p_e) + A_{Py}^2(p_s - p_n) \tag{7.19}$$

7.4.2.4 Implementation of the QUICK Scheme

As discussed in Chapter 5, the implementation of QUICK scheme in its full form may give rise to stability problems and unbounded solutions under some flow conditions. To avoid that, troublesome negative coefficients can be placed in source terms to retain positive main coefficients so that we treat the main coefficients implicitly and source terms explicitly; this is known as *deferred correction*. The scheme then becomes stable and fast converging variant. In this case, for the east face we can write

$$F_e\emptyset_e = \left[\emptyset_P + \frac{1}{8}(3\emptyset_E - 2\emptyset_P - \emptyset_W)\right][\![F_e, 0]\!] - \left[\emptyset_E + \frac{1}{8}(3\emptyset_P - 2\emptyset_E - \emptyset_{EE})\right][\![-F_e, 0]\!].$$

Similarly, equations can be written for other faces.

Combining all the terms, the total transport equation, Eq. (7.6) for steady state becomes

$$\begin{aligned} a_P^*\emptyset_P &= a_w\emptyset_W + a_e\emptyset_E + a_s\emptyset_S + a_n\emptyset_N + \bar{b}_P \\ a_P^* &= a_w + a_e + a_n + a_s - S_P + [(F_e - F_w) + (F_n - F_s)] \\ a_w &= [\![F_w, 0]\!] + D_w; a_e = [\![-F_e, 0]\!] + D_e \\ a_s &= [\![F_s, 0]\!] + D_s; a_n = [\![-F_n, 0]\!] + D_n \\ \bar{b}_P &= b_P + b_{no} + b_{qu} \end{aligned} \quad (7.20)$$

$$\begin{aligned} b_{QU} = &- \left[\frac{1}{8}(3\emptyset_E - 2\emptyset_P - \emptyset_W)\right][\![F_e, 0]\!] + \left[\frac{1}{8}(3\emptyset_P - 2\emptyset_E - \emptyset_{EE})\right][\![-F_e, 0]\!] \\ &+ \left[\frac{1}{8}(3\emptyset_P - 2\emptyset_W - \emptyset_{WW})\right][\![F_w, 0]\!] - \left[\frac{1}{8}(3\emptyset_W - 2\emptyset_P - \emptyset_E)\right][\![-F_w, 0]\!] \\ &- \left[\frac{1}{8}(3\emptyset_N - 2\emptyset_P - \emptyset_S)\right][\![F_n, 0]\!] + \left[\frac{1}{8}(3\emptyset_P - 2\emptyset_N - \emptyset_{NN})\right][\![-F_n, 0]\!] \\ &+ \left[\frac{1}{8}(3\emptyset_P - 2\emptyset_S - \emptyset_{SS})\right][\![F_s, 0]\!] - \left[\frac{1}{8}(3\emptyset_S - 2\emptyset_P - \emptyset_N)\right][\![-F_s, 0]\!] \end{aligned} \quad (7.21)$$

In general, the discretized transport equation is used in an under-relaxed form, which can be expressed as

$$a_P\emptyset_P = \sum_{nb} a_{nb}\emptyset_{nb} + (1 - \alpha_\emptyset)a_P\emptyset_P^0 + \bar{b}_P \quad (7.22)$$

where

$$\sum_{nb} a_{nb}\emptyset_{nb} = a_w\emptyset_W + a_e\emptyset_E + a_s\emptyset_S + a_n\emptyset_N$$

$a_P = a_P^*/\alpha_\emptyset$; $\emptyset_P^0$ is the previous iterate value and $\alpha_\emptyset$ is under-relaxation factor.

7.5 SOLUTION OF THE PRESSURE FIELD

7.5.1 Derivation of Pressure Correction and Velocity Correction Equations

The implementation of the SIMPLE algorithm is discussed here.

The momentum equation given in Eq. (7.22) can be written as

$$a_P \emptyset_P = \sum_{nb} a_{nb} \emptyset_{nb} + \delta P_i + (1 - \alpha_\emptyset) a_P \emptyset_P^0 + \bar{b}_P \tag{7.23}$$

where

$\delta P_i = \delta P_x$for an X – momentum and δP_y for a Y – momentum equation.

and

$$\emptyset = u, v.$$

Eq. (7.23) can be rewritten as

$$\emptyset_P = \frac{\sum a_{nb} \emptyset_{nb} + \bar{b}_P}{a_P} + \frac{\delta P_i}{a_P} + (1 - \alpha_\emptyset) \emptyset_P^0 \tag{7.24}$$

or

$$\emptyset_P = H_{nb} + \frac{\delta P_i}{a_P} + (1 - \alpha_\emptyset) \emptyset_P^0,$$

where

$$H_{nb} = \frac{\sum a_{nb} \emptyset_{nb} + \bar{b}_P}{a_P}.$$

The algorithm can be implemented as follows:

Predictor Step:

In the SIMPLE algorithm, for a guess at pressure field P, the predicted velocities of velocity field u^*, v^* are obtained by solving this equation given below.:

$$\emptyset_P^* = (H_{nb})_P^* + \left(\frac{\delta P_i}{a_P}\right)_P^* + (1 - \alpha_\emptyset) \emptyset_P^0 \tag{7.25}$$

Since these predicted values do not satisfy the continuity equation, the corrected pressure and velocity fields are obtained from the satisfaction of the continuity equation.

Corrector step:

$$\emptyset_P^{**} = (H_{nb})_P^{**} + \left(\frac{\delta P_i}{a_P}\right)_P^{**} + (1 - \alpha_\emptyset) \emptyset_P^0 \tag{7.26}$$

Subtracting Eq. (7.25) from Eq. (7.26), we get

$$\emptyset_P^{**} = \emptyset_P^* + \left[(H_{nb})_P^{**} - (H_{nb})_P^*\right] + \left(\frac{\delta P_i}{a_P}\right)_P' \tag{7.27}$$

where the velocity correction $\emptyset_P^{**} = \emptyset_P^{**} - \emptyset_P^*$, and pressure correction $p' = p^* - p^0$.

In SIMPLE algorithm, the second term in the right-hand side of Eq. (7.27) is neglected resulting in

$$\emptyset_P^{**} = \emptyset_P^* + \left(\frac{\delta P_i}{a_P}\right)_P' \tag{7.28}$$

Now, the continuity equation can be written as

$$F_e^{**} - F_w^{**} + F_n^{**} - F_s^{**} = 0 \tag{7.29}$$

or

$$\begin{aligned} &\left[\langle\rho u^{**}\rangle_e A_{ex}^1 + \langle\rho v^{**}\rangle_e A_{ey}^1\right] - \left[\langle\rho u^{**}\rangle_w A_{wx}^1 + \langle\rho v^{**}\rangle_w A_{wy}^1\right] \\ &+ \left[\langle\rho u^{**}\rangle_n A_{nx}^2 + \langle\rho v^{**}\rangle_n A_{ny}^2\right] - \left[\langle\rho u^{**}\rangle_s A_{sx}^2 + \langle\rho v^{**}\rangle_s A_{sy}^2\right] = 0 \end{aligned} \tag{7.30}$$

7.5.2 Implementation of Momentum Interpolation

The evaluation of convective fluxes requires the calculation of cell face velocity components. These velocities must be interpolated from the nodal values for co-located grid arrangement. As discussed earlier, linear interpolation leads to the occurrence of non-physical oscillations of the pressure and/or velocity field. This is due to the de-coupling of pressure and the velocity field, resulting in a checkerboard pressure distribution, as shown by Patankar [1980]. Hence, a special interpolation technique, like that of Rhie and Chow [1983], is required for the success of the scheme. Various interpolation techniques are available in the literature. Here, we discuss the pressure weighted interpolation method-corrected, as suggested by Kobayashi and Pereira [1991], to avoid the dependence of the solution field on the under-relaxation factor and convergence rate.

In this method, the momentum equation for node P is expressed as

$$\emptyset_P = (H_{nb})_P + \left(\frac{\delta P_i}{a_P}\right)_P + (1 - \alpha_\emptyset)\emptyset_P^0,$$

where

$$H_{nb} = \frac{\sum a_{nb}\emptyset_{nb} + \bar{b}_P}{a_P}.$$

Similarly, for node E, one can write

$$\emptyset_E = (H_{nb})_E + \left(\frac{\delta P_i}{a_P}\right)_E + (1 - \alpha_\emptyset)\emptyset_E^0.$$

The velocity at the east face of control volume is obtained from equations for nodes P and E using linear interpolation of $\emptyset$ terms and the pressure between the nodes of each side of the interface resulting

$$\emptyset_e = (\bar{H}_{nb})_e + \left(\frac{\bar{1}}{a_P}\right)_e (\delta P_i)_e + (1 - \alpha_\emptyset)\emptyset_e^0 \tag{7.31}$$

where

$$(\bar{H}_{nb})_e = f(H_{nb})_P + (1 - f)(H_{nb})_E$$

$$\left(\frac{\bar{1}}{a_P}\right)_e = f\left(\frac{1}{a_P}\right)_P + (1 - f)\left(\frac{1}{a_P}\right)_E,$$

and f is the interpolation factor.

Comparing Eqs. (7.18) and (7.19), we get

$$(\delta P_i)_e = \left[A_{ei}^1(p_P - p_E) + A_{ei}^2(p_{se} - p_{ne})\right].$$

The last term in the right-hand side of Eq. (7.31) is kept explicit to avoid the dependence of solution field on under-relaxation factor. Hence, the corrected velocities are given by

$$\begin{aligned}
\emptyset_P^{**} &= \emptyset_P^* + \left(\frac{1}{a_P}\right)_P \left[A_{Pj}^1(p'_w - p'_e) + A_{Pj}^2(p'_s - p'_n)\right] \\
\emptyset_e^{**} &= \emptyset_e^* + \left(\frac{\bar{1}}{a_P}\right)_e \left[A_{ej}^1(p'_P - p'_E) + \underbrace{A_{ej}^2(p'_{se} - p'_{ne})}\right] \\
\emptyset_w^{**} &= \emptyset_w^* + \left(\frac{\bar{1}}{a_P}\right)_w \left[A_{wj}^1(p'_W - p'_P) + \underbrace{A_{wj}^2(p'_{sw} - p'_{nw})}\right] \\
\emptyset_n^{**} &= \emptyset_n^* + \left(\frac{\bar{1}}{a_P}\right)_n \left[\underbrace{A_{nj}^1(p'_{nw} - p'_{ne})} + A_{nj}^2(p'_P - p'_N)\right] \\
\emptyset_s^{**} &= \emptyset_s^* + \left(\frac{\bar{1}}{a_P}\right)_s \left[\underbrace{A_{sj}^1(p'_{sw} - p'_{se})} + A_{sj}^2(p'_S - p'_P)\right]
\end{aligned} \tag{7.32}$$

The underlined terms in the right-hand side of Eq. (7.32) arise due to non-orthogonality of the grid. The Eq. (7.32) can be expanded as

$$u_e^{**} = u_e^* + \left(\frac{1}{a_P}\right)_e \left[A_{ex}^1(p'_P - p'_E) + \underbrace{A_{ex}^2(p'_{se} - p'_{ne})}\right]$$

$$v_e^{**} = v_e^* + \left(\frac{1}{a_P}\right)_e \left[A_{ey}^1(p'_P - p'_E) + \underbrace{A_{ey}^2(p'_{se} - p'_{ne})}\right]$$

$$u_w^{**} = u_w^* + \left(\frac{1}{a_P}\right)_w \left[A_{wx}^1(p'_W - p'_P) + \underbrace{A_{wx}^2(p'_{sw} - p'_{nw})}\right]$$

$$v_w^{**} = v_w^* + \left(\frac{1}{a_P}\right)_w \left[A_{wy}^1(p'_W - p'_P) + \underbrace{A_{wy}^2(p'_{sw} - p'_{nw})}\right]$$

$$u_n^{**} = u_n^* + \left(\frac{1}{a_P}\right)_n \left[\underbrace{A_{nx}^1(p'_{nw} - p'_{ne})} + A_{nx}^2(p'_P - p'_N)\right]$$

$$v_n^{**} = v_n^* + \left(\frac{1}{a_P}\right)_n \left[\underbrace{A_{ny}^1(p'_{nw} - p'_{ne})} + A_{ny}^2(p'_P - p'_N)\right]$$

$$u_s^{**} = u_s^* + \left(\frac{1}{a_P}\right)_s \left[\underbrace{A_{sx}^1(p'_{sw} - p'_{se})} + A_{sx}^2(p'_S - p'_P)\right]$$

$$v_s^{**} = v_s^* + \left(\frac{1}{a_P}\right)_s \left[\underbrace{A_{sy}^1(p'_{sw} - p'_{se})} + A_{sy}^2(p'_S - p'_P)\right].$$

By substituting the cell face velocities in the continuity equation, a pressure correction equation can be derived in the form

$$a_P^P p'_P = \sum a_{nb}^P p'_{nb} + b_P + b_{Pno} \tag{7.33}$$

where

$$a_e^P = \left(\frac{\bar{1}}{a_P}\right)_e \left[\left(\rho_e A_{ex}^1\right) A_{ex}^1 + \left(\rho_e A_{ey}^1\right) A_{ey}^1\right]$$
$$a_w^P = \left(\frac{\bar{1}}{a_P}\right)_w \left[\left(\rho_w A_{wx}^1\right) A_{wx}^1 + \left(\rho_w A_{wy}^1\right) A_{wy}^1\right]$$
$$a_n^P = \left(\frac{\bar{1}}{a_P}\right)_n \left[\left(\rho_n A_{nx}^2\right) A_{nx}^2 + \left(\rho_n A_{ny}^2\right) A_{ny}^2\right] \quad (7.34)$$
$$a_s^P = \left(\frac{\bar{1}}{a_P}\right)_s \left[\left(\rho_s A_{sx}^2\right) A_{sx}^2 + \left(\rho_s A_{sy}^2\right) A_{sy}^2\right]$$
$$a_P^P = a_e^P + a_w^P + a_n^P + a_s^P$$

Here, b_{Pno} represents the contribution arising due to non-orthogonality of the grid for the pressure correction equation and b_P is the source term for the pressure correction equation, which is the residue of the continuity equation, i.e.

$$b_P = -\left(F_e^* - F_w^* + F_n^* - F_s^*\right) \quad (7.35)$$

The contribution due to non-orthogonal terms can be neglected altogether in the pressure correction equation, in order to achieve a faster rate of convergence. Hence, the final form of pressure correction equation becomes

$$a_P^P p'_P = \sum a_{nb}^P p'_{nb} + b_P \quad (7.36)$$

The SIMPLE algorithm can be summarized as

1. $\emptyset$ and p fields are guessed.
2. Coefficients of momentum equations are calculated from Eq. (7.20) and Eq. (7.23) and solved to obtain $u^* and\ v^*$.
3. Cell face velocities $\emptyset_e^*, \emptyset_w^*$ *etc.* are obtained from Eq. (7.31).
4. Coefficients of pressure correction equations are calculated from Eq. (7.34a), source term from Eq. (7.35) and finally Eq. (7.36) is solved to obtain p'.
5. The velocities $u^{**} and\ v^{**}$ are updated from Eq. (7.32) using the pressure corrections calculated.
6. The pressure field is updated from $p = p^* + \alpha_P p'$.
7. Other scalar equations are solved, if required.
8. Steps 2 to 7 are repeated till convergence.

7.6 EXTENSION TO THREE DIMENSION

As shown earlier, extension from two dimensions to three dimensions is straightforward. Here, we shall discuss only the form of discretized equations and the coefficients involved.

7.6.1 Discretization of Continuity Equations

$$\sum \langle \rho u\rangle_k A_{kx} + \langle \rho v\rangle_k A_{ky} + \langle \rho w\rangle_k A_{kz} = 0$$

or

$$\left[\langle \rho u\rangle_e A^1_{ex} + \langle \rho v\rangle_e A^1_{ey} + \langle \rho w\rangle_e A^1_{ez}\right] - \left[\langle \rho u\rangle_w A^1_{wx} + \langle \rho v\rangle_w A^1_{wy} + \langle \rho w\rangle_w A^1_{wz}\right]$$
$$+\left[\langle \rho u\rangle_n A^2_{nx} + \langle \rho v\rangle_n A^2_{ny} + \langle \rho w\rangle_n A^2_{nz}\right] - \left[\langle \rho u\rangle_s A^2_{sx} + \langle \rho v\rangle_s A^2_{sy} + \langle \rho w\rangle_s A^2_{sz}\right]$$
$$+\left[\langle \rho u\rangle_t A^3_{tx} + \langle \rho v\rangle_t A^3_{ty} + \langle \rho w\rangle_t A^3_{tz}\right] - \left[\langle \rho u\rangle_b A^3_{bx} + \langle \rho v\rangle_b A^3_{by} + \langle \rho w\rangle_b A^3_{bz}\right] = 0$$

or

$$F_e - F_w + F_n - F_s + F_t - F_b = 0 \tag{7.37}$$

where $F_k = \sum \rho_k \emptyset_k A_{kj};\ \emptyset \Rightarrow u, v, w;\ k \Rightarrow e, w, n, s, t, b;\ j \Rightarrow x, y, z$

7.6.2 Discretization of Convective Flux

The net convective flux through the control volume shown becomes

$$J_C = \sum \langle \rho u\rangle_k \emptyset_k A_{kx} + \sum \langle \rho v\rangle_k \emptyset_k A_{ky} + \sum \langle \rho w\rangle_k \emptyset_k A_{kz}$$

or

$$\begin{aligned} J_c = & \left[\langle \rho u\rangle_e A^1_{ex} + \langle \rho v\rangle_e A^1_{ey} + \langle \rho w\rangle_e A^1_{ez}\right]\emptyset_e \\ & - \left[\langle \rho u\rangle_w A^1_{wx} + \langle \rho v\rangle_w A^1_{wy} + \langle \rho w\rangle_w A^1_{wz}\right]\emptyset_w \\ & + \left[\langle \rho u\rangle_n A^2_{nx} + \langle \rho v\rangle_n A^2_{ny} + \langle \rho w\rangle_n A^2_{nz}\right]\emptyset_n \\ & - \left[\langle \rho u\rangle_s A^2_{sx} + \langle \rho v\rangle_s A^2_{sy} + \langle \rho w\rangle_s A^2_{sz}\right]\emptyset_s \\ & + \left[\langle \rho u\rangle_t A^3_{tx} + \langle \rho v\rangle_t A^3_{ty} + \langle \rho w\rangle_t A^3_{tz}\right]\emptyset_t \\ & - \left[\langle \rho u\rangle_b A^3_{bx} + \langle \rho v\rangle_b A^3_{by} + \langle \rho w\rangle_b A^3_{bz}\right]\emptyset_b \end{aligned}$$

or

$$J_C = F_e\emptyset_e - F_w\emptyset_w + F_n\emptyset_n - F_s\emptyset_s + F_t\emptyset_t - F_b\emptyset_b \tag{7.38}$$

These statements can be written in compact form as

$$\begin{aligned} F_e\emptyset_e &= [\![F_e,0]\!]\emptyset_P - [\![-F_e,0]\!]\emptyset_E \\ F_w\emptyset &= -[\![-F_w,0]\!]\emptyset_P + [\![F_w,0]\!]\emptyset_W \\ F_n\emptyset_n &= [\![F_n,0]\!]\emptyset_P - [\![-F_n,0]\!]\emptyset_N \\ F_s\emptyset_s &= -[\![-F_s,0]\!]\emptyset_P + [\![F_s,0]\!]\emptyset_S \\ F_t\emptyset_t &= [\![F_t,0]\!]\emptyset_P - [\![-F_t,0]\!]\emptyset_T \\ F_b\emptyset_b &= -[\![-F_b,0]\!]\emptyset_P + [\![F_b,0]\!]\emptyset_B \end{aligned} \tag{7.39}$$

Hence, the total convective flux can be written as

$$\begin{aligned} J_C =& [[\![F_e,0]\!] + [\![-F_w,0]\!] + [\![F_n,0]\!] + [\![-F_s,0]\!] + [\![F_t,0]\!] + [\![-F_b,0]\!]]\emptyset_P - [\![-F_e,0]\!]\emptyset_E \\ &- [\![F_w,0]\!]\emptyset_W - [\![-F_n,0]\!]\emptyset_N - [\![F_s,0]\!]\emptyset_S - [\![-F_t,0]\!]\emptyset_T - [\![F_b,0]\!]\emptyset_B \end{aligned} \tag{7.40}$$

7.6.3 Discretization of Diffusive Flux

The net diffusive flux is given by

$$J_D = -\sum\langle\Gamma\frac{\partial\emptyset}{\partial x}\rangle_k A_{kx} - \sum\langle\Gamma\frac{\partial\emptyset}{\partial y}\rangle_k A_{ky} - \sum\langle\Gamma\frac{\partial\emptyset}{\partial z}\rangle_k A_{kz}$$

or

$$\begin{aligned} J_D =& \left[\left\{\langle\Gamma\frac{\partial\emptyset}{\partial x}\rangle_e A^1_{ex} + \langle\Gamma\frac{\partial\emptyset}{\partial y}\rangle_e A^1_{ey} + \langle\Gamma\frac{\partial\emptyset}{\partial z}\rangle_e A^1_{ez}\right\} - \left\{\langle\Gamma\frac{\partial\emptyset}{\partial x}\rangle_w A^1_{wx} + \langle\Gamma\frac{\partial\emptyset}{\partial y}\rangle_w A^1_{wy} + \langle\Gamma\frac{\partial\emptyset}{\partial z}\rangle_w A^1_{wz}\right\}\right. \\ &+ \left\{\langle\Gamma\frac{\partial\emptyset}{\partial x}\rangle_n A^2_{nx} + \langle\Gamma\frac{\partial\emptyset}{\partial y}\rangle_n A^2_{ny} + \langle\Gamma\frac{\partial\emptyset}{\partial z}\rangle_n A^2_{nz}\right\} - \left\{\langle\Gamma\frac{\partial\emptyset}{\partial x}\rangle_s A^2_{sx} + \langle\Gamma\frac{\partial\emptyset}{\partial y}\rangle_s A^2_{sy} + \langle\Gamma\frac{\partial\emptyset}{\partial z}\rangle_s A^2_{sz}\right\} \\ &\left.+ \left\{\langle\Gamma\frac{\partial\emptyset}{\partial x}\rangle_t A^3_{tx} + \langle\Gamma\frac{\partial\emptyset}{\partial y}\rangle_t A^3_{ty} + \langle\Gamma\frac{\partial\emptyset}{\partial z}\rangle_t A^3_{tz}\right\} - \left\{\langle\Gamma\frac{\partial\emptyset}{\partial x}\rangle_b A^3_{bx} + \langle\Gamma\frac{\partial\emptyset}{\partial y}\rangle_b A^3_{by} + \langle\Gamma\frac{\partial\emptyset}{\partial z}\rangle_b A^3_{bz}\right\}\right] \end{aligned} \tag{7.41}$$

Now the diffusive flux through east face can be written as

$$J_{De} = \Gamma\frac{\partial\emptyset}{\partial x_e} A^1_{ex} + \Gamma\frac{\partial\emptyset}{\partial y_e} A^1_{ey} + \Gamma\frac{\partial\emptyset}{\partial z_e} A^1_{ez}$$

$$\begin{aligned} J_{De} =& \frac{\Gamma_e A^1_{ex}}{V_e}\left[A^1_{ex}(\emptyset_E - \emptyset_P) + A^2_{ex}(\emptyset_{ne} - \emptyset_{se}) + A^3_{ex}(\emptyset_{te} - \emptyset_{be})\right] \\ &+ \frac{\Gamma_e A^1_{ey}}{V_e}\left[A^1_{ey}(\emptyset_E - \emptyset_P) + A^2_{ey}(\emptyset_{ne} - \emptyset_{se}) + A^3_{ey}(\emptyset_{te} - \emptyset_{be})\right] \\ &+ \frac{\Gamma_e A^1_{ez}}{V_e}\left[A^1_{ez}(\emptyset_E - \emptyset_P) + A^2_{ez}(\emptyset_{ne} - \emptyset_{se}) + A^3_{ez}(\emptyset_{te} - \emptyset_{be})\right] \end{aligned}$$

$$= \frac{\Gamma_e}{V_e}\left[A^1_{ex}A^1_{ex} + A^1_{ey}A^1_{ey} + A^1_{ez}A^1_{ez}\right](\emptyset_E - \emptyset_P)$$
$$+ \frac{\Gamma_e}{V_e}\left[A^1_{ex}A^2_{ex} + A^1_{ey}A^2_{ey} + A^1_{ez}A^2_{ez}\right](\emptyset_{ne} - \emptyset_{se})$$
$$+ \frac{\Gamma_e}{V_e}\left[A^1_{ex}A^3_{ex} + A^1_{ey}A^3_{ey} + A^1_{ez}A^3_{ez}\right](\emptyset_{te} - \emptyset_{be})$$

or

$$J_{De} = d^1_e(\emptyset_E - \emptyset_P) + d^2_e(\emptyset_{ne} - \emptyset_{se}) + d^3_e(\emptyset_{te} - \emptyset_{be}).$$

Similarly, we get

$$J_{Dw} = d^1_w(\emptyset_P - \emptyset_W) + d^2_w(\emptyset_{nw} - \emptyset_{sw}) + d^3_w(\emptyset_{tw} - \emptyset_{bw})$$

$$J_{Dn} = d^1_n(\emptyset_N - \emptyset_P) + d^2_n(\emptyset_{ne} - \emptyset_{nw}) + d^3_n(\emptyset_{tn} - \emptyset_{bn})$$

$$J_{Ds} = d^1_s(\emptyset_P - \emptyset_S) + d^2_s(\emptyset_{se} - \emptyset_{sw}) + d^3_s(\emptyset_{ts} - \emptyset_{bs})$$

$$J_{Dt} = d^1_t(\emptyset_T - \emptyset_P) + d^2_t(\emptyset_{te} - \emptyset_{tw}) + d^3_t(\emptyset_{tn} - \emptyset_{ts})$$

$$J_{Db} = d^1_b(\emptyset_P - \emptyset_B) + d^2_b(\emptyset_{be} - \emptyset_{bw}) + d^3_b(\emptyset_{bn} - \emptyset_{bs}),$$

where

$$d^1_e = \frac{\Gamma_e}{V_e}\left[A^1_{ex}A^1_{ex} + A^1_{ey}A^1_{ey} + A^1_{ez}A^1_{ez}\right]$$

$$d^1_w = \frac{\Gamma_w}{V_w}\left[A^1_{wx}A^1_{wx} + A^1_{wy}A^1_{wy} + A^1_{wz}A^1_{wz}\right]$$

$$d^1_n = \frac{\Gamma_n}{V_n}\left[A^2_{nx}A^2_{nx} + A^2_{ny}A^2_{ny} + A^2_{nz}A^2_{nz}\right]$$

$$d^1_s = \frac{\Gamma_s}{V_s}\left[A^2_{sx}A^2_{sx} + A^2_{sy}A^2_{sy} + A^2_{sz}A^2_{sz}\right]$$

$$d^1_t = \frac{\Gamma_t}{V_t}\left[A^3_{tx}A^3_{tx} + A^3_{ty}A^3_{ty} + A^3_{tz}A^3_{tz}\right]$$

$$d^1_b = \frac{\Gamma_b}{V_b}\left[A^3_{bx}A^3_{bx} + A^3_{by}A^3_{by} + A^3_{bz}A^3_{bz}\right]$$

are the orthogonal parts of the diffusive fluxes, and

$$d_e^2 = \frac{\Gamma_e}{V_e}\left[A_{ex}^1 A_{ex}^2 + A_{ey}^1 A_{ey}^2 + A_{ez}^1 A_{ez}^2\right]$$

$$d_e^3 = \frac{\Gamma_e}{V_e}\left[A_{ex}^1 A_{ex}^3 + A_{ey}^1 A_{ey}^3 + A_{ez}^1 A_{ez}^3\right]$$

$$d_w^2 = \frac{\Gamma_w}{V_w}\left[A_{wx}^1 A_{wx}^2 + A_{wy}^1 A_{wy}^2 + A_{wz}^1 A_{wz}^2\right]$$

$$d_w^3 = \frac{\Gamma_w}{V_w}\left[A_{wx}^1 A_{wx}^3 + A_{wy}^1 A_{wy}^3 + A_{wz}^1 A_{wz}^3\right]$$

$$d_n^2 = \frac{\Gamma_n}{V_n}\left[A_{nx}^1 A_{nx}^2 + A_{ny}^1 A_{ny}^2 + A_{nz}^1 A_{nz}^2\right]$$

$$d_n^3 = \frac{\Gamma_n}{V_n}\left[A_{nx}^2 A_{nx}^3 + A_{ny}^2 A_{ny}^3 + A_{nz}^2 A_{nz}^3\right]$$

$$d_s^2 = \frac{\Gamma_s}{V_s}\left[A_{sx}^1 A_{sx}^2 + A_{sy}^1 A_{sy}^2 + A_{sz}^1 A_{sz}^2\right]$$

$$d_s^3 = \frac{\Gamma_s}{V_s}\left[A_{sx}^2 A_{sx}^3 + A_{sy}^2 A_{sy}^3 + A_{sz}^2 A_{sz}^3\right]$$

$$d_t^2 = \frac{\Gamma_t}{V_t}\left[A_{tx}^1 A_{tx}^3 + A_{ty}^1 A_{ty}^3 + A_{tz}^1 A_{tz}^3\right]$$

$$d_t^3 = \frac{\Gamma_t}{V_t}\left[A_{tx}^2 A_{tx}^3 + A_{ty}^2 A_{ty}^3 + A_{tz}^2 A_{tz}^3\right]$$

$$d_b^2 = \frac{\Gamma_b}{V_b}\left[A_{bx}^1 A_{bx}^3 + A_{by}^1 A_{by}^3 + A_{bz}^1 A_{bz}^3\right]$$

$$d_b^3 = \frac{\Gamma_b}{V_b}\left[A_{bx}^2 A_{bx}^3 + A_{by}^2 A_{by}^3 + A_{bz}^2 A_{bz}^3\right]$$

are the non-orthogonal parts of the diffusive flux (i.e. extra terms arising from the non-orthogonality of the grid). The detailed calculation of area vectors and volumes is given in Appendix 1.

The net diffusive flux contribution for control volume is

$$J_D = -[J_{De} - J_{Dw} + J_{Dn} - J_{Ds} + J_{Dt} - J_{Db}]$$

or

$$\begin{aligned}J_D = -\ &[d_e^1\emptyset_E + d_w^1\emptyset_W + d_n^1\emptyset_N + d_s^1\emptyset_S + d_t^1\emptyset_T + d_b^1\emptyset_B \\ &- (d_e^1 + d_w^1 + d_n^1 + d_s^1 + d_t^1 + d_b^1)\emptyset_P \\ &+ \{(d_e^2 + d_n^2)\emptyset_{ne} - (d_w^2 + d_n^2)\emptyset_{nw} - (d_e^2 + d_s^2)\emptyset_{se} + (d_w^2 + d_s^2)\emptyset_{sw} \\ &+ (d_e^3 + d_t^2)\emptyset_{te} - (d_w^3 + d_t^2)\emptyset_{tw} + (d_n^3 + d_t^3)\emptyset_{tn} - (d_s^3 + d_t^3)\emptyset_{ts} \\ &- (d_e^3 + d_b^2)\emptyset_{be} + (d_w^3 + d_b^2)\emptyset_{bw} - (d_n^3 + d_b^3)\emptyset_{bn} + (d_s^3 + d_b^3)\emptyset_{bs}\}]\end{aligned}$$

or

$$\begin{aligned}J_D = -\ &[d_e^1\emptyset_E + d_w^1\emptyset_W + d_n^1\emptyset_N + d_s^1\emptyset_S + d_t^1\emptyset_T + d_b^1\emptyset_B \\ &-(d_e^1 + d_w^1 + d_n^1 + d_s^1 + d_t^1 + d_b^1)\emptyset_P + (b_{no})]\end{aligned} \tag{7.42}$$

where

$$\begin{aligned}b_{no} = &(d_e^2 + d_n^2)\emptyset_{ne} - (d_w^2 + d_n^2)\emptyset_{nw} - (d_e^2 + d_s^2)\emptyset_{se} + (d_w^2 + d_s^2)\emptyset_{sw} \\ &+ (d_e^3 + d_t^2)\emptyset_{te} - (d_w^3 + d_t^2)\emptyset_{tw} + (d_n^3 + d_t^3)\emptyset_{tn} - (d_s^3 + d_t^3)\emptyset_{ts} \\ &- (d_e^3 + d_b^2)\emptyset_{be} + (d_w^3 + d_b^2)\emptyset_{bw} - (d_n^3 + d_b^3)\emptyset_{bn} + (d_s^3 + d_b^3)\emptyset_{bs}\end{aligned} \tag{7.43}$$

7.6.4 Discretization of the Pressure Term

The pressure contribution for an X-momentum equation in discretized form is

$$\delta P_x = A_{Px}^1(p_w - p_e) + A_{Px}^2(p_s - p_n) + A_{Px}^3(p_b - p_t).$$

The pressure contribution for Y- and Z-momentum equations are similarly given by

$$\delta P_y = A_{Py}^1(p_w - p_e) + A_{Py}^2(p_s - p_n) + A_{Py}^3(p_b - p_t)$$

$$\delta P_z = A_{Pz}^1(p_w - p_e) + A_{Pz}^2(p_s - p_n) + A_{Pz}^3(p_b - p_t).$$

7.6.5 Implementation of the QUICK Scheme

Extension of the QUICK scheme as deferred correction to three dimensions is straightforward. In this case, the coefficients become

$$a_P\emptyset_P - \sum_{nb} a_{nb}\emptyset_{nb} = \bar{b}_P \tag{7.44}$$

$$a_w = [\![F_w, 0]\!] + D_w; a_e = [\![-F_e, 0]\!] + D_e$$

$$a_s = [\![F_s, 0]\!] + D_s; a_n = [\![-F_n, 0]\!] + D_n$$

$$a_b = [\![F_b, 0]\!] + D_b; a_t = [\![-F_t, 0]\!] + D_t$$

$$a_P = a_w + a_e + a_n + a_s + a_t + a_b + [(F_e - F_w) + (F_n - F_s) + (F_t - F_b)] - S_P$$

$$\bar{b}_P = b_P + b_{QU}$$

$$\begin{aligned} b_{QU} = & -\left[\frac{1}{8}(3\emptyset_E - 2\emptyset_P - \emptyset_W)\right][\![F_e, 0]\!] + \left[\frac{1}{8}(3\emptyset_P - 2\emptyset_E - \emptyset_{EE})\right][\![-F_e, 0]\!] \\ & + \left[\frac{1}{8}(3\emptyset_P - 2\emptyset_W - \emptyset_{WW})\right][\![F_w, 0]\!] - \left[\frac{1}{8}(3\emptyset_W - 2\emptyset_P - \emptyset_E)\right][\![-F_w, 0]\!] \\ & - \left[\frac{1}{8}(3\emptyset_N - 2\emptyset_P - \emptyset_S)\right][\![F_n, 0]\!] + \left[\frac{1}{8}(3\emptyset_P - 2\emptyset_N - \emptyset_{NN})\right][\![-F_n, 0]\!] \\ & + \left[\frac{1}{8}(3\emptyset_P - 2\emptyset_S - \emptyset_{SS})\right][\![F_s, 0]\!] - \left[\frac{1}{8}(3\emptyset_S - 2\emptyset_P - \emptyset_N)\right][\![-F_s, 0]\!] \\ & - \left[\frac{1}{8}(3\emptyset_T - 2\emptyset_P - \emptyset_B)\right][\![F_t, 0]\!] + \left[\frac{1}{8}(3\emptyset_P - 2\emptyset_T - \emptyset_{TT})\right][\![-F_t, 0]\!] \\ & + \left[\frac{1}{8}(3\emptyset_P - 2\emptyset_B - \emptyset_{BB})\right][\![F_b, 0]\!] - \left[\frac{1}{8}(3\emptyset_B - 2\emptyset_P - \emptyset_T)\right][\![-F_b, 0]\!] \end{aligned}$$

7.6.6 Implementation of the SIMPLE Algorithm

The steps involved in two dimensions were discussed in Section 7.5. We mentioned that for deriving the pressure correction equation, the contribution from the non-orthogonal terms can be neglected altogether to achieve a faster rate of convergence. Hence, the final form of the equations becomes

$$u_e^{**} = u_e^* + \left(\frac{\bar{1}}{a_P}\right)_e [A_{ex}^1(p'_P - p'_E)]$$

$$v_e^{**} = v_e^* + \left(\frac{\bar{1}}{a_P}\right)_e \left[A_{ey}^1(p'_P - p'_E)\right]$$

$$w_e^{**} = w_e^* + \left(\frac{\bar{1}}{a_P}\right)_e [A_{ez}^1(p'_P - p'_E)]$$

$$u_w^{**} = u_w^* + \left(\frac{\bar{1}}{a_P}\right)_w [A_{wx}^1(p'_W - p'_P)+]$$

$$v_w^{**} = v_w^* + \left(\frac{\bar{1}}{a_P}\right)_w \left[A_{wy}^1(p'_W - p'_P)\right]$$

$$w_w^{**} = w_w^* + \left(\frac{1}{a_P}\right)_w \left[A_{wz}^1(p'_W - p'_P)\right]$$

$$u_n^{**} = u_n^* + \left(\frac{1}{a_P}\right)_n \left[A_{nx}^2(p'_P - p'_N)\right]$$

$$v_n^{**} = v_n^* + \left(\frac{1}{a_P}\right)_n \left[A_{ny}^2(p'_P - p'_N)\right]$$

$$w_n^{**} = w_n^* + \left(\frac{1}{a_P}\right)_n \left[A_{nz}^2(p'_P - p'_N)\right]$$

$$u_s^{**} = u_s^* + \left(\frac{1}{a_P}\right)_s \left[A_{sx}^2(p'_S - p'_P)\right]$$

$$v_s^{**} = v_s^* + \left(\frac{1}{a_P}\right)_s \left[A_{sy}^2(p'_S - p'_P)\right]$$

$$w_s^{**} = w_s^* + \left(\frac{1}{a_P}\right)_s \left[A_{sz}^2(p'_S - p'_P)\right]$$

$$u_t^{**} = u_t^* + \left(\frac{1}{a_P}\right)_t \left[A_{tx}^3(p'_P - p'_T)\right]$$

$$v_t^{**} = v_t^* + \left(\frac{1}{a_P}\right)_t \left[A_{ty}^3(p'_P - p'_T)\right]$$

$$w_t^{**} = w_t^* + \left(\frac{1}{a_P}\right)_t \left[A_{tz}^3(p'_P - p'_T)\right]$$

$$u_b^{**} = u_b^* + \left(\frac{1}{a_P}\right)_b \left[A_{bx}^3(p'_B - p'_P)\right]$$

$$v_b^{**} = v_b^* + \left(\frac{1}{a_P}\right)_b \left[A_{by}^3(p'_B - p'_P)\right]$$

$$w_b^{**} = w_b^* + \left(\frac{1}{a_P}\right)_b \left[A_{bz}^3(p'_B - p'_P)\right].$$

By substituting the cell face velocities in the continuity equation, a pressure correction equation can be derived in the form

$$a_P^P p'_P = \sum a_{nb}^P p'_{nb} + b_P,$$

where

$$a_e^P = \left(\frac{\bar{1}}{a_P}\right)_e \left[\left(\rho_e A_{ex}^1\right) A_{ex}^1 + \left(\rho_e A_{ey}^1\right) A_{ey}^1 + \left(\rho_e A_{ez}^1\right) A_{ez}^1\right]$$

$$a_w^P = \left(\frac{\bar{1}}{a_P}\right)_w \left[\left(\rho_w A_{wx}^1\right) A_{wx}^1 + \left(\rho_w A_{wy}^1\right) A_{wy}^1 + \left(\rho_w A_{wz}^1\right) A_{wz}^1\right]$$

$$a_n^P = \left(\frac{\bar{1}}{a_P}\right)_n \left[\left(\rho_n A_{nx}^2\right) A_{nx}^2 + \left(\rho_n A_{ny}^2\right) A_{ny}^2 + \left(\rho_n A_{nz}^2\right) A_{nz}^2\right]$$

$$a_s^P = \left(\frac{\bar{1}}{a_P}\right)_s \left[\left(\rho_s A_{sx}^2\right) A_{sx}^2 + \left(\rho_s A_{sy}^2\right) A_{sy}^2 + \left(\rho_s A_{sz}^2\right) A_{sz}^2\right]$$

$$a_t^P = \left(\frac{\bar{1}}{a_P}\right)_t \left[\left(\rho_t A_{tx}^3\right) A_{tx}^3 + \left(\rho_t A_{ty}^3\right) A_{ty}^3 + \left(\rho_t A_{tz}^3\right) A_{tz}^3\right]$$

$$a_b^P = \left(\frac{\bar{1}}{a_P}\right)_b \left[\left(\rho_b A_{bx}^3\right) A_{bx}^3 + \left(\rho_b A_{by}^3\right) A_{by}^3 + \left(\rho_b A_{bz}^3\right) A_{bz}^3\right]$$

$$a_P^P = a_e^P + a_w^P + a_n^P + a_s^P + a_t^P + a_b^P$$

$$b_P = -\left(F_e^* - F_w^* + F_n^* - F_s^* + F_t^* - F_b^*\right)$$

The SIMPLE algorithm can be summarized as

1. $\emptyset$ and p fields are guessed.
2. Coefficients of momentum equations are calculated, and momentum equations are solved to obtain $u^*, v^* and\ w^*$.
3. Cell face velocities $\emptyset_e^*, \emptyset_w^*$ *etc.* are obtained like two-dimensional case.
4. Coefficients of pressure correction equations and source term are calculated and finally pressure correction equation is solved to obtain p'.
5. The velocities $u^{**}, v^{**} and\ w^{**}$ are then updated using the pressure corrections calculated.
6. The pressure field is updated from $p = p^* + \alpha_P p'$.
7. Other scalar equations are solved, if required.
8. Steps 2 to 7 are repeated till convergence.

It may be noted that here we have used QUICK as the deferred correction for the convective scheme for both two- and three-dimensional cases. However, higher-order/high-resolution convective schemes, which are discussed in Chapter 5, can also be easily implemented.

7.7 DISCRETIZATION METHOD FOR THE CARTESIAN STRUCTURED GRID

We shall now concentrate on the rectangular domain in Cartesian coordinates as shown in Figure 7.7 and develop the flow algorithm for this geometry with co-located grids.

7.7.1 The Continuity Equation

Discretizing the continuity equation on the control volume shown in Figure 7.8, we get

$$\sum \langle\rho u\rangle_k A_{kx} + \langle\rho v\rangle_k A_{ky} = 0 \tag{7.45}$$

where k represents the faces e, w, n and s.

or

$$[\langle\rho u\rangle_e A_e] - [\langle\rho u\rangle_w A_w] + [\langle\rho v\rangle_n A_n] - [\langle\rho v\rangle_s A_s] = 0$$

or

$$F_e - F_w + F_n - F_s = 0 \tag{7.46}$$

where

$$F_k = \sum \rho_k \emptyset_k A_k; \quad \emptyset \Rightarrow u, v; \quad k \Rightarrow e, w, n, s$$

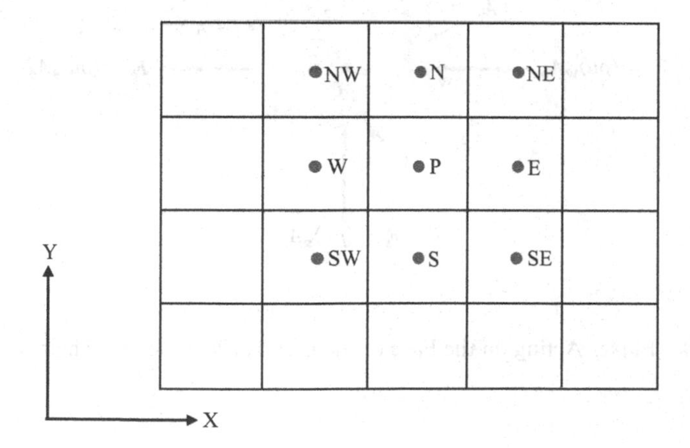

FIGURE 7.7 A Cartesian Grid.

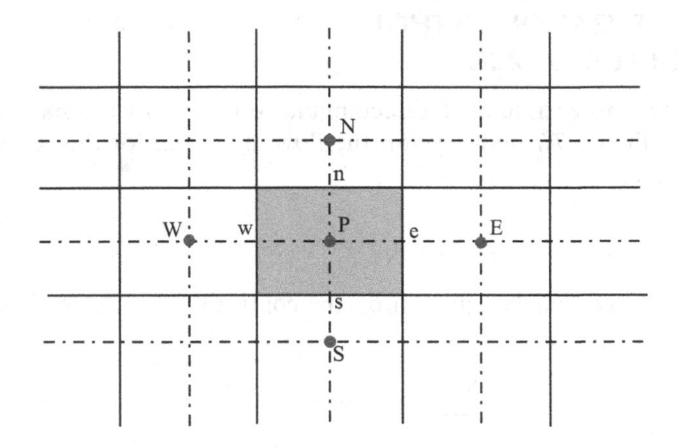

FIGURE 7.8 Typical Control Volume for the Cartesian Grid.

and e, w, n, s indicate the control volume faces in east, west, north and south directions, respectively. Here, F_k represents the total mass flux entering or leaving the control volume through normal to face k as shown in Figure 7.9. Also, A_k represents the projected area in the x or y direction for the kth face of the control volume as shown in Figure 7.9.

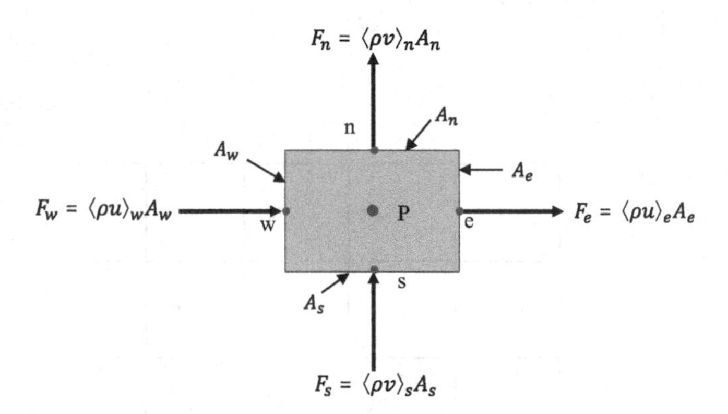

FIGURE 7.9 Fluxes Acting on the Face of the Control Volume and Their Area.

7.7.2 The Transport Equation

In general, the governing equation for the transport equation can be expressed as

$$\frac{\partial}{\partial t}\langle\rho\emptyset\rangle_P V_P + \sum\langle J\rangle_k = \langle S_\emptyset\rangle_P$$

where

$$\emptyset = u, v;\ \Gamma = \mu\ for\ momentum\ equation;$$

$$J_k = Total\ flux\ crossing\ the\ boundary = J_C + J_D;$$

$$J_C = Convective\ flux = \langle\rho u_j\emptyset\rangle_k A_k;$$

$$J_D = Diffusive\ flux = -\langle\Gamma\frac{\partial\emptyset}{\partial x_j}\rangle_k A_k;$$

$$S_\emptyset = Source\ term = body\ forces, s_i V_p + \delta p_i\ for\ momentum\ equation;$$

V_p being the volume of the cell connected with node P.

7.7.2.1 Discretization of Convective Flux

The net convective flux through the control volume shown in Figure 7.9 becomes

$$J_C = \left[\langle\rho u\rangle_e A_e\right]\emptyset_e - \left[\langle\rho u\rangle_w A_w\right]\emptyset_w + \left[\langle\rho v\rangle_n A_n\right]\emptyset_n - \langle\rho v\rangle_s A_s\emptyset_s$$

or

$$J_C = F_e\emptyset_e - F_w\emptyset_w + F_n\emptyset_n - F_s\emptyset_s \tag{7.48}$$

As discussed, for the convective terms it is required to assume the value of $\emptyset_k$ at the interface. The value of the convective property $\emptyset_k$ at the interface can be treated as described in earlier sections.

We already know that for UDS

and
$$\begin{aligned} \emptyset_e &= \emptyset_P \quad \text{if } F_e < 0 \\ \emptyset_e &= \emptyset_E \quad \text{if } F_e > 0 \end{aligned}$$

The above statements can be written in compact form as

$$F_w\emptyset_w = \emptyset_W[\![F_w, 0]\!] - \emptyset_P[\![-F_w, 0]\!]$$

$$F_e\emptyset_e = \emptyset_P[\![F_e, 0]\!] - \emptyset_E[\![-F_e, 0]\!]$$

$$F_s\emptyset_s = \emptyset_S[\![F_s, 0]\!] - \emptyset_P[\![-F_s, 0]\!]$$

$$F_n\emptyset_n = \emptyset_P[\![F_n, 0]\!] - \emptyset_N[\![-F_n, 0]\!]$$

Hence, the total convective flux can be written as

$$J_C = [[\![F_e, 0]\!] + [\![-F_w, 0]\!] + [\![F_n, 0]\!] + [\![-F_s, 0]\!]]\emptyset_P - [\![-F_e, 0]\!]\emptyset_E - [\![F_w, 0]\!]\emptyset_W - [\![-F_n, 0]\!]\emptyset_N - [\![F_s, 0]\!]\emptyset_S \tag{7.49}$$

7.7.2.2 Discretization of Diffusive Flux

Similarly, the net diffusive flux is given by

$$J_D = -\sum\left\langle \Gamma\frac{\partial\emptyset}{\partial x}\right\rangle_k A_{kx} - \sum\left\langle \Gamma\frac{\partial\emptyset}{\partial y}\right\rangle_k A_{ky}$$

or

$$J_D = -\left[\left\{\left\langle \Gamma\frac{\partial\emptyset}{\partial x}\right\rangle_e A_e\right\} - \left\{\left\langle \Gamma\frac{\partial\emptyset}{\partial x}\right\rangle_w A_w\right\} + \left\{\left\langle \Gamma\frac{\partial\emptyset}{\partial x}\right\rangle_n A_n\right\} - \left\{\left\langle \Gamma\frac{\partial\emptyset}{\partial x}\right\rangle_s A_s\right\}\right] \tag{7.50}$$

Consider an auxiliary cell centered at face e as shown in Figure 7.10.

The expression for the diffusive flux through the east face can be obtained in terms of projected areas and neighboring node values of the transport quantity $\emptyset$ as

$$J_{De} = \left\langle \Gamma\frac{\partial\emptyset}{\partial x}\right\rangle_e A_e = \frac{\Gamma_e A_e}{V_e}[A_e(\emptyset_E - \emptyset_P)]$$

or

$$J_{De} = d_e(\emptyset_E - \emptyset_P) \tag{7.51a}$$

Similarly, we get

$$J_{Dw} = d_w(\emptyset_P - \emptyset_W) \tag{7.51b}$$

$$J_{Dn} = d_n(\emptyset_N - \emptyset_P) \tag{7.51c}$$

$$J_{Ds} = d_s(\emptyset_P - \emptyset_S) \tag{7.51d}$$

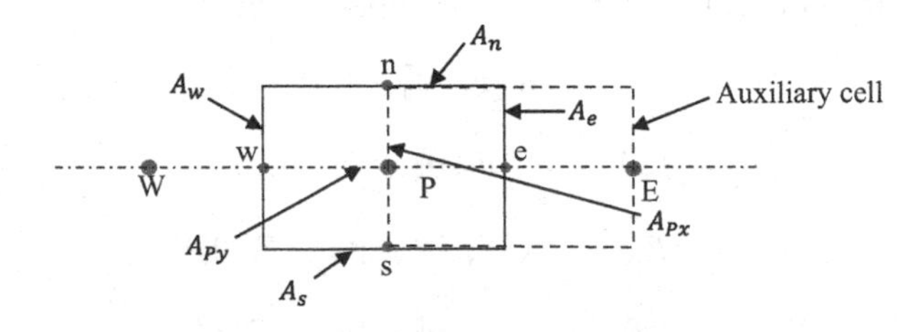

FIGURE 7.10 Auxiliary Cell Centered at Face e for the Discretization of Diffusive Flux.

where

$$d_e = \frac{\Gamma_e}{V_e}[A_e A_e]$$

$$d_w = \frac{\Gamma_w}{V_w}[A_w A_w]$$

$$d_n = \frac{\Gamma_n}{V_n}[A_n A_n]$$

$$d_s = \frac{\Gamma_s}{V_s}[A_s A_s]$$

$$J_D = -[J_{De} - J_{Dw} + J_{Dn} - J_{Ds}].$$

or

$$J_D = -[d_e \emptyset_E + d_w \emptyset_W + d_n \emptyset_N + d_s \emptyset_S - (d_e + d_w + d_n + d_s)\emptyset_P]$$

7.7.2.3 Discretization of the Pressure Term

The pressure contribution for the X-momentum equation in discretized form is

$$\delta P_x = A_{Px}(p_w - p_e)$$

The pressure contribution for Y-momentum equation is similarly given by

$$\delta P_y = A_{Py}(p_s - p_n)$$

7.7.2.4 Implementation of the QUICK Scheme

Here, the QUICK scheme is implemented as *deferred correction*, which is already outlined in Section 7.4.2.4.

In this case, for the east face we can write

$$F_e \emptyset_e = \left[\emptyset_P + \frac{1}{8}(3\emptyset_E - 2\emptyset_P - \emptyset_W)\right][\![F_e, 0]\!] - \left[\emptyset_E + \frac{1}{8}(3\emptyset_P - 2\emptyset_E - \emptyset_{EE})\right][\![-F_e, 0]\!]$$

Similarly, equations can be written for other faces.

Combining all the terms, the total transport equation, Eq. (7.47) for steady state becomes

$$\begin{aligned}
a_P^*\emptyset_P &= a_w\emptyset_W + a_e\emptyset_E + a_s\emptyset_S + a_n\emptyset_N + \bar{b}_P \\
a_P^* &= a_w + a_e + a_n + a_s - S_P + [(F_e - F_w) + (F_n - F_s)] \\
a_w &= [\![F_w, 0]\!] + D_w; a_e = [\![-F_e, 0]\!] + D_e \\
a_s &= [\![F_s, 0]\!] + D_s; a_n = [\![-F_n, 0]\!] + D_n \\
\bar{b}_P &= b_P + b_{qu}
\end{aligned} \tag{7.52}$$

$$\begin{aligned}
b_{QU} = &- \left[\frac{1}{8}(3\emptyset_E - 2\emptyset_P - \emptyset_W)\right][\![F_e, 0]\!] + \left[\frac{1}{8}(3\emptyset_P - 2\emptyset_E - \emptyset_{EE})\right][\![-F_e, 0]\!] \\
&+ \left[\frac{1}{8}(3\emptyset_P - 2\emptyset_W - \emptyset_{WW})\right][\![F_w, 0]\!] - \left[\frac{1}{8}(3\emptyset_W - 2\emptyset_P - \emptyset_E)\right][\![-F_w, 0]\!] \\
&- \left[\frac{1}{8}(3\emptyset_N - 2\emptyset_P - \emptyset_S)\right][\![F_n, 0]\!] + \left[\frac{1}{8}(3\emptyset_P - 2\emptyset_N - \emptyset_{NN})\right][\![-F_n, 0]\!] \\
&+ \left[\frac{1}{8}(3\emptyset_P - 2\emptyset_S - \emptyset_{SS})\right][\![F_s, 0]\!] - \left[\frac{1}{8}(3\emptyset_S - 2\emptyset_P - \emptyset_N)\right][\![-F_s, 0]\!]
\end{aligned}$$

In general, the discretized transport equation is used in an under-relaxed form, which can be expressed as

$$a_P\emptyset_P = \sum_{nb} a_{nb}\emptyset_{nb} + (1 - \alpha_\emptyset)a_P\emptyset_P^0 + \bar{b}_P \tag{7.53}$$

where,

$$\sum_{nb} a_{nb}\emptyset_{nb} = a_w\emptyset_W + a_e\emptyset_E + a_s\emptyset_S + a_n\emptyset_N$$

$a_P = {a_P^*}/{\alpha_\emptyset}$; $\emptyset_P^0$ is the previous iterate value and $\alpha_\emptyset$ is an under-relaxation factor.

7.7.2.5 Derivation of Pressure Correction and Velocity Correction Equations

The implementation of the SIMPLE algorithm is discussed here.

The momentum equation given in Eq. (7.53) can be written as

$$a_P\emptyset_P = \sum_{nb} a_{nb}\emptyset_{nb} + \delta P_i + (1 - \alpha_\emptyset)a_P\emptyset_P^0 + \bar{b}_P \tag{7.54}$$

where, $\delta P_i = \delta P_x$ for X-momentum and δP_y for Y-momentum,

and $\emptyset = u, v.$

Eq. (7.54) can be rewritten as

$$\emptyset_P = \frac{\sum a_{nb}\emptyset_{nb} + \bar{b}_P}{a_P} + \frac{\delta P_i}{a_P} + (1 - \alpha_\emptyset)\emptyset_P^0 \tag{7.55}$$

or

$$\emptyset_P = H_{nb} + \frac{\delta P_i}{a_P} + (1 - \alpha_\emptyset)\emptyset_P^0$$

where

$$H_{nb} = \frac{\sum a_{nb}\emptyset_{nb} + \bar{b}_P}{a_P}$$

The algorithm can be implemented as follows:

Predictor Step:

In the SIMPLE algorithm, for a guess pressure field P, the predicted velocities of velocity field u^*, v^* are obtained by solving

$$\emptyset_P^* = (H_{nb})_P^* + \left(\frac{\delta P_i}{a_P}\right)_P^* + (1 - \alpha_\emptyset)\emptyset_P^0 \tag{7.56}$$

Since these predicted values do not satisfy the continuity equation, the corrected pressure and velocity fields are obtained from the satisfaction of the continuity equation.

Corrector step:

$$\emptyset_P^{**} = (H_{nb})_P^{**} + \left(\frac{\delta P_i}{a_P}\right)_P^{**} + (1 - \alpha_\emptyset)\emptyset_P^0 \tag{7.57}$$

Subtracting Eq. (7.56) from (7.57) we get

$$\emptyset_P^{**} = \emptyset_P^* + \left[(H_{nb})_P^{**} - (H_{nb})_P^*\right] + \left(\frac{\delta P_i}{a_P}\right)_P' \tag{7.58}$$

where the velocity correction $\emptyset_p' = \emptyset_P^{**} - \emptyset_P^*$, and pressure correction $p' = p^* - p^0$.
In SIMPLE algorithm, the second term in Eq. (7.58) is neglected resulting

$$\emptyset_P^{**} = \emptyset_P^* + \left(\frac{\delta P_i}{a_P}\right)_P' \tag{7.59}$$

Now, the continuity equation can be written as

$$F_e^{**} - F_w^{**} + F_n^{**} - F_s^{**} = 0 \tag{7.60}$$

or

$$\left[(\rho u^{**})_e A_e\right] - \left[(\rho u^{**})_w A_w\right] + \left[(\rho v^{**})_n A_n\right] - \left[(\rho v^{**})_s A_s\right] = 0 \tag{7.61}$$

7.7.2.6 Implementation of the Momentum Interpolation

The momentum equation for node P is expressed as

$$\emptyset_P = (H_{nb})_P + \left(\frac{\delta P_i}{a_P}\right)_P + (1-\alpha_\emptyset)\emptyset_P^0$$

where

$$H_{nb} = \frac{\sum a_{nb}\emptyset_{nb} + \bar{b}_P}{a_P}$$

For example, for X-momentum the above equation becomes

$$u_P = (H_{nb})_P + \frac{A_{Px}(p_w - p_e)}{(a_P)_P} + (1-\alpha_\emptyset)u_P^0$$

$$H_{nb} = \frac{\sum a_{nb}u_{nb} + \bar{b}_P}{a_P}$$

Similarly, for node E, we can write

$$u_E = (H_{nb})_E + \left(\frac{\delta P_x}{a_P}\right)_E + (1-\alpha_\emptyset)u_E^0$$

The velocity at the east face of control volume is obtained from equations for nodes P and E using linear interpolation of $\emptyset$ terms and the pressure between the nodes of each side of the interface resulting

$$u_e = (\bar{H}_{nb})_e + \left(\frac{\bar{1}}{a_P}\right)_e (\delta P_x)_e + (1-\alpha_\emptyset)u_e^0 \tag{7.62}$$

where

$$(\bar{H}_{nb})_e = f(H_{nb})_P + (1-f)(H_{nb})_E$$

$$\left(\frac{\bar{1}}{a_P}\right)_e = f\left(\frac{1}{a_P}\right)_P + (1-f)\left(\frac{1}{a_P}\right)_E$$

and f is the interpolation factor.

Now

$$(\delta P_x)_e = [A_e(p_P - p_E)]$$

Hence, the corrected velocities are given by

$$u_P^{**} = u_P^* + \left(\frac{1}{a_P}\right)_P [A_{Px}(p'_W - p'_e)]$$
$$v_P^{**} = v_P^* + \left(\frac{1}{a_P}\right)_P [A_{Py}(p'_s - p'_n)]$$
$$u_e^{**} = u_e^* + \left(\frac{\bar{1}}{a_P}\right)_e [A_e(p'_P - p'_E)]$$
$$u_w^{**} = u_w^* + \left(\frac{\bar{1}}{a_P}\right)_w [A_w(p'_W - p'_P)] \tag{7.63}$$
$$v_n^{**} = v_n^* + \left(\frac{\bar{1}}{a_P}\right)_n [A_n(p'_P - p'_N)]$$
$$v_s^{**} = v_s^* + \left(\frac{\bar{1}}{a_P}\right)_s [A_s(p'_S - p'_P)]$$

By substituting the cell face velocities in the continuity equation, a pressure correction equation can be derived in the form

$$a_P^P p'_P = \sum a_{nb}^P p'_{nb} + b_P \tag{7.64}$$

where

$$a_e^P = \left(\frac{\bar{1}}{a_P}\right)_e [(\rho_e A_e) A_e]; \; a_w^P = \left(\frac{\bar{1}}{a_P}\right)_w [(\rho_w A_w) A_w]$$
$$a_n^P = \left(\frac{\bar{1}}{a_P}\right)_n [(\rho_n A_n) A_n]; \; a_s^P = \left(\frac{\bar{1}}{a_P}\right)_s [(\rho_s A_s) A_s] \tag{7.65}$$
$$a_P^P = a_e^P + a_w^P + a_n^P + a_s^P$$

Here, b_P is the source term for the pressure correction equation, which is the residue of the continuity equation, i.e.

$$b_P = -\left(F_e^* - F_w^* + F_n^* - F_s^*\right) \tag{7.66}$$

Here, the contribution due to non-orthogonal terms is absent in the pressure correction equation due to Cartesian grid. Hence, the final form of pressure correction equation becomes

$$a_P^P p'_P = \sum a_{nb}^P p'_{nb} + b_P \tag{7.67}$$

The SIMPLE algorithm can be summarized as

1. $\emptyset$ and p fields are guessed.
2. Coefficients of momentum equations are calculated, and it is solved to obtain u^* *and* v^*.
3. Cell face velocities $\emptyset_e^*, \emptyset_w^*$ *etc.* are then obtained.

4. Coefficients of pressure correction equations are calculated from Eq. (7.65) and source term from Eq. (7.66) and finally Eq. (7.67) is solved to obtain p'.
5. The velocities $u^{**} and\ v^{**}$ are updated from the Eq. (7.63) using the pressure corrections calculated.
6. The pressure field is updated from $p = p^* + \alpha_P p'$.
7. Other scalar equations are solved, if required.
8. Steps 2 to 7 are repeated till convergence.

7.8 DISCRETIZATION METHOD FOR THE NON-ORTHOGONAL UNSTRUCTURED GRID

For complex geometries, grid generation with structured grids requires more computational effort than flow computations. The use of unstructured grids can considerably reduce computational effort for grid generation. For local mesh refining or adaptive meshes, use of unstructured grids is very useful. Hence, most commercial computer codes adapt grid generation by unstructured grids by default.

7.8.1 THE CONTINUITY EQUATION

$$\sum \langle \rho u \rangle_k A_{kx} + \langle \rho v \rangle_k A_{ky} = 0 \tag{7.68}$$

where k represents the faces.

or

$$\begin{aligned} &\left[\langle \rho u \rangle_{k1} A^1_{k1x} + \langle \rho v \rangle_{k1} A^1_{k1y}\right] + \left[\langle \rho u \rangle_{k2} A^1_{k2x} + \langle \rho v \rangle_{k2} A^1_{k2y}\right] \\ &+ \left[\langle \rho u \rangle_{k3} A^1_{k3x} + \langle \rho v \rangle_{k3} A^1_{k3y}\right] = 0 \end{aligned} \tag{7.69}$$

or

$$F_{k1} + F_{k2} + F_{k3} = 0 \tag{7.70}$$

Here, F_k represents total mass flux entering or leaving the control volume through normal to face k as shown in Figure 7.11.

In these expressions, A_{kj} represents the projected area in the x or y direction for the kth face of the control volume as shown in Figure 7.12.

7.8.2 THE TRANSPORT EQUATION

In general, the governing equation for the transport equation can be expressed as

$$\frac{\partial}{\partial t} \langle \rho \emptyset \rangle_P V_P + \sum \langle J \rangle_k = \langle S_\emptyset \rangle_P \tag{7.71}$$

where $\emptyset = u, v$; $\Gamma = \mu$ for momentum equation;

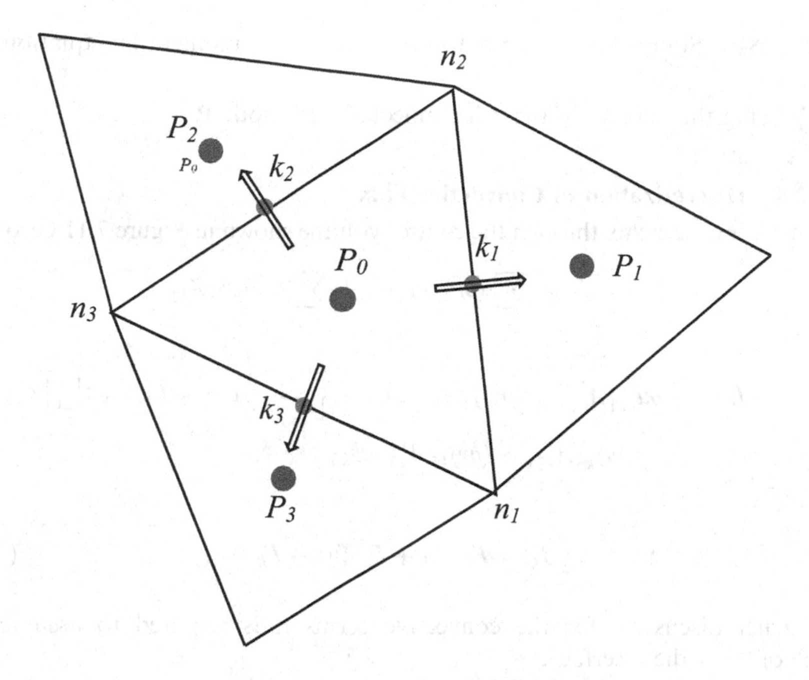

FIGURE 7.11 Control Volume Used for the Unstructured Grid.

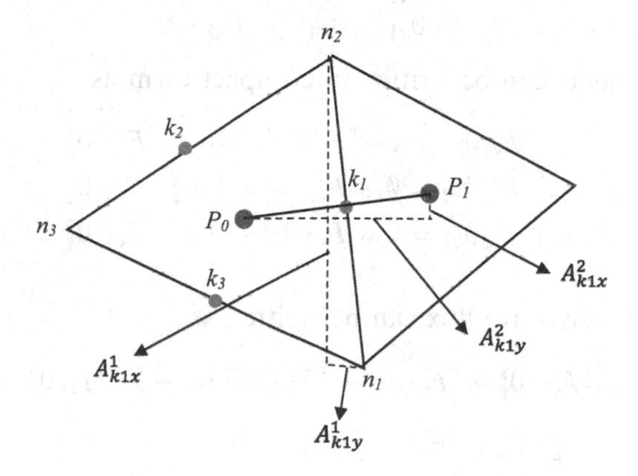

FIGURE 7.12 Projected Area for the k_1 Face for the Unstructured Grid.

$$J_k = Total\ flux\ crossing\ the\ boundary = J_C + J_D;$$

$$J_C = Convective\ flux = \langle \rho u \rangle_j \emptyset_k A_{kj};$$

$$J_D = Diffusive\ flux = -\langle \Gamma \frac{\partial \emptyset}{\partial x_{j_k}} \rangle A_{kj};$$

$$S_\emptyset = \text{Source term} = \text{body forces} = s_i V_p + \delta p_i \text{ momentum equation;}$$

V_p being the volume of the cell connected with node P.

7.8.2.1 Discretization of Convective Flux

The net convective flux through the control volume shown in Figure 7.11 becomes

$$J_C = \sum \langle \rho u \rangle_k \emptyset_k A_{kx} + \sum \langle \rho v \rangle_k \emptyset_k A_{ky}$$

or

$$J_C = \left[\langle \rho u \rangle_{k1} A^1_{k1x} + \langle \rho v \rangle_{k1} A^1_{k1y} \right] \emptyset_{k1} + \left[\langle \rho u \rangle_{k2} A^1_{k2x} + \langle \rho v \rangle_{k2} A^1_{k2y} \right] \emptyset_{k2} + \left[\langle \rho u \rangle_{k3} A^1_{k3x} + \langle \rho v \rangle_{k3} A^1_{k3y} \right] \emptyset_{k3}$$

or

$$J_C = F_{k1}\emptyset_{k1} + F_{k2}\emptyset_{k2} + F_{k3}\emptyset_{k3}. \tag{7.72}$$

As earlier discussed, for the convective terms it is required to assume the value of $\emptyset_k$ at the interface.

We already know that for UDS

$$\emptyset_{k1} = \emptyset_{P0} \quad \text{if } F_{k1} > 0$$

and

$$\emptyset_{k1} = \emptyset_{P1} \quad \text{if } F_{k1} < 0$$

The above statements can be written in compact form as

$$\begin{aligned} F_{k1}\emptyset_{k1} &= \emptyset_{P0}[\![F_{k1}, 0]\!] - \emptyset_{P1}[\![-F_{k1}, 0]\!] \\ F_{k2}\emptyset_{k2} &= \emptyset_{P0}[\![F_{k2}, 0]\!] - \emptyset_{P2}[\![-F_{k2}, 0]\!] \\ F_{k3}\emptyset_{k3} &= \emptyset_{P0}[\![F_{k3}, 0]\!] - \emptyset_{P3}[\![-F_{k3}, 0]\!] \end{aligned} \tag{7.73}$$

Hence, the total convective flux can be written as

$$\begin{aligned} J_C = {} & [[\![F_{k1}, 0]\!] + [\![F_{k2}, 0]\!] + [\![F_{k3}, 0]\!]]\emptyset_{P0} - [\![-F_{k1}, 0]\!]\emptyset_{P1} \\ & - [\![-F_{k2}, 0]\!]\emptyset_{P2} - [\![-F_{k3}, 0]\!]\emptyset_{P3} \end{aligned} \tag{7.74}$$

7.8.2.2 Discretization of Diffusive Flux

Similarly, the net diffusive flux is given by

$$J_D = -\sum \left\langle \Gamma \frac{\partial \emptyset}{\partial x} \right\rangle_k A_{kx} - \sum \left\langle \Gamma \frac{\partial \emptyset}{\partial y} \right\rangle_k A_{ky} \tag{7.75}$$

or

$$J_D = -\left[\left\{\left\langle\Gamma\frac{\partial\emptyset}{\partial x}\right\rangle_{k1}A^1_{k1x} + \left\langle\Gamma\frac{\partial\emptyset}{\partial y}\right\rangle_{k1}A^1_{k1y}\right\} + \left\{\left\langle\Gamma\frac{\partial\emptyset}{\partial x}\right\rangle_{k2}A^1_{k2x} + \left\langle\Gamma\frac{\partial\emptyset}{\partial y}\right\rangle_{k2}A^1_{k2y}\right\}\right.$$
$$\left. + \left\{\left\langle\Gamma\frac{\partial\emptyset}{\partial x}\right\rangle_{k3}A^1_{k3x} + \left\langle\Gamma\frac{\partial\emptyset}{\partial y}\right\rangle_{k3}A^1_{k3y}\right\}\right] \tag{7.76}$$

Consider an auxiliary cell centered at face $k_{1,}$ as shown in Figure 7.13. Referring to Figure 7.12, the expression for the diffusive flux through the east face can be obtained in terms of projected areas and neighboring node values for the transport quantity $\langle\emptyset\rangle$ as

$$J_{Dk1} = \left\langle\Gamma\frac{\partial\emptyset}{\partial x}\right\rangle_{k1}A^1_{k1x} + \left\langle\Gamma\frac{\partial\emptyset}{\partial y}\right\rangle_{k1}A^1_{k1y}$$

$$\begin{aligned} J_{Dk1} = {} & \frac{\Gamma_{k1}A^1_{k1x}}{V_{k1}}\left[A^1_{k1x}(\emptyset_{P1} - \emptyset_{P0}) + A^2_{k1x}(\emptyset_{n2} - \emptyset_{n1})\right] \\ & + \frac{\Gamma_{k1}A^1_{ey}}{V_{k1}}\left[A^1_{k1y}(\emptyset_{P1} - \emptyset_{P0}) + A^2_{k1y}(\emptyset_{n2} - \emptyset_{n1})\right] \end{aligned}$$

$$= \frac{\Gamma_{k1}}{V_{k1}}\left[A^1_{k1x}A^1_{k1x} + A^1_{k1y}A^1_{k1y}\right](\emptyset_{P1} - \emptyset_{P0}) + \frac{\Gamma_{k1}}{V_{k1}}\left[A^1_{k1x}A^2_{k1x} + A^1_{k1y}A^2_{k1y}\right](\emptyset_{n2} - \emptyset_{n1})$$

or
similarly, we get

$$\begin{aligned} J_{Dk1} &= d^1_{k1}(\emptyset_{P1} - \emptyset_{P0}) + d^2_{k1}(\emptyset_{n2} - \emptyset_{n1}) \\ J_{Dk2} &= d^1_{k2}(\emptyset_{P2} - \emptyset_{P0}) + d^2_{k2}(\emptyset_{n3} - \emptyset_{n2}) \\ J_{Dk3} &= d^1_{k3}(\emptyset_{P3} - \emptyset_{P0}) + d^2_{k3}(\emptyset_{n1} - \emptyset_{n3}) \end{aligned} \tag{7.77}$$

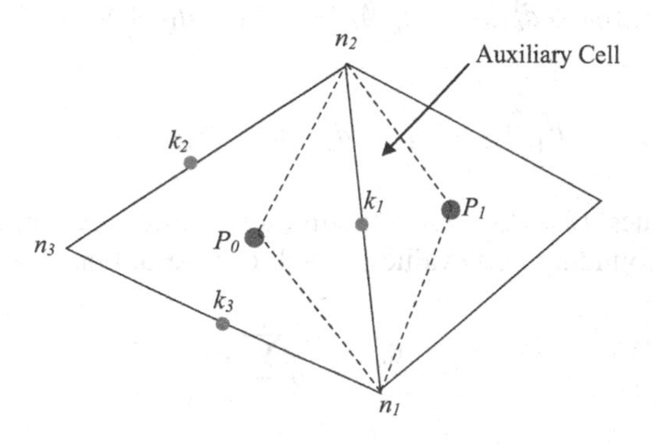

FIGURE 7.13 Auxiliary Cell for the k_1 Face for the Unstructured Grid.

where

$$d_{k1}^{2} = \frac{\Gamma_{k1}}{V_{k1}}\left[A_{k1x}^{1}A_{k1x}^{2} + A_{k1y}^{1}A_{k1y}^{2}\right]$$

$$d_{k2}^{2} = \frac{\Gamma_{k2}}{V_{k2}}\left[A_{k2x}^{1}A_{k2x}^{2} + A_{k2y}^{1}A_{k2y}^{2}\right]$$

$$d_{k3}^{2} = \frac{\Gamma_{k3}}{V_{k3}}\left[A_{k3x}^{1}A_{k3x}^{2} + A_{k3y}^{1}A_{k3y}^{2}\right]$$

are the orthogonal parts of the diffusive fluxes, and

$$d_{k1}^{2} = \frac{\Gamma_{k1}}{V_{k1}}\left[A_{k1x}^{1}A_{k1x}^{2} + A_{k1y}^{1}A_{k1y}^{2}\right]$$

$$d_{k2}^{2} = \frac{\Gamma_{k2}}{V_{k2}}\left[A_{k2x}^{1}A_{k2x}^{2} + A_{k2y}^{1}A_{k2y}^{2}\right]$$

$$d_{k3}^{2} = \frac{\Gamma_{k3}}{V_{k3}}\left[A_{k3x}^{1}A_{k3x}^{2} + A_{k3y}^{1}A_{k3y}^{2}\right]$$

are the non-orthogonal parts of the diffusive flux (i.e. extra terms arising from the non-orthogonality of the grid). Areas $A^{1}$and A^{2} are defined in Figure 7.12.

The net diffusive flux contribution for control volume is

$$J_D = -[J_{Dk1} + J_{Dk2} + J_{Dk3}]$$

or

$$J_D = -\left[d_{k1}^{1}\emptyset_{P1} + d_{k2}^{1}\emptyset_{P2} + d_{k3}^{1}\emptyset_{P3} - \left(d_{k1}^{1} + d_{k2}^{1} + d_{k3}^{1}\right)\emptyset_{P0} + \left\{d_{k1}^{2}(\emptyset_{n2} - \emptyset_{n1}) + d_{k2}^{2}(\emptyset_{n3} - \emptyset_{n2}) + d_{k3}^{2}(\emptyset_{n1} - \emptyset_{n3})\right\}\right]$$

or

$$J_D = -\left[d_{k1}^{1}\emptyset_{P1} + d_{k2}^{1}\emptyset_{P2} + d_{k3}^{1}\emptyset_{P3} - \left(d_{k1}^{1} + d_{k2}^{1} + d_{k3}^{1}\right)\emptyset_{P0} + (b_{no})\right] \tag{7.78}$$

where

$$b_{no} = d_{k1}^{2}(\emptyset_{n2} - \emptyset_{n1}) + d_{k2}^{2}(\emptyset_{n3} - \emptyset_{n2}) + d_{k3}^{2}(\emptyset_{n1} - \emptyset_{n3})$$

The corner values ($\emptyset_{n1}, \emptyset_{n2}$ *etc.*) are unknown and are approximated using neighboring surrounding node values, which can be obtained from

$$\emptyset_{ni} = \frac{1}{nb}\sum_{i=1}^{nb}\emptyset_{Pi} \tag{7.79}$$

The non-orthogonal diffusive flux terms, b_{no}, can be treated either implicitly or explicitly. As discussed earlier, the non-orthogonal contributions are lumped in the source term, which is handled explicitly.

7.8.2.3 Discretization of the Pressure Term

The pressure contribution for the X-momentum equation in the discretized form is

$$\delta P_x = -\sum \langle p \rangle_k A_{kx} = -[p_{k1}A_{k1x} + p_{k2}A_{k2x} + p_{k3}A_{k3x}] \tag{7.80}$$

The pressure contribution for Y-momentum equation is similarly given by

$$\delta P_y = -[p_{k1}A_{k1y} + p_{k2}A_{k2y} + p_{k3}A_{k3y}] \tag{7.81}$$

In general, the discretized transport equation is used in under-relaxed form, which can be expressed as

$$a_{P0}\emptyset_{P0} = \sum_{nb} a_{nb}\emptyset_{nb} + (1-\alpha_\emptyset)a_P\emptyset_P^0 + \bar{b}_P \tag{7.82}$$

where,

$$\sum\nolimits_{nb} a_{nb}\emptyset_{nb} = a_{k1}\emptyset_{k1} + a_{k2}\emptyset_{k2} + a_{k3}\emptyset_{k3}$$

$a_P = a_P^*/\alpha_\emptyset$; $\emptyset_P^0$ is the previous iterate value and $\alpha_\emptyset$ is an under-relaxation factor.

and

$$a_{k1} = [\![-F_{k1}, 0]\!] + d_{k1}^1\ ;\ a_{k2} = [\![-F_{k2}, 0]\!] + d_{k2}^1\ ;\ a_{k3} = [\![-F_{k3}, 0]\!] + d_{k3}^1$$

$$a_{P0} = \sum_{nb} a_{nb}\emptyset_{nb} - S_P + (F_{k1} + F_{k2} + F_{k3}) \tag{7.83}$$

$$\bar{b}_P = b_P + b_{no}$$

7.9 SOLUTION OF THE PRESSURE FIELD

7.9.1 Derivation of Pressure Correction and Velocity Correction Equations

The implementation of the SIMPLE algorithm for the unstructured grid is discussed here.

The momentum equation given in Eq. (7.82) can be written as

$$a_{P0}\emptyset_{P0} = \sum_{nb} a_{nb}\emptyset_{nb} + \delta P_i + (1-\alpha_\emptyset)a_{P0}\emptyset_{P0}^0 + \bar{b}_P \tag{7.84}$$

where

$\delta P_i = \delta P_x$ for X-momentum and δP_y for Y-momentum,

and

$$\emptyset = u, v.$$

Eq. (7.84) can be rewritten as

$$\emptyset_{P0} = \frac{\sum a_{nb}\emptyset_{nb} + \bar{b}_P}{a_{P0}} + \frac{\delta P_i}{a_{P0}} + (1 - \alpha_\emptyset)\emptyset^0_{P0} \tag{7.85}$$

or

$$\emptyset_{P0} = H_{nb} + \frac{\delta P_i}{a_{P0}} + (1 - \alpha_\emptyset)\emptyset^0_{P0}$$

Where

$$H_{nb} = \frac{\sum a_{nb}\emptyset_{nb} + \bar{b}_P}{a_{P0}}$$

The algorithm can be implemented as follows:

Predictor Step:

In the SIMPLE algorithm, for a guess pressure field P, the predicted velocities of velocity field u^*, v^* are obtained by solving this equation:

$$\emptyset^*_{P0} = (H_{nb})^*_{P0} + \left(\frac{\delta P_i}{a_{P0}}\right)^*_{P0} + (1 - \alpha_\emptyset)\emptyset^0_{P0} \tag{7.86}$$

Since these predicted values do not satisfy the continuity equation, the corrected pressure and velocity fields are obtained from the satisfaction of the continuity equation.

Corrector step:

$$\emptyset^{**}_{P0} = (H_{nb})^{**}_{P0} + \left(\frac{\delta P_i}{a_{P0}}\right)^{**}_{P0} + (1 - \alpha_\emptyset)\emptyset^0_{P0} \tag{7.87}$$

Subtracting Eq. (7.86) from (7.87) we get

$$\emptyset^{**}_{P0} = \emptyset^*_{P0} + \left[(H_{nb})^{**}_{P0} - (H_{nb})^*_{P0}\right] + \left(\frac{\delta P_i}{a_{P0}}\right)'_{P0} \tag{7.88}$$

where the velocity correction $\emptyset'_{P0} = \emptyset^{**}_{P0} - \emptyset^*_{P0}$, and pressure correction $p' = p^* - p^0$.

In SIMPLE algorithm, the second term in Eq. (7.88) is neglected resulting

$$\emptyset^{**}_{P0} = \emptyset^*_{P0} + \left(\frac{\delta P_i}{a_{P0}}\right)'_{P0} \tag{7.89}$$

Now, the continuity equation can be written as

$$F_{k1}^{**} + F_{k2}^{**} + F_{k3}^{**} = 0 \tag{7.90}$$

or

$$\begin{aligned}\left[\langle\rho u^{**}\rangle_{k1}A_{k1x}^{1} + \langle\rho v^{**}\rangle_{k1}A_{k1y}^{1}\right] + \left[\langle\rho u^{**}\rangle_{k2}A_{k2x}^{1} + \langle\rho v^{**}\rangle_{k2}A_{k2y}^{1}\right] \\ + \left[\langle\rho u^{**}\rangle_{k3}A_{k3x}^{1} + \langle\rho v^{**}\rangle_{k3}A_{k3y}^{1}\right] = 0\end{aligned} \tag{7.91}$$

7.9.2 Implementation of the Momentum Interpolation

The evaluation of convective fluxes requires the calculation of the cell face velocity components. These velocities must be interpolated from the nodal values for co-located grid arrangement as follows:

The momentum equation for node P_0 is expressed as

$$\emptyset_{P0} = (H_{nb})_{P0} + \left(\frac{\delta P_i}{a_P}\right)_{P0} + (1 - \alpha_{\emptyset})\emptyset_{P0}^{0}$$

where

$$H_{nb} = \frac{\sum a_{nb}\emptyset_{nb} + \bar{b}_P}{a_{P0}}$$

Similarly, for node P_0, one can write

$$\emptyset_{P1} = (H_{nb})_{P1} + \left(\frac{\delta P_i}{a_P}\right)_{P1} + (1 - \alpha_{\emptyset})\emptyset_{P1}^{0}$$

The velocity at the k_1 face of control volume is obtained from equations for nodes P_0 and P_1 using linear interpolation of $\emptyset$ terms and the pressure between the nodes of each side of the interface resulting

$$\emptyset_{k1} = (\bar{H}_{nb})_{k1} + \left(\frac{\bar{1}}{a_P}\right)_{k1}(\delta P_i)_{k1} + (1 - \alpha_{\emptyset})\emptyset_{k1}^{0} \tag{7.92}$$

where

$$(\bar{H}_{nb})_{k1} = f(H_{nb})_{P0} + (1 - f)(H_{nb})_{P1}$$

$$\left(\frac{\bar{1}}{a_P}\right)_{k1} = f\left(\frac{1}{a_P}\right)_{P0} + (1 - f)\left(\frac{1}{a_P}\right)_{P1}$$

and f is the interpolation factor.

The last term in the right-hand side of Eq. (7.92) can be kept explicit as discussed earlier.

Hence, the corrected velocities are given by

$$\emptyset^{**}_{k1} = \emptyset^{*}_{k1} + \left(\frac{\bar{1}}{a_P}\right)_{k1} \left[A^{1}_{k1j}(p'_{P0} - p'_{P1}) + \underbrace{A^{2}_{k1j}(p'_{n1} - p'_{n2})} \right]$$

$$\emptyset^{**}_{k2} = \emptyset^{*}_{k2} + \left(\frac{\bar{1}}{a_P}\right)_{k2} \left[A^{1}_{k2j}(p'_{P2} - p'_{P0}) + \underbrace{A^{2}_{k2j}(p'_{n3} - p'_{n2})} \right] \qquad (7.93)$$

$$\emptyset^{**}_{k3} = \emptyset^{*}_{k3} + \left(\frac{\bar{1}}{a_P}\right)_{k3} \left[A^{1}_{k3j}(p'_{P0} - p'_{P3}) + \underbrace{A^{2}_{k3j}(p'_{n1} - p'_{n3})} \right]$$

The underlined terms in the right-hand side of Eq. (7.93) arise from the non-orthogonality of the grid. Eq. (7.93) can be expanded as

$$u^{**}_{k1} = u^{*}_{k1} + \left(\frac{\bar{1}}{a_P}\right)_{k1} \left[A^{1}_{k1x}(p'_{P0} - p'_{P1}) + A^{2}_{k1x}(p'_{n1} - p'_{n2}) \right]$$

$$v^{**}_{k1} = v^{*}_{k1} + \left(\frac{\bar{1}}{a_P}\right)_{k1} \left[A^{1}_{k1y}(p'_{P0} - p'_{P1}) + A^{2}_{k1y}(p'_{n1} - p'_{n2}) \right]$$

$$u^{**}_{k2} = u^{*}_{k2} + \left(\frac{\bar{1}}{a_P}\right)_{k2} \left[A^{1}_{k2x}(p'_{P2} - p'_{P0}) + A^{2}_{k2x}(p'_{Pn3} - p'_{n2}) \right]$$

$$v^{**}_{k2} = v^{*}_{k2} + \left(\frac{\bar{1}}{a_P}\right)_{k2} \left[A^{1}_{k2y}(p'_{P2} - p'_{P0}) + A^{2}_{k2y}(p'_{n3} - p'_{n2}) \right]$$

$$u^{**}_{k3} = u^{*}_{k3} + \left(\frac{\bar{1}}{a_P}\right)_{k3} \left[A^{1}_{k3x}(p'_{P0} - p'_{P3}) + A^{2}_{k3x}(p'_{n1} - p'_{n3}) \right]$$

$$v^{**}_{k3} = v^{*}_{k3} + \left(\frac{\bar{1}}{a_P}\right)_{k3} \left[A^{1}_{k3y}(p'_{P0} - p'_{P3}) + A^{2}_{k3y}(p'_{n1} - p'_{n3}) \right]$$

By substituting the cell face velocities in the continuity equation, a pressure correction equation can be derived in the form

$$a^{P}_{P0}p'_{nb} = \sum a^{P}_{nb}p'_{nb} + b_P + b_{Pno} \qquad (7.94)$$

where

$$a^{P}_{P1} = \left(\frac{\bar{1}}{a_P}\right)_{k1} \left[(\rho_{k1}A^{1}_{k1x})A^{1}_{k1x} + \left(\rho_{k1}A^{1}_{k1y}\right)A^{1}_{k1y} \right]$$

$$a_{P2}^{P} = \left(\frac{1}{a_P}\right)_{k2}\left[\left(\rho_{k2}A_{k2x}^{1}\right)A_{k2x}^{1} + \left(\rho_{k2}A_{k2y}^{1}\right)A_{k2y}^{1}\right] \quad (7.95)$$

$$a_{P3}^{P} = \left(\frac{1}{a_P}\right)_{k3}\left[\left(\rho_{k3}A_{k3x}^{1}\right)A_{k3x}^{1} + \left(\rho_{k3}A_{k3y}^{1}\right)A_{k3y}^{1}\right]$$

Here, b_{Pno} represents the contribution arising due to non-orthogonality of the grid for pressure correction equation and b_P is the source term for the pressure correction equation, which is the residue of the continuity equation, i.e.

$$b_P = -\left(F_{k1}^{*} + F_{k2}^{*} + F_{k3}^{*}\right) \quad (7.96)$$

The contribution from the non-orthogonal terms, as explained earlier, can be neglected altogether in the pressure correction equation. Hence, the final form of the pressure correction equation becomes

$$a_{P0}^{P}p'_{P0} = \sum a_{nb}^{P}p'_{nb} + b_P \quad (7.97)$$

The SIMPLE algorithm can be summarized as

1. $\emptyset$ and p fields are guessed.
2. Coefficients of the momentum equations are calculated and Eq. (7.86) is solved to obtain u^{*} *and* v^{*}.
3. Cell face velocities $\emptyset_{k1}^{*}, \emptyset_{k2}^{*}$ *etc.* are obtained from Eq. (7.92).
4. Coefficients of pressure correction equations are calculated from Eq. (7.95), and source term from Eq. (7.96) and finally Eq. (7.97) is solved to obtain p'.
5. The velocities u^{**} *and* v^{**} are updated from the Eq. (7.93) using the pressure corrections calculated.
6. The pressure field is updated from $p = p^{*} + \alpha_P p'$
7. Other scalar equations are solved, if required.
8. Steps 2 to 7 are repeated till convergence.

7.9.3 Implementation of Higher-Order Schemes

In Chapter 5, we saw that Higher-Order (HO) schemes involve the nodes' relations to *UU, U* and *D*. Although the nodes *U* and *D* are available, for unstructured grids defining the UU node may not be straightforward as can be seen in Figure 7.14.

However, one way is to define a pseudo *UU* node. Assume *UU* node lies on the line joining *U* and *D* such that *U* is mid-point of the line joining *UU* and *D*. Then

$$\emptyset_{UU} = \emptyset_D - 2\nabla\emptyset_U . d_{UD}$$

where d_{UD} is the vector between *U* and *D*.

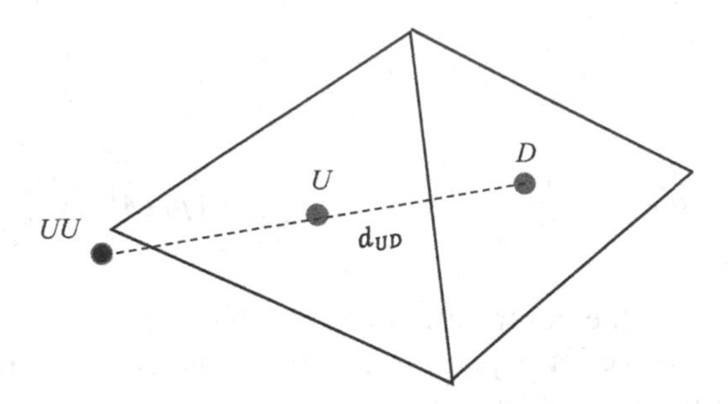

FIGURE 7.14 The *UU* Node of HO Schemes for an Unstructured Grid.

$\emptyset_{UU}$ can be found by using various HR schemes discussed in Chapter 5. For further details, readers may refer Moukalled et al. [2015].

7.10 SUMMARY

- The use of Cartesian coordinate is not suitable for complex geometries.
- Non-orthogonal grids provide the flexibility of mapping an irregular boundary.
- The use of a staggered grid system may increase the complexity in flow problems requiring a three-dimensional generalized co-ordinate systems.
- Co-located grid arrangement is attractive due to its simplicity for the location of variables.
- However, this arrangement leads to decoupling between pressure and the velocity field, if linear interpolation is used.
- Hence, a special interpolation technique is required for the success of the scheme.
- Non-orthogonal terms arising from the discretization of diffusive fluxes can be lumped into the source term and handled explicitly to avoid solving a nine-diagonal matrix for two-dimensional and 19 diagonal matrices for three-dimensional problems.
- The contribution of non-orthogonal terms can be neglected altogether in a pressure-correction equation, to achieve a faster rate of convergence.

QUESTIONS

1) A non-Cartesian orthogonal structure grid is shown in Figure 7.15 Discretize the transport equation and formulate the SIMPLE procedure to solve it. Consider only UDS as convective scheme.

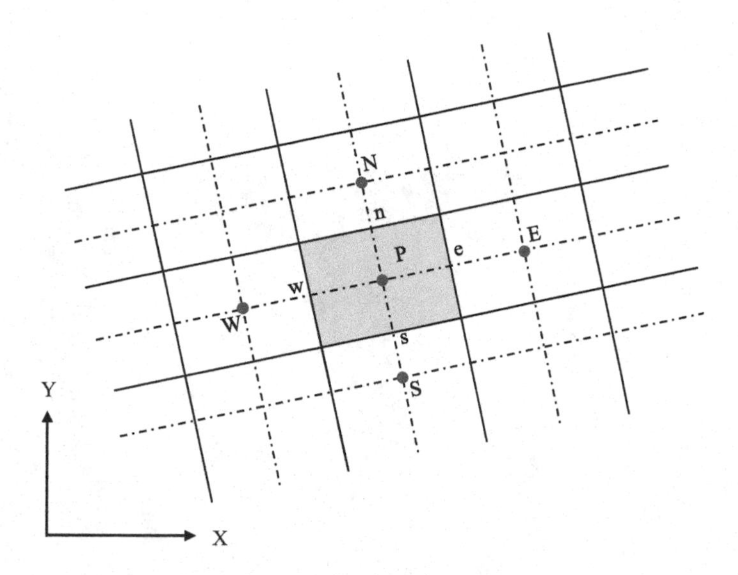

FIGURE 7.15 Non-Cartesian Orthogonal Structured Grid.

2) Write the discretized equations using the various convective schemes discussed in Section 5.3.5.4 for the two-dimensional non-orthogonal collocated structured grid shown in Figure 7.15.

8 Solution of Algebraic Equations

8.1 INTRODUCTION

In earlier chapters, finite difference/finite volume discretization methods of various equations are discussed. We have seen that during the discretization process, we obtain a system of algebraic equations, which can be written in matrix form as

$$A\emptyset = b \tag{8.1}$$

where **A** is the coefficient matrix, $\emptyset$ is an unknown vector and **b** is the known vector. Eq. (8.1) can be linear or non-linear. Iterative methods are used to solve a non-linear system of equations. It should be noted that the nature of coefficient matrix A is always sparse where non-zero elements lie on a small number of well-defined diagonals. There are two types of methods available, i) direct methods and ii) iterative methods.

8.2 DIRECT METHODS

8.2.1 Gauss Elimination

Gauss elimination is a basic method for solving linear systems of algebraic equations. It is very useful for small problems. However, this method is not so fast for a large system of equations, and accuracy deteriorates when the number of equations is large.

Let us consider the following equations:

$$a_{11}\emptyset_1 + a_{12}\emptyset_2 + a_{13}\emptyset_3 + \ldots + a_{1n}\emptyset_n = b_1$$

$$a_{21}\emptyset_1 + a_{22}\emptyset_2 + a_{23}\emptyset_3 + \ldots + a_{2n}\emptyset_n = b_2$$

$$a_{31}\emptyset_1 + a_{32}\emptyset_2 + a_{33}\emptyset_3 + \ldots + a_{3n}\emptyset_n = b_3$$

$$\vdots \qquad\qquad\qquad\qquad \vdots$$

$$a_{n1}\emptyset_1 + a_{n2}\emptyset_2 + a_{n3}\emptyset_3 + \ldots + a_{nn}\emptyset_n = b_n$$

The above equations can be expressed in matrix form as

$$\begin{bmatrix} a_{11} & a_{12} & a_{13} \cdots\cdots & a_{1n} \\ a_{21} & a_{22} & a_{23} \cdots\cdots & a_{2n} \\ a_{31} & a_{32} & a_{33} \cdots\cdots & a_{3n} \\ & & \vdots & \\ a_{n1} & a_{n2} & a_{n3} \cdots\cdots & a_{nn} \end{bmatrix} \begin{bmatrix} \emptyset_1 \\ \emptyset_2 \\ \emptyset_3 \\ \vdots \\ \emptyset_n \end{bmatrix} = \begin{bmatrix} b_1 \\ b_2 \\ b_3 \\ \vdots \\ b_n \end{bmatrix} \tag{8.2}$$

The objective is to transform the system into an upper diagonal triangular array by eliminating some of the unknowns from some of the equations by algebraic operations. This is known as *forward elimination.* For example, eliminate $\emptyset_1$ from 2nd to nth equations to get

$$\begin{bmatrix} a_{11} & a_{12} & a_{13} \cdots\cdots & a_{1n} \\ 0 & a'_{22} & a'_{23} \cdots\cdots & a'_{2n} \\ 0 & a'_{32} & a'_{33} \cdots\cdots & a'_{3n} \\ & & \vdots & \\ 0 & a'_{n2} & a'_{n3} \cdots\cdots & a'_{nn} \end{bmatrix} \begin{bmatrix} \emptyset_1 \\ \emptyset_2 \\ \emptyset_3 \\ \vdots \\ \emptyset_n \end{bmatrix} = \begin{bmatrix} b_1 \\ b'_2 \\ b'_3 \\ \vdots \\ b'_n \end{bmatrix} \text{ where } ' \text{ represents modified value}$$

In the next step, we eliminate $\emptyset_2$ from the 3rd to the nth equations to get

$$\begin{bmatrix} a_{11} & a_{12} & a_{13} \cdots\cdots & a_{1n} \\ 0 & a'_{22} & a'_{23} \cdots\cdots & a'_{2n} \\ 0 & 0 & a''_{33} \cdots\cdots & a''_{3n} \\ & & \vdots & \\ 0 & 0 & a''_{n3} \cdots\cdots & a''_{nn} \end{bmatrix} \begin{bmatrix} \emptyset_1 \\ \emptyset_2 \\ \emptyset_3 \\ \vdots \\ \emptyset_n \end{bmatrix} = \begin{bmatrix} b_1 \\ b'_2 \\ b''_3 \\ \vdots \\ b''_n \end{bmatrix} \text{ where } '' \text{ represents modified value}$$

Finally, we get (primes are removed for clarity)

$$\begin{bmatrix} a_{11} & a_{12} & a_{13} \cdots\cdots & a_{1n} \\ 0 & a_{22} & a_{23} \cdots\cdots & a_{2n} \\ 0 & 0 & a_{33} \cdots\cdots & a_{3n} \\ & & \vdots & \\ 0 & 0 & 0 \cdots\cdots & a_{nn} \end{bmatrix} \begin{bmatrix} \emptyset_1 \\ \emptyset_2 \\ \emptyset_3 \\ \vdots \\ \emptyset_n \end{bmatrix} = \begin{bmatrix} b_1 \\ b_2 \\ b_3 \\ \vdots \\ b_n \end{bmatrix} \tag{8.3}$$

Now, by *back substitution*, we can find the unknowns starting from the last equation.

Hence, the algorithm for Gauss elimination is as follows:

Forward elimination:

For $k = 1$ *to* $N - 1$

{

For $i = k + 1$ *to* n

```
    {
        Factor = a_ik / a_kk
        {
            For j = k + 1 to n
                a_ij = a_ij − Factor * a_kj
        }
                b_i = b_i − Factor * b_k
}
```

Backward Substitution:

```
∅_n = b_n / a_nn
For i = n − 1 to 1
{
    Sum = 0
    {
        For j = i + 1 to n
        Sum = Sum + a_ij * ∅_j
    }
        ∅_i = (∅_i − Sum) / a_ii
}
```

In Gauss Elimination, a division by zero occurs if the pivot element is zero. Zero pivot elements may be created during the elimination step even if they are not present in the original matrix. To avoid this problem, pivoting is used, by which we interchange rows and columns at each step to put the coefficient with the largest magnitude on the diagonal. Normally, partial pivoting is used; this uses only row interchanges. However, by this elimination process, the non-zero elements in the upper diagonal matrix are filled.

8.2.2 LU Decomposition

Another direct method is to decompose the ***A*** matrix into ***L*** and ***U*** by modified Crout algorithm, where ***L*** and ***U*** is lower and upper triangular matrix respectively, such that

$$LU = A \tag{8.4}$$

While factorization, pivoting process as described above is also carried out. Hence, we get

$$\begin{bmatrix} 1 & 0 & 0 \ldots\ldots\ldots & 0 \\ l_{21} & 1 & 0 \ldots\ldots\ldots & 0 \\ l_{31} & l_{32} & \ldots\ldots\ldots & 0 \\ & & \vdots & \\ l_{n1} & l_{n2} & \ldots\ldots\ldots & 1 \end{bmatrix} \begin{bmatrix} u_{11}u_{12}u_{13} & \ldots\ldots\ldots u_{1n} \\ 0 \;\; u_{22}u_{23} & \ldots\ldots\ldots u_{2n} \\ 0 \;\; 0 \;\; u_{33} & \ldots\ldots\ldots u_{3n} \\ \vdots & \\ 0 \;\; 0 \;\; 0 & \ldots\ldots\ldots u_{nn} \end{bmatrix} = \begin{bmatrix} a_{11}a_{12}a_{13} & \ldots\ldots\ldots a_{1n} \\ a_{21}a_{22}a_{23} & \ldots\ldots\ldots a_{2n} \\ a_{31}a_{32}a_{33} & \ldots\ldots\ldots a_{3n} \\ \vdots & \\ a_{n1}a_{n2}a_{n3} & \ldots\ldots\ldots a_{nn} \end{bmatrix} \tag{8.5}$$

The algorithm for $\boldsymbol{LU}$ decomposition is as follows:

$u_{1j} = a_{1j} \quad j = 1 \ to \ n$

$l_{i1} = a_{i1}/u_{11} \quad i = 2 \ to \ n$

$For \ i = 2 \ to \ n$

{

$u_{ij} = a_{ij} - \sum_{k=1}^{i-1} l_{ik}u_{kj} \qquad j = i, i+1, \ldots.., n$

$l_{ki} = \left(a_{ki} - \sum_{j=1}^{i-1} l_{kj}u_{ji}\right)/u_{ii} \qquad k = i+1, i+2, \ldots, n$

}

$$u_{nn} = a_{nn} - \sum_{i-1}^{n-1} l_{ni}u_{in}$$

8.2.3 Tri-Diagonal Matrix Algorithm (See Tannehill et al., [1997])

This is a direct method for solving the *tri-diagonal* matrix, also known as the Thomas algorithm. Here we transform the system of equations into an upper diagonal triangular array. This method maintains the sparsity of the matrix and requires less computational time.

Let us take a one-dimensional diffusion equation:

$$\frac{\partial u}{\partial t} = D\frac{\partial^2 u}{\partial x^2} \tag{8.6}$$

The finite difference discretization of fully Implicit scheme is given by

$$\frac{u_k^{n+1} - u_k^n}{\Delta t} = D\left[\frac{u_{k+1}^{n+1} - 2u_k^{n+1} + u_{k-1}^{n+1}}{\Delta x^2}\right]$$

or

$$-\lambda u_{k-1}^{n+1} + (1 + 2\lambda)u_k^{n+1} - \lambda u_{k+1}^{n+1} = u_k^n \tag{8.7}$$

Eq. (8.7) can also be written as

$$bu_{k-1}^{n+1} + du_k^{n+1} + au_{k+1}^{n+1} = c$$

where

$$a = b = -\lambda; d = (1 + 2\lambda) \text{ and } c = u_k^n$$

Hence, we get the matrix like:

$$\begin{bmatrix} d_1 & a_1 & \cdots\cdots\cdots & & \cdots \\ b_2 & d_2 & a_2 \cdots & \cdots & \cdots \\ \cdots & b_3 & d_3 \; a_3 & \cdots & \cdots \\ & & \vdots & & \\ & & \vdots & & \\ \cdots & \cdots & \cdots\cdots\cdots & b_{NK} & d_{NK} \end{bmatrix} \begin{bmatrix} u_1^{n+1} \\ u_2^{n+1} \\ \vdots \\ \vdots \\ \vdots \\ u_{NK}^{n+1} \end{bmatrix} = \begin{bmatrix} c_1 \\ c_2 \\ \vdots \\ \vdots \\ \vdots \\ c \end{bmatrix} \tag{8.8}$$

where

$$c_1 = u_1^n - b_1 u_0^{n+1}; c_{Nk} = u_{NK}^n - a_{NK} u_{NK+1}^{n+1}$$

Algorithm:

Compute

$$d_k = d_k - \frac{b_k}{d_{k-1}} a_{k-1}; \; k = 2, 3, \ldots\ldots, NK$$

and new c_k by

$$c_k = c_k - \frac{b_k}{d_{k-1}} c_{k-1}; \; k = 2, 3, \ldots\ldots, NK$$

Then we compute unknowns by back substitution according to

$$u_{NK} = \frac{c_{NK}}{d_{NK}}$$

$$u_k = \frac{[c_k - a_k u_{k+1}]}{d_k}; \; k = NK - 1, NK - 2, \ldots\ldots\ldots 1$$

8.3 ITERATIVE METHODS

In the iterative methods, we make an initial guess at the solution vector and iteratively go on updating the solution vector until we achieve convergence (i.e. we reach the pre-specified tolerance value and round-off errors are not an issue).

The convergence is achieved when

$$Residual \leq \varepsilon_s \tag{8.9a}$$

where ε_s is the pre-specified tolerance value, normally varies from 10^{-3} to 10^{-6}.

The residual may be calculated in various ways. One way is

$$Residual = \left| \frac{\emptyset_i^{k+1} - \emptyset_i^k}{\emptyset_i^{k+1}} \right| \tag{8.9b}$$

where k denotes the iteration number.

Another way to is to calculate for equation $a_P\emptyset_P - \sum_{nb} a_{nb}\emptyset_{nb} = b_P$ is

$$Residual = \frac{\sum \left| \left(a_P\emptyset_P - \sum_{nb} a_{nb}\emptyset_{nb} - b_P\right)^{k+1} \right|}{\sum \left| \left(a_P\emptyset_P - \sum_{nb} a_{nb}\emptyset_{nb} - b_P\right)^{k} \right|} \tag{8.9c}$$

8.3.1 THE JACOBI METHOD

In the Jacobi method, a dependable variable is solved at each grid point using initial guess values of neighboring points or previously computed values, so

$$\begin{aligned}
\emptyset_1^{k+1} &= \frac{1}{a_{11}}\left(b_1 - a_{12}\emptyset_2^k - a_{13}\emptyset_3^k - \ldots - a_{1n}\emptyset_n^k\right) \\
\emptyset_2^{k+1} &= \frac{1}{a_{22}}\left(b_2 - a_{21}\emptyset_1^k - a_{23}\emptyset_3^k - \ldots - a_{2n}\emptyset_n^k\right) \\
\emptyset_3^{k+1} &= \frac{1}{a_{33}}\left(b_3 - a_{31}\emptyset_1^k - a_{32}\emptyset_2^k - \ldots - a_{3n}\emptyset_n^k\right) \\
\emptyset_4^{k+1} &= \frac{1}{a_{44}}\left(b_4 - a_{41}\emptyset_1^k - a_{42}\emptyset_2^k - \ldots - a_{4n}\emptyset_n^k\right) \\
\emptyset_n^{k+1} &= \frac{1}{a_{nn}}\left(b_n - a_{n1}\emptyset_1^k - a_{n2}\emptyset_2^k - \ldots - a_{n,n-1}\emptyset_{n-1}^k\right)
\end{aligned} \tag{8.10}$$

This method is very slow and rarely used.

Algorithm:

Initialize: b, A, kmax, $\emptyset^{(0)}$, error(ε), number of variables (n)

k = 0

WHILE (residue < ε) and (k < kmax)
FOR i = 1 : n

$$r_i^{(k+1)} = b_i - \sum_{j=1,\, j\neq i}^{n} a_{ij} x_j^{(k)}$$

$$x_i^{(k+1)} = \frac{r_i^{(k+1)}}{a_{ii}}$$

END FOR

$$residue = \max\left|\frac{x_i^{(k+1)} - x_i^{(k)}}{x_i^{(k+1)}}\right|$$

k = k + 1
END WHILE

8.3.2 The Point Gauss-Seidel Method

In this method, current values of dependent variables are used to compute neighboring points as soon as they are available. By this process, the convergence rate increases. The method is as follows:

$$\begin{aligned}
\emptyset_1^{k+1} &= \frac{1}{a_{11}}\left(b_1 - a_{12}\emptyset_2^k - a_{13}\emptyset_3^k - \ldots - a_{1n}\emptyset_n^k\right)\\
\emptyset_2^{k+1} &= \frac{1}{a_{22}}\left(b_2 - a_{21}\emptyset_1^{k+1} - a_{23}\emptyset_3^k - \ldots - a_{2n}\emptyset_n^k\right)\\
\emptyset_3^{k+1} &= \frac{1}{a_{33}}\left(b_3 - a_{31}\emptyset_1^{k+1} - a_{32}\emptyset_2^{k+1} - \ldots - a_{3n}\emptyset_n^k\right)\\
\emptyset_4^{k+1} &= \frac{1}{a_{44}}\left(b_4 - a_{41}\emptyset_1^{k+1} - a_{42}\emptyset_2^{k+1} - \ldots - a_{4n}\emptyset_n^k\right)\\
\emptyset_n^{k+1} &= \frac{1}{a_{nn}}\left(b_n - a_{n1}\emptyset_1^{k+1} - a_{n2}\emptyset_2^{k+1} - \ldots - a_{n,n-1}\emptyset_{n-1}^{k+1}\right)
\end{aligned} \tag{8.11}$$

Algorithm:
Initialize: b, A, kmax, $\emptyset^{(0)}$, error(ε), number of variables (n)
k = 0
WHILE (residue < ε) and (k < kmax)
FOR i = 1 : n

$$r_i = b_i - \sum_{j=1,\, j\neq i}^{n} a_{ij} x_j$$

$$x_i = \frac{r_i}{a_{ii}}$$

END FOR

$$residue = \max\left|\frac{x_i^{(k+1)} - x_i^{(k)}}{x_i^{(k+1)}}\right|$$

k = k + 1
END WHILE

8.3.3 Point Successive Over-Relaxation Method

To further improve the convergence rate of Gauss-Seidel method, relaxation method can be used. In this method, after computing new value of $\emptyset$, this value is modified as

$$\emptyset^{k+1} = \alpha\emptyset^{k+1} + (1-\alpha)\emptyset^{k}; 0<\alpha<2$$

where α is the relaxation factor.

If $\alpha = 1$; no relaxation, we get original Guess-Seidel method.

If $0<\alpha<1$; under relaxation, this is used to make a diverging system converge.

If $1<\alpha<2$; over relaxation, this is used to speed up the convergence of an already converging system.

It should be remembered that the choice of relaxation factor,α, is problem specific.

8.3.4 The Line Gauss-Seidel Method

The Jacobi, Point Gauss-Seidel and Point Successive Over-Relaxation methods can be used for one-dimensional problems. However, for two-dimensional problems, especially steady-state problems, they are inefficient. Instead, the line Gauss-Seidel method, which is more efficient, can be used. The method is illustrated in Figure 8.1.

When we are computing the k+1 iteration, values of $\emptyset$ at j location, the values of $\emptyset$ at j-1 location for k+1 iteration level is already known whereas for j+1 location only k iteration value are known. For example, let us consider the Laplace equation

$$\left(\frac{\partial^2 u}{\partial x^2} + \frac{\partial^2 u}{\partial y^2}\right) = 0 \tag{8.12a}$$

The finite difference discretization is given by

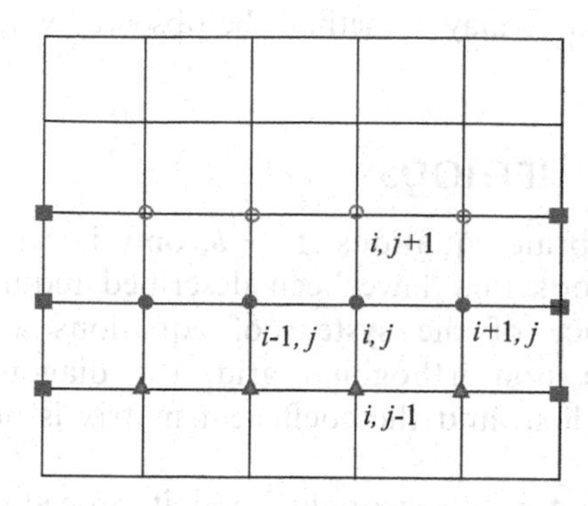

FIGURE 8.1 Line Gauss-Seidel method.

$$\frac{u_{i+1,j} - 2u_{i,j} + u_{i-1,j}}{(\Delta x)^2} + \frac{u_{i,j+1} - 2u_{i,j} + u_{i,j-1}}{(\Delta y)^2} = 0$$

or

$$u_{i+1,j} - 2u_{i,j} + u_{i-1,j} + \beta^2 (u_{i,j+1} - 2u_{i,j} + u_{i,j-1}) = 0 \quad (8.12b)$$

This can be written for line Gauss-Seidel form as

$$-u_{i-1,j}^{k+1} + 2(1+\beta^2)u_{i,j}^{k+1} - u_{i-1,j}^{k+1} = \beta^2 \left(u_{i,j+1}^{k} + u_{i,j-1}^{k+1}\right) \quad (8.13)$$

This equation can be solved iteratively by the Thomas algorithm until convergence is achieved. The convergence is faster than for the point Gauss-Seidel method.

8.3.5 Convergence of the Iterative Methods

It is to be noted that the iterative methods may converge or diverge. The Scarborough criteria gives the sufficient (not necessary) condition for convergence.

For all rows:

$$a_{ii} \geq \sum_{j=1;j\neq i}^{n} |a_{ij}|$$

and at least for one row

$$a_{ii} > \sum_{j=1;j\neq i}^{n} |a_{ij}|$$

Sufficient condition means that convergence may sometimes be observed when these conditions are not met.

8.4 CONJUGATE GRADIENT (CG) METHODS

For solving very large systems of algebraic equations $\boldsymbol{A}\emptyset = \boldsymbol{b}$, only iterative methods are used. The iterative methods that have been described require properties such as diagonal dominance of the system of equations and sometimes symmetry. However, for a non-orthogonal grid, the diagonal dominance of the coefficient matrix is lost, and the coefficient matrix is not symmetric.

Such restrictions are not necessary for the strongly implicit procedure (SIP) as proposed by Stone [1968]. The acceleration parameter used in SIP is less critical than the relaxation parameter for Successive Over-Relaxation (SOR). However, the main drawback of SIP is that the efficiency of the method depends on the efficient exploitation of the sparsity of matrix A and therefore demands new algorithms for different sparsity patterns.

The use of the CG method for solving linear algebraic equations has shown encouraging improvements in speed and efficiency. However, the method is restricted to symmetric and positive definite matrices. Hence, pre-conditioning the original non-symmetric system is recommended. An alternative is the bi-conjugate gradient (BCG) method. For BCG, the amount of computation involved is greater than for the CG, and the convergence rate is slower. However, the BCG solver is very robust. On the other hand, in the conjugate gradient square (CGS) method the rate of convergence is faster, and the amount of computational effort is nearly equal to that of the CG method. In addition, the CGS method is also very robust. Both BCG and CGS become very efficient when a pre-conditioner is applied to the original non-symmetric matrix. Here, we shall discuss pre-conditioned BCG and CGS methods.

Here, the coefficient matrix $\boldsymbol{A}$ of the algebraic equation $\boldsymbol{A}\emptyset = \boldsymbol{b}$ is preconditioned by incomplete $\boldsymbol{L}$-$\boldsymbol{U}$ (ILU) decomposition by Cholesky's method, resulting in the modified matrix equation

$$\boldsymbol{M}\overline{\emptyset} = \overline{\boldsymbol{b}} \tag{18.14a}$$

where

$$\boldsymbol{M} = \boldsymbol{L}^{-1}\boldsymbol{A}\boldsymbol{U}^{-1}; \overline{\emptyset} = \boldsymbol{U}\emptyset \text{ and } \overline{\boldsymbol{b}} = \boldsymbol{L}^{-1}\boldsymbol{b} \tag{8.14b}$$

The details of incomplete $\boldsymbol{L}$-$\boldsymbol{U}$ decomposition are given in next section. The pre-conditioned BCG and CGS methods as proposed by Joly and Eymard [1990] are outlined here.

8.4.1 THE PRE-CONDITIONED BCG METHOD

- Choice of initial estimate of solution $\bar{\emptyset}_0$
- Calculation of initial residual $r_0 = \bar{b} - M\bar{\emptyset}_0$
- Setting first direction vector p and bi-direction vector q as

$$s_0 = q_0 = p_0 = r_0$$

Iterative solution procedure:
In the iterative procedure, the following recurrence relations are employed till convergence.

i) Calculation of coefficient

$$\gamma_k = \frac{(s_k, r_k)}{(q_k, Mp_k)}$$

where (. , .) represents the scalar product.
ii) Calculation of the updated estimate

$$\bar{\emptyset}_{k+1} = \bar{\emptyset}_k + \gamma_k p_k$$

iii) Determination of new residual

$$r_{k+1} = r_k - \gamma_k M p_k$$

$$s_{k+1} = s_k - \gamma_k M^T p_k$$

$$\beta_{k+1} = \frac{(s_{k+1}, r_{k+1})}{(s_k, r_k)}$$

iv) Setting of new direction vector and bi-direction vector as

$$p_{k+1} = r_{k+1} + \beta_{k+1} p_k$$

$$q_{k+1} = s_{k+1} + \beta_{k+1} q_k$$

These iterations are continued until the flow variables satisfy specified convergence criteria.

8.4.2 THE PRE-CONDITIONED CGS METHOD

Initialization of iterate variables:

- Choice of initial estimate of solution $\bar{\emptyset}_0$
- Calculation of initial residual $r_0 = \bar{b} - M\bar{\emptyset}_0$

- Setting first direction vector $\boldsymbol{p}$ and bi-direction vector $\boldsymbol{q}$ as

$$\boldsymbol{q}_0 = \boldsymbol{p}_0 = \boldsymbol{r}_0$$

Iterative solution procedure:
In the iterative procedure, the following recurrence relations are employed till convergence.

i) Calculation of coefficient

$$\gamma_k = \frac{(\boldsymbol{r}_0, \boldsymbol{r}_k)}{(\boldsymbol{r}_0, \boldsymbol{M}\boldsymbol{q}_k)}$$

where (. , .) represents the scalar product.

ii) Calculation of the updated estimate

$$\overline{\emptyset}_{k+1} = \overline{\emptyset}_k + \gamma_k(\boldsymbol{p}_k + \boldsymbol{v}_k)$$

$$where\ \boldsymbol{v}_k = \boldsymbol{p}_k - \gamma_k \boldsymbol{M}\boldsymbol{q}_k$$

iii) Determination of new residual

$$\boldsymbol{r}_{k+1} = \boldsymbol{r}_k - \gamma_k \boldsymbol{M}(\boldsymbol{p}_k + \nu_k)$$

$$\beta_{k+1} = \frac{(\boldsymbol{r}_0, \boldsymbol{r}_{k+1})}{(\boldsymbol{r}_0, \boldsymbol{r}_k)}$$

iv) Setting of new direction vector and bi-direction vector as

$$\boldsymbol{p}_{k+1} = \boldsymbol{r}_{k+1} + \beta_{k+1}\boldsymbol{v}_k$$

$$\boldsymbol{q}_{k+1} = \boldsymbol{p}_{k+1} + \beta_{k+1}(\beta_{k+1}\boldsymbol{q}_k + \boldsymbol{v}_k)$$

These iterations are continued until flow variables satisfy specified convergence criteria.

8.5 THE INCOMPLETE L-U DECOMPOSITION METHOD

8.5.1 INTRODUCTION

The pre-conditioned conjugate gradient algorithm or its variant have become the favored iterative method for solving large sparse linear systems of equations of the form

$$A\emptyset = \boldsymbol{b} \tag{8.15}$$

In Section 8.4, the CG methods along with pre-conditioning are described. The technique of pre-conditioning is done by applying the method to the problem

$$\tilde{A}^{-1}A\emptyset = \tilde{A}^{-1}b \tag{8.16}$$

where $\tilde{A}$ approximates A such that the system $\tilde{A}z = t$ can be solved easily.

One way to derive such an approximation is incomplete LU decomposition. This leads to L, U such that

$$\tilde{A} = LU \cong A$$

A slight variation of preconditioning of Eq. (8.16) is obtained by solving

$$L^{-1}AU^{-1}\overline{\emptyset} = L^{-1}b \tag{8.17}$$

and then putting

$$\emptyset = U^{-1}\overline{\emptyset}$$

8.5.2 Pre-Conditioning by L-U Decomposition

As discussed, good pre-conditioning is of prime importance for the faster convergence. Generally, it is desirable that the matrix A should have a small spectral radius, and preferably its eigenvalues are clustered for a faster convergence rate. In fluid-flow computations, the matrix A is generally non-symmetric. It tends to have large spectral radii and fairly uniform distributions of eigenvalues. So, the CGS method converges very slowly if implemented directly. Hence, the coefficient matrix A of the algebraic equation $A\emptyset = b$ is pre-conditioned by incomplete L-U (ILU) decomposition by Cholesky's method, resulting in the modified matrix equation. The idea is that the sparsity pattern, which may be same as that of matrix A, is prescribed and then L-U decomposition is performed, except that anywhere an entry of L or U falls outside the sparsity pattern is neglected. This pre-conditioning is denoted by ILU. For ILU, the matrices L and U satisfy the following rules:

1) Diag. $(L) = I$.
2) The non-zero structure of the matrix $L+U$ is identical to the non-zero structure of A.
3) If $A_{ij} \neq 0$, then $(LU)_{ij} = A_{ij}$.

The algorithm for ILU decomposition for the general sparse matrix is as follows:

For $i = 2$ *to* n

{

For $k = 1$ *to* $i - 1$ *and if* $a_{ik} \neq 0$ *and* $a_{kk} \neq 0$ *Do*:

{

$a_{ik} = {}^{a_{ik}}/_{a_{kk}}$ (*l values*)

For $j = k + 1$ *to* n *and if* $a_{ij} \neq 0$ *Do*:

{

$a_{ij} = a_{ij} - a_{ik}a_{kj}$ (*u values*)

}

}

}

However, the sparsity pattern of matrix A for two and three dimensions arising from the discretization procedure results in a banded sparse matrix and is shown in Figures 8.2 and 8.3, respectively, while that of $\boldsymbol{L}$ and $\boldsymbol{U}$ for two dimensions is shown in Figure 8.4. Here we store coefficients as column vectors.

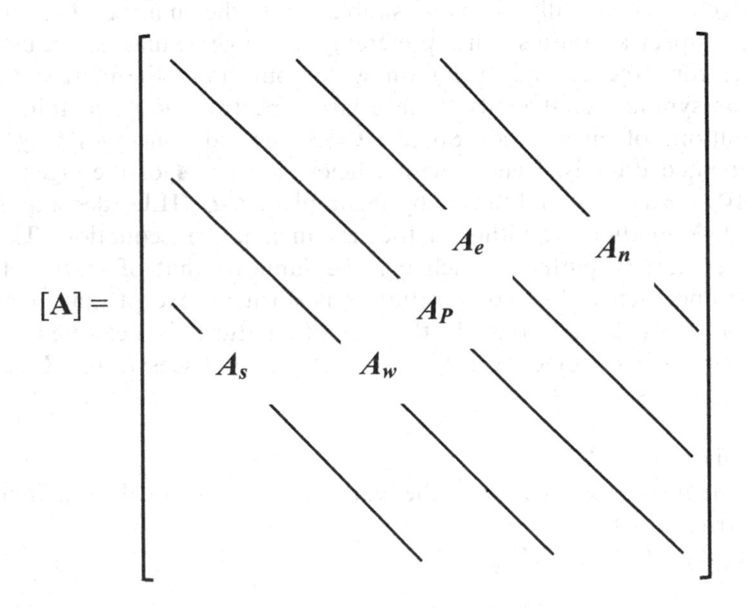

FIGURE 8.2 Sparsity Pattern of Matrix A in Two Dimensions.

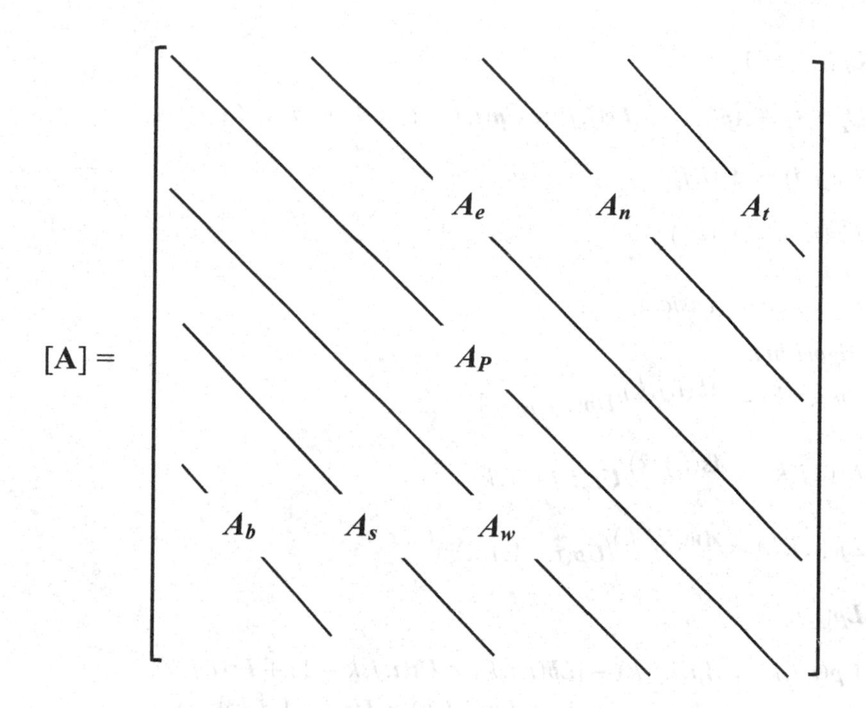

FIGURE 8.3 Sparsity Pattern of Matrix A in Three Dimensions.

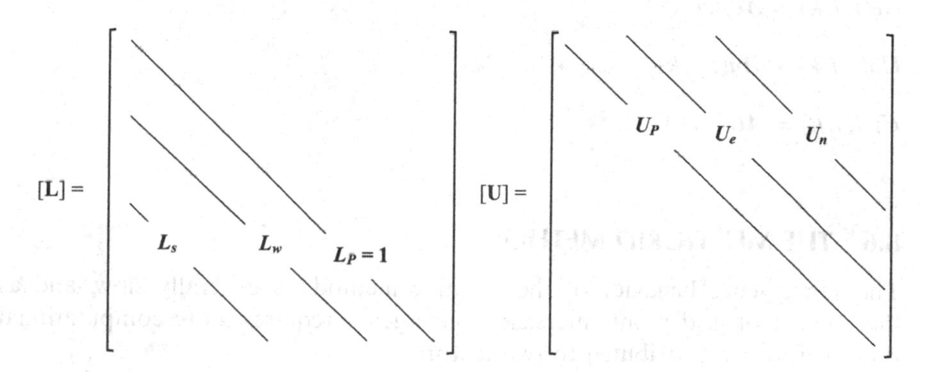

FIGURE 8.4 Sparsity Pattern of Matrices L and U in Two Dimensions.

For two dimensions

Algorithm:

$$Ls(i,j) = {As(i,j)}/{Up(i,j-1)}$$

$$Lw(i,j) = {Aw(i,j)}/{Up(i-1,j)}$$

$$\boldsymbol{Lp}(i,j) = 1$$

$$\boldsymbol{Up}(i,j) = \boldsymbol{Ap}(i,j) - [\boldsymbol{Ls}(i,j) \times \boldsymbol{Un}(i,j-1) + \boldsymbol{Lw}(i,j) \times \boldsymbol{Ue}(i-1,j)]$$

$$\boldsymbol{Ue}(i,j) = \boldsymbol{Ae}(i,j)$$

$$\boldsymbol{Un}(i,j) = \boldsymbol{An}(i,j)$$

For three dimensions

Algorithm:

$$\boldsymbol{Lb}(i,j,k) = {}^{\boldsymbol{Ab}(i,j,k)}/_{\boldsymbol{Up}(i,j,k-1)}$$

$$\boldsymbol{Ls}(i,j,k) = {}^{\boldsymbol{As}(i,j,k)}/_{\boldsymbol{Up}(i,j-1,k)}$$

$$\boldsymbol{Lw}(i,j,k) = {}^{\boldsymbol{Aw}(i,j,k)}/_{\boldsymbol{Up}(i-1,j,k)}$$

$$\boldsymbol{Lp}(i,j,k) = 1$$

$$\begin{aligned}\boldsymbol{Up}(i,j,k) = \boldsymbol{Ap}(i,j,k) - [&\boldsymbol{Lb}(i,j,k) \times \boldsymbol{Ut}(i,j,k-1) + \boldsymbol{Ls}(i,j,k) \\ &\times \boldsymbol{Un}(i,j-1,k) + \boldsymbol{Lw}(i,j,k) \times \boldsymbol{Ue}(i-1,j,k)]\end{aligned}$$

$$\boldsymbol{Ue}(i,j,k) = \boldsymbol{Ae}(i,j,k)$$

$$\boldsymbol{Un}(i,j,k) = \boldsymbol{An}(i,j,k)$$

$$\boldsymbol{Ut}(i,j,k) = \boldsymbol{At}(i,j,k)$$

8.6 THE MULTIGRID METHOD

The convergence behavior of the iterative methods is generally slow, and as the number of grid points increase, convergence requires more computational time. This can be attributed to two factors.

Firstly, as we have seen during Von Neumann's stability analysis, the error function can be represented using a Fourier series using the collection of sine waves of different frequency and amplitudes. In the iterative process, higher frequency components quickly smooth out. However, lower frequency components extend to a large number of grid points and require more iterations to be dissipated. This behavior can be clearly observed when we plot residual history during the iteration process. In the first few iterations, the convergence rate is faster and then slows down.

Secondly, when there is a large number of grid points, the boundary information propagation is slower and requires many iterations to reach the centre. With the increase of the number of grid points, this process further deteriorates and the number of iterations increases because higher frequency waves are removed rapidly and lower frequency waves remain.

In the multigrid method, a series of coarse grids is used to approximate the solution of the equations because lower frequency waves become higher frequency waves in coarser grids, which smooth out at a faster rate. Also, boundary information travels faster to the centre in a coarser grid. Hence, the convergence rate is faster.

The multigrid method gained popularity through the pioneering work of Brandt [1977]. The basic steps of multigrid algorithm follow.

8.6.1 Coarsening Step

The first step in the multigrid process is to generate coarse and fine grid levels. Initially, we generate the fine grid, which is used for solving the problems. Then, coarse grid levels are generated by the agglomeration of fine grids, as shown in Figure 8.5.

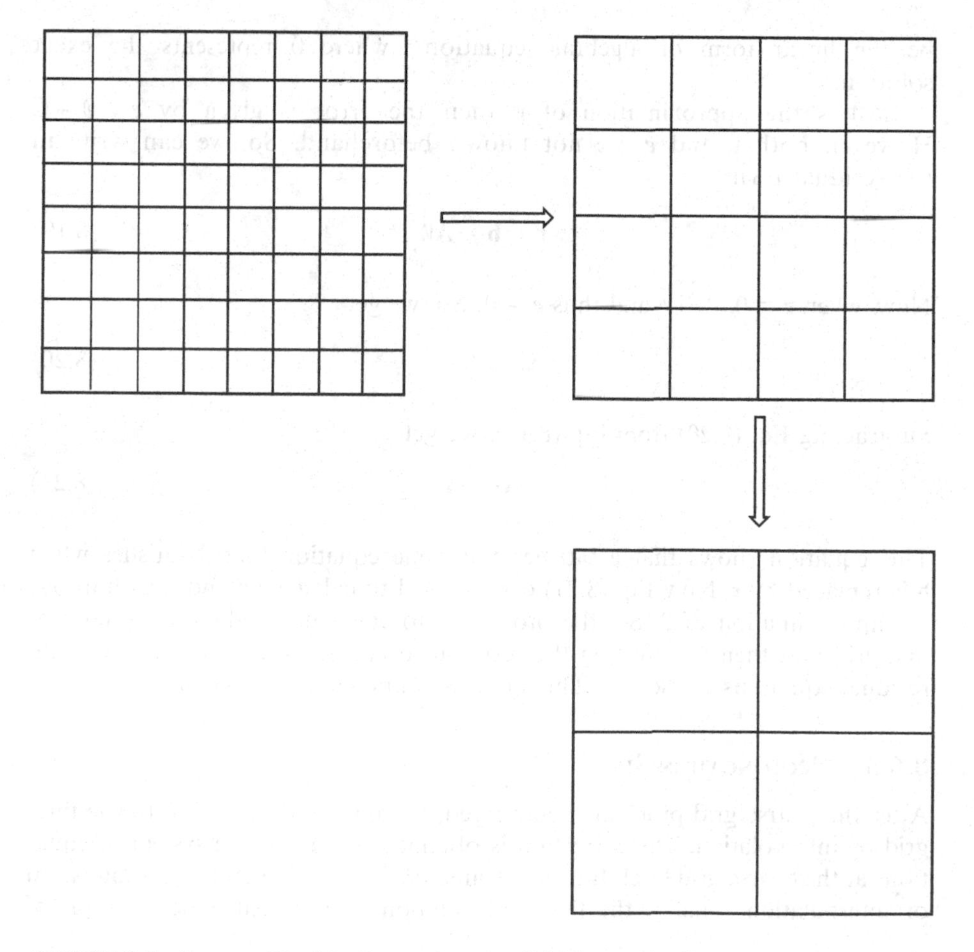

FIGURE 8.5 Strategy Used in Generating the Grid System in the Multigrid Approach.

As we can see from Figure 8.5, coarse cells are obtained by joining four cells of finer mesh. Hence, the centre of coarse cells will not coincide with the centre of fine mesh. This cell-centre approach is easy to implement in the multigrid procedure and is suitable for adaptive mesh. This process is continued until the coarsest grid contains fewer interior cells.

8.6.2 Restriction Step

Initially, the solution starts at the finest grid. However, the same form of equations being solved in the fine grid cannot be applied to the coarse grid. Instead, the equations in the coarse grid are rewritten by considering the amount by which the fine grid solutions are not satisfied. That is the residual form of equations solved in the course grid.

Let

$$\mathbf{A}\emptyset = \mathbf{b} \tag{8.18}$$

be the linear form of algebraic equations, where $\emptyset$ represents the exact solution.

If $\tilde{\emptyset}$ is the approximation of $\emptyset$ then the error is given by $e = \emptyset - \tilde{\emptyset}$. However, both $\emptyset$ and e are not known beforehand. So, we can write in the residual form

$$\mathbf{r} = \mathbf{b} - \mathbf{A}\tilde{\emptyset} \tag{8.19}$$

Now, when $\mathbf{r} = 0$, $\emptyset = \tilde{\emptyset}$ and thus $e = 0$. So, we get

$$\mathbf{A}\tilde{\emptyset} = \mathbf{b} - \mathbf{r} \tag{8.20}$$

Subtracting Eq. (8.20) from Eq. (8.18), we get

$$\mathbf{A}e = \mathbf{r} \tag{8.21}$$

This equation shows that **e** satisfies the same equation that $\emptyset$ satisfies when **b** is replaced by **r**. Now Eq. (8.21) can be used to calculate e and thus improve the approximation of $\tilde{\emptyset}$. So, the process is to calculate residual using current fine grid and then transferring the residual to coarse grid. In the coarse grid, residual equations are solved. This process is known as *restriction*.

8.6.3 Prolongation Step

After the coarse grid problem is converged, we apply the correction to the finer grid by interpolation. The correction is obtained by solving the system of equations at the coarse grid level. It is then transferred to the fine grid by prolongation or interpolation. Finally, the fine grid solution is corrected. One iteration of multigrid method consists of finest grid to the coarser grids and back to the finest grid. This whole process is known as a *cycle.*

8.6.4 Cycling Strategy

There are various methods for moving from one grid to another. Normally, after updating a solution on the finer grid, few iterations need to be done in the finer grid to smooth high-frequency error. Figure 8.6 shows various cycling strategies adopted in the multigrid process: a) V-cycle, b) F-cycle and c) W-cycle.

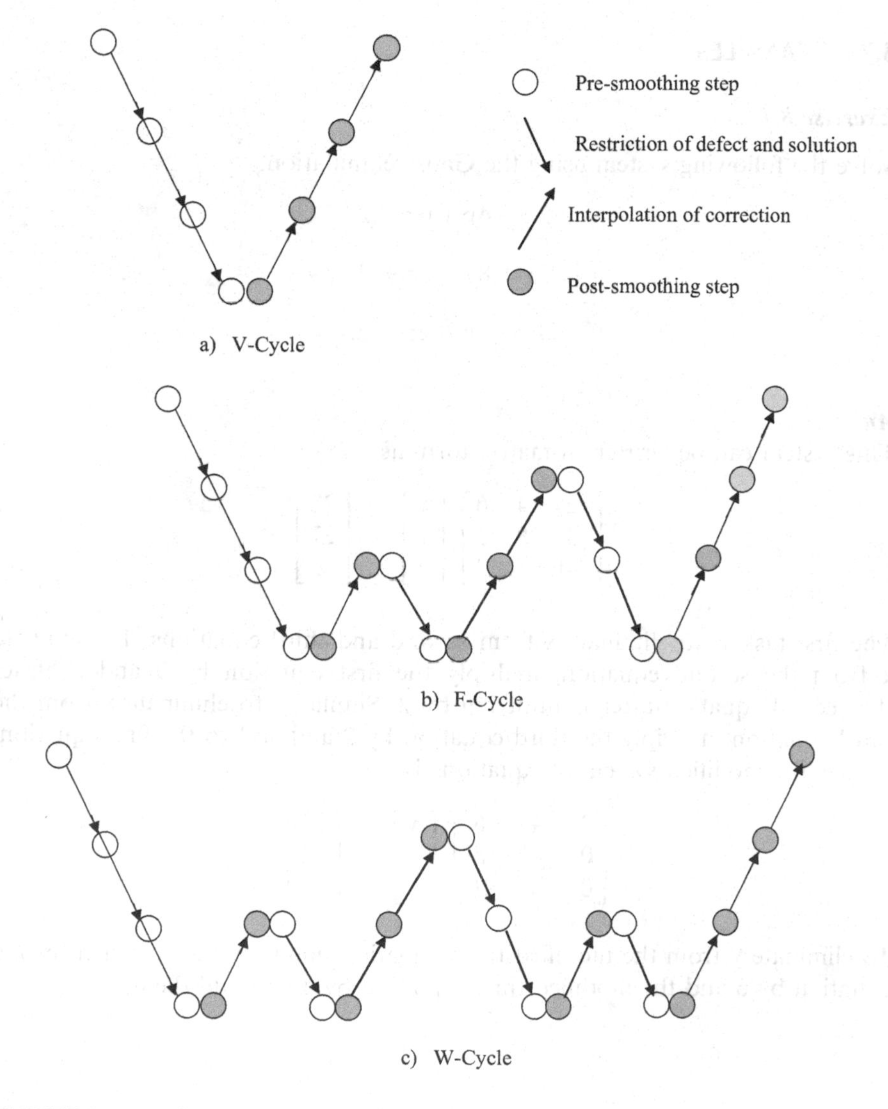

FIGURE 8.6 Various Cycling Strategies.

The simplest of all cycles is V-cycle in which one needs to visit each grid level only once. Normally, few iterations are carried out in the restriction phase and then the residual equations are solved in the coarse grid. The same procedure is continued until the coarsest grid is reached. For a stiff system, more iterations are required in the coarse grid level. The W-cycle can be thought of as applying smaller V-cycles at each visited coarse grid level. The F-cycle can be thought of as splitting the W-cycle in half. Hence, the F-cycle requires fewer coarse level sweeps than the W-cycle but more than the V-cycle. Normally, F-cycle and W-cycle iterations take more time than the V-cycle.

8.7 EXAMPLES

Exercise 8.1

Solve the following system using the Gauss elimination.

$$2x + 4y + 6z = 22$$

$$3x + 8y + 5z = 27$$

$$-x + y + 2z = 2.$$

Answer:
This system can be written in matrix form as

$$\begin{bmatrix} 2 & 4 & 6 \\ 3 & 8 & 5 \\ -1 & 1 & 2 \end{bmatrix} \begin{bmatrix} x \\ y \\ z \end{bmatrix} = \begin{bmatrix} 22 \\ 27 \\ 2 \end{bmatrix}.$$

The first task is to eliminate x from second and third equations. To eliminate x from the second equation, multiply the first equation by 3 and subtract the second equation after multiplying by 2. Similarly, to eliminate x from the third equation, multiply the third equation by 2 and add to the first equation. Hence, the modified system of equations is

$$\begin{bmatrix} 2 & 4 & 6 \\ 0 & -4 & 8 \\ 0 & 6 & 10 \end{bmatrix} \begin{bmatrix} x \\ y \\ z \end{bmatrix} = \begin{bmatrix} 22 \\ 12 \\ 26 \end{bmatrix}.$$

To eliminate y from the modified third equation, multiply the modified second equation by 6 and the modified third equation by 4 and add them.

Hence, we get the modified system of equations as

$$\begin{bmatrix} 2 & 4 & 6 \\ 0 & -4 & 8 \\ 0 & 0 & 88 \end{bmatrix} \begin{bmatrix} x \\ y \\ z \end{bmatrix} = \begin{bmatrix} 22 \\ 12 \\ 176 \end{bmatrix}$$

and by back substitution we get

$$x = 3; y = 1; z = 2$$

As a check, you can put the values in any of the equations and verify.

Exercise 8.2

Solve the system of equation given in Exercise 8.1 using the Jacobi iteration starting with an initial guess of x = y = z =1.

Answer:

The system of equations can be written as

$$x^{(k+1)} = \left(22 - 4y^k - 6z^k\right)/2$$

$$y^{(k+1)} = \left(27 - 3x^k - 5z^k\right)/8$$

$$z^{(k+1)} = \left(2 + x^k - y^k\right)/2.$$

where k is the iteration number.

Initial guess: x = y = z =1.

First Iteration:

$$x^1 = \left(22 - 4y^0 - 6z^0\right)/2 = (22 - 4 - 6)/2 = 6$$

$$y^1 = \left(27 - 3x^0 - 5z^0\right)/8 = (27 - 3 - 5)/8 = 2.375$$

Second Iteration:

$$y^2 = \left(27 - 3x^1 - 5z^1\right)/8 = (27 - 3 \times 6 - 5 \times 1)/8 = 0.5$$

$$z^2 = \left(2 + x^1 - y^1\right)/2 = (2 + 6 - 2.375)/2 = 2.8125$$

The values of x, y and z with the residual against number of iterations are given in Table 8.1.

If the error specified is 1%, then we can terminate the calculation after 9 iterations. For 0.1% error, 16 iterations are required to get the converged solution.

TABLE 8.1
Values of x, y and z against Number of Iterations

Iteration	Values of					
	x	Residue	y	Residue	z	Residue
1	6	0.833	2.375	0.833	1	0
2	3.25	0.846	0.5	3.750	2.8125	0.644
3	1.563	1.079	0.398	0.256	2.375	0.184
4	3.078	0.492	1.305	0.695	1.582	0.501
5	3.645	0.156	1.232	0.059	1.887	0.162
6	2.876	0.267	0.829	0.486	2.206	0.145
7	2.723	0.056	0.918	0.097	2.023	0.090
8	3.095	0.120	1.089	0.157	1.903	0.063
9	3.113	0.006	1.025	0.062	2.003	0.050
10	2.941	0.058	0.956	0.072	2.044	0.020
11	2.956	0.005	0.994	0.038	1.993	0.026
12	3.033	0.025	1.021	0.026	1.981	0.006
13	3.015	0.006	1.000	0.021	2.006	0.012
14	2.983	0.011	0.990	0.010	2.008	0.001
15	2.995	0.004	1.001	0.011	1.996	0.006
16	3.008	0.004	1.004	0.003	1.997	0.001
17	3.001	0.002	0.999	0.005	2.002	0.002

Exercise 8.3

Solve the following system of equations with Jacobi and Gauss-Seidel iterations starting with an initial guess of x = y = z =40.

$$2x + y + z = 180$$

$$x + 3y + 2z = 300$$

$$2x + y + 2z = 240.$$

Answer:

a) For the Gauss-Seidel iterative method, the system of equations can be written as

$$x^{(k+1)} = \left(180 - y^k - z^k\right)/2$$

$$y^{(k+1)} = \left(300 - x^{(k+1)} - 2z^k\right)/3$$

$$z^{(k+1)} = \left(240 - 2x^{(k+1)} - y^{(k+1)}\right)/2.$$

where k is the iteration number.

Initial guess: x = y = z =40.

The values of x, y and z are shown in Table 8.2.

If the error specified is 1%, then we can terminate the calculation after 8 iterations. For 0.1% error, 13 iterations are required to get the converged solution.

b) For the Jacobi iterative method, the above system of equations can be written as

$$x^{(k+1)} = (180 - y^k - z^k)/2$$

$$y^{(k+1)} = (300 - x^k - 2z^k)/3$$

$$z^{(k+1)} = (240 - 2x^k - y^k)/2$$

Initial guess: x = y = z =40.

TABLE 8.2
Values of x, y, z against the Iteration Number

	Values of					
Iteration	**x**	**Residue**	**y**	**Residue**	**z**	**Residue**
1	50.0000	0.2	56.6667	0.2941	41.6667	0.04
2	40.8333	0.2245	58.6111	0.0332	49.8611	0.1643
3	35.7639	0.1417	54.8380	0.0688	56.8171	0.1224
4	34.1725	0.0466	50.7311	0.0810	60.4620	0.0603
5	34.4035	0.0067	48.2242	0.0520	61.4845	0.0166
6	35.1457	0.0211	47.2951	0.0196	61.2068	0.0045
7	35.7491	0.0169	47.2791	0.0003	60.6114	0.0098
8	36.0547	0.0085	47.5742	0.0062	60.1582	0.0075
9	36.1338	0.0022	47.8499	0.0058	59.9412	0.0036
10	36.1044	0.0008	48.0044	0.0032	59.8934	0.0008
11	36.0511	0.0015	48.0540	0.0010	59.9219	0.0005
12	36.0120	0.0011	48.0481	0.0001	59.9639	0.0007
13	35.9940	0.0005	48.0261	0.0005	59.9930	0.0005

Solving, after 23 iterations, we get x = 36.1885, y = 48.1515, and z = 59.7357. Corresponding errors are 0.0044, 0.0043 and 0.0010, respectively, for x, y and z, which is less than 1%. If the error criterion is 0.1%, then we need 27 iterations to get the converge results. The exact answer is x = 36, y = 48, z = 60.

8.8 SUMMARY

- Direct methods are not as fast for a large system of equations. The accuracy also deteriorates when the number of equations is large.
- For solving very large systems of algebraic equations, only iterative methods are used.
- Given an error criterion, the Gauss-Seidel iterative method converges faster than the Jacobi method. This is because the Gauss-Seidel method uses updated values of the variable as soon as they are available, whereas the Jacobi method uses old values until the next iteration.
- The convergence rate can be further accelerated by using a relaxation parameter.
- As the error criterion is made stricter, the number of iterations increases and with them the computational cost. However, the error criterion is problem specific, and the user must specify the value carefully to maintain solution accuracy.
- The iterative methods require properties such as diagonal dominance of the system of equations and sometimes symmetry. However, for a non-orthogonal grid, the diagonal dominance of the coefficient matrix is lost, and the coefficient matrix is not symmetric.
- The use of CG for solving linear algebraic equations has shown encouraging improvements in speed and efficiency.
- Both BCG and CGS solvers are very robust. However, for CGS, the rate of convergence is faster, and the amount of computational effort is less than for BCG.
- Both BCG and CGS methods become very efficient when a preconditioner is applied to the original non-symmetric matrix.
- The convergence behavior of the iterative methods is generally slow, and as the number of grid point increases, convergence requires more computational time.
- In the iterative process, higher frequency components quickly smooth out. However, lower frequency components extend to a large number of grid points and require more iterations to be dissipated.
- For a large number of grid points, the boundary information propagation is slower and requires many iterations for information to reach the centre.
- In the multigrid method, a series of coarse grids is used to approximate the solution of the equations.

- The lower frequency waves become higher frequency waves in coarser grids and are smoothed out at a faster rate.
- The boundary information travels more quickly to the centre in a coarser grid. Hence, the convergence rate also becomes faster.
- The multigrid method is very popular and is incorporated in most commercial codes.

QUESTIONS:

1) Define Scarborough criteria for solving algebraic equations.
2) Solve the following system of equations with Jacobi and Gauss-Seidel iteration starting with an initial guess of x = y = z =1.

$$3x + 2y + 5z = 13$$

$$-x + 6y + z = -15$$

$$5x - 3y - 3z = 23.$$

Comment on the accuracy of your solution and the relative efficiency of the two methods.

3) Solve the following system using the Gauss elimination method.

$$2x_1 + 3x_2 + 5x_3 - 2x_4 = 8$$

$$6x_1 - 8x_2 + 3x_3 - 7x_4 = 9$$

$$5x_1 + 2x_2 - 7x_3 + 3x_4 = -13$$

$$x_1 + 5x_2 - 2x_3 + 4x_4 = 1$$

4) Solve the following system using the Gauss Elimination method.

$$2y + 3z = 7$$

$$3x + 6y - 12z = -3$$

$$5x - 2y + 2z = -7.$$

5) Solve the following system of equations with the Jacobi and Gauss-Seidel iterations for error ≤. 1%.

$$8x - 2y + 3z = 23$$

$$-2x + 4y + z = 3$$

$$x + 2y - 3z = -5.$$

6) An insulated rod of length 2.0m and 2.0cm × 2.0cm square cross-section has its temperature fixed at 150^0C at one end and 750^0C at the other end. Assuming steady-state heat conduction in one dimension, find the temperature at each node using the Thomas algorithm. Divide the rod in ten equal parts and take conductivity of the material k = 1000W/mK.
7) Solve the same problem by dividing the rod in 15 equal parts. Comment on the accuracy of your solution. In both cases, compare the results with the analytical solution.

9 Turbulence Modeling

9.1 INTRODUCTION

I am an old man now, and when I die and go to heaven there are two matters on which I hope for enlightenment. One is quantum electrodynamics, and the other is the turbulent motion of fluids. And about the former I am rather optimistic.

- *Sir Horace Lamb* FRS (1849–1934) 2nd Wrangler Trinity College

Turbulence is the last great unsolved problem in classical physics.

- Richard Feynman (Nobel Prize in Physics – quantum electrodynamics)

Most of the flows encountered in nature or in industrial applications are turbulent in nature. Turbulence plays a major part in the design of industrial equipment. Thus, modeling of turbulent flow is one of the important and challenging areas in CFD. In this chapter, the physical nature of turbulence and mathematical framework of turbulence modeling will be discussed, as well as the limitations of various models.

It is very difficult to define turbulence. According to Hinze [1975], turbulence can be described as: *"Turbulent Fluid motion is an irregular condition of flow in which the various quantities show a random variation with time and space coordinates, so that statistically distinct average values can be discerned."*

9.2 WHAT IS TURBULENCE?

Defining turbulence is not easy, but some important features can be described. Turbulence is a three-dimensional, highly non-linear, irregular, disorderly, irreversible stochastic phenomenon. It is governed by unsteady Navier-Stokes equations. Turbulent flow takes place at high Reynolds numbers.

9.2.1 Characteristics of Turbulent Flows

Characteristics of turbulent flow are as follows:

- The flow is random (i.e. the flow is disorderly and non-repetitive in nature).
- It is non-linear and three dimensional.
- Larger eddies break down into smaller eddies with a range of frequencies.
- Broken-down eddies cause transfer of energy from larger eddies to smaller eddies. The energy transfer process is irreversible and dissipative.
- This causes an increased transfer of momentum, energy, species, etc.
- Turbulence can only occupy parts of the flow domain.

- It increases drag, mixing and energy diffusion. In engineering machinery, this is sometimes welcome and sometimes detrimental to performance.

A useful way of looking at turbulence is through vortex dynamics. Large-scale vortices or eddies are created by the flow. Through the process of vortex stretching, vortices are broken up into smaller vortices. This moves the energy from large eddies to smaller eddies. At the smallest scales, turbulent kinetic energy is dissipated into heat due to viscous stresses. Large-scale vortices are influenced by the shape of flow domain and global flow field and contain the highest part of energy. Large-scale turbulence is problematic; it is difficult to decide what is a coherent structure and what is actually turbulence. This is described by the *Taylor scale.* Small vortices contain a low proportion of overall energy but contribute most in dissipation. This is also the smallest relevant scale in turbulent flows, characterized by the *Kolmogorov micro-scale.*

9.2.2 Task of Turbulence Modeling

We try to find approximate simplified solutions for the Navier-Stokes (N-S) equations in a manner that either describes turbulence in terms of mean properties or limits the spatial/temporal resolution requirements associated with the full model. Turbulence modeling is therefore about employing N-S equations in such a manner that we can simulate turbulence interaction in a simpler form. For example, a set of equations describing mean properties would allow us to perform steady-state simulations when only mean properties are of interest.

We shall here examine three modeling frameworks:

- Direct numerical simulation (DNS)
- Reynolds-averaged Navier-Stokes (RANS) equations, including eddy viscosity models and higher moment closure
- Large eddy simulation (LES)

9.3 DIRECT NUMERICAL SIMULATION (DNS)

DNS is, strictly speaking, not a turbulence model at all. Here, we simulate all scales of interest in both space and time in a well-resolved transient mode. Thus, computer-resource requirements are immense, and we can only handle relatively modest Reynolds numbers and very simple geometry. However, this is the best way to gather detailed information on turbulent interaction and mean properties in full fields. The method is unrealistically expensive for flows of engineering interest, but it still has a role (e.g. gathering data for model evaluation and fundamental understanding of turbulent interaction). DNS is much more reliable than experiments, and complete data sets can be obtained, including correlations and visualization. Today, DNS includes scalar mixing, heat transfer, buoyancy effects, etc.

9.4 REYNOLDS AVERAGING

Turbulent flow is three dimensional and unsteady in nature, but in turbulence modeling, we are mainly interested in mean flow solutions (e.g. velocity, pressure, lift, drag). Looking at turbulent flow, mean flow properties may be steady despite turbulent fluctuations. If this is so, we can derive equations for mean properties directly and reduce the cost of computation by orders of magnitude because:

- It is no longer necessary to perform unsteady simulations as we are solving for average properties directly.
- We can tackle high Reynolds numbers and determine resolution based on the required engineering accuracy.

Reynolds decomposition of any flow variable, as shown in Figure 9.1, gives

$$\emptyset = \overline{\emptyset} + \emptyset'$$

where $\emptyset$ = instantaneous parameter (velocity, pressure etc.) at any particular position and time

$\overline{\emptyset}$ = Mean value or time averaged value

$$= \lim_{T \to \infty} \frac{1}{T} \int_0^T \emptyset dt$$

$\emptyset'$ = fluctuating component due to turbulence.

Time scales of ensemble-averaged flow are separate from scales from turbulent fluctuations.

9.4.1 Reynolds-Averaged Navier-Stokes (RANS) Equations

In N-S equations, the flow parameters (velocities, pressure, temperature, species etc.) are instantaneous. This instantaneous parameter ($\emptyset$) can be expressed by

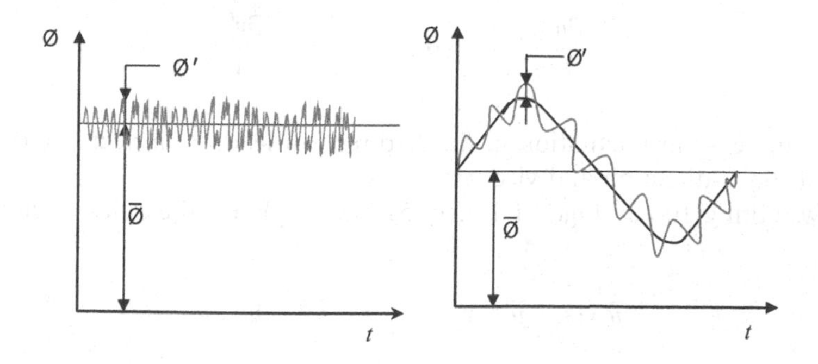

FIGURE 9.1 Turbulence Notations.

superimposing fluctuating component, denoted as $\emptyset'$, on the mean component ($\overline{\emptyset}$) as shown in Figure 9.1. *Reynolds averaging* involves the process of taking mean or average of flow parameters.

Let us decompose velocities and pressure into a mean and fluctuating component:

$$u = \bar{u} + u'; \; v = \bar{v} + v' \; and \; p = \bar{p} + p' \tag{9.1}$$

Let us consider two-dimensional incompressible flow. We have
Continuity:

$$\frac{\partial u}{\partial x} + \frac{\partial v}{\partial y} = 0 \tag{9.2}$$

X –Momentum

$$\frac{\partial}{\partial t}(\rho u) + \frac{\partial}{\partial x}(\rho uu) + \frac{\partial}{\partial y}(\rho uv) = -\frac{\partial p}{\partial x} + \frac{\partial}{\partial x}\left(\mu\frac{\partial u}{\partial x}\right) + \frac{\partial}{\partial y}\left(\mu\frac{\partial u}{\partial y}\right) \tag{9.3}$$

Similarly, Y-Momentum:

$$\frac{\partial}{\partial t}(\rho v) + \frac{\partial}{\partial x}(\rho uv) + \frac{\partial}{\partial y}(\rho vv) = -\frac{\partial p}{\partial y} + \frac{\partial}{\partial x}\left(\mu\frac{\partial v}{\partial x}\right) + \frac{\partial}{\partial y}\left(\mu\frac{\partial v}{\partial y}\right) \tag{9.4}$$

Now, putting Eq. (9.1) in Eq. (9.2), we get

$$\frac{\partial \bar{u}}{\partial x} + \frac{\partial \bar{v}}{\partial y} + \frac{\partial u'}{\partial x} + \frac{\partial v'}{\partial y} = 0 \tag{9.5}$$

Hence,

$$\frac{\partial \bar{u}}{\partial x} + \frac{\partial \bar{v}}{\partial y} = 0 \; and \; \frac{\partial u'}{\partial x} + \frac{\partial v'}{\partial y} = 0$$

Let us consider

$$\frac{\partial u'}{\partial x} + \frac{\partial v'}{\partial y} = 0 \; i.e. \; \frac{\partial u'}{\partial x} = -\frac{\partial v'}{\partial y}$$

So, any increase in fluctuation in the X-direction leads to a decrease of fluctuation in the Y-direction and vice versa.

Now, taking the average of Eq. (9.5) and applying the rules of averaging, that is

$$\overline{\bar{u}} = \bar{u} \; ; \; \overline{u + v} = \bar{u} + \bar{v} \; ; \; \overline{\bar{u}\bar{v}} = \bar{u}\bar{v} \; etc.$$

and

$$\overline{u'} = 0 \ ; \ \overline{u' + v'} = \overline{u'} + \overline{v'} = 0 \ ; \ but \ \overline{u'v'} \neq 0$$

Hence,

$$\overline{\frac{\partial \bar{u}}{\partial x} + \frac{\partial \bar{v}}{\partial y} + \frac{\partial u'}{\partial x} + \frac{\partial v'}{\partial y}} = \frac{\partial \bar{u}}{\partial x} + \frac{\partial \bar{v}}{\partial y} + \frac{\partial \overline{u'}}{\partial x} + \frac{\partial \overline{v'}}{\partial y} = \frac{\partial \bar{u}}{\partial x} + \frac{\partial \bar{v}}{\partial y}$$

Let us now take X-momentum equation

$$\frac{\partial}{\partial t}(\rho u) + \frac{\partial}{\partial x}(\rho uu) + \frac{\partial}{\partial y}(\rho uv) = -\frac{\partial p}{\partial x} + \frac{\partial}{\partial x}\left(\mu \frac{\partial u}{\partial x}\right) + \frac{\partial}{\partial y}\left(\mu \frac{\partial u}{\partial y}\right)$$

Now putting $u = \bar{u} + u'; v = \bar{v} + v'$ *and* $p = \bar{p} + p'$ and averaging we get,

$$\overline{\frac{\partial}{\partial t}(\rho u)} = \frac{\partial}{\partial t}\left\{\overline{\rho(\bar{u} + u')}\right\} = \frac{\partial}{\partial t}\left\{\rho(\bar{u} + \overline{u'})\right\} = \frac{\partial}{\partial t}(\rho \bar{u})$$

Similarly, we get

$$\overline{\frac{\partial p}{\partial x}} = \frac{\partial \bar{p}}{\partial x} \ and \ \overline{\mu \frac{\partial^2 u}{\partial x^2}} = \mu \frac{\partial^2 \bar{u}}{\partial x^2}$$

For convective term

$$\frac{\partial}{\partial x}\overline{(\rho uu)} = \frac{\partial}{\partial x}\left\{\rho\left(\overline{\bar{u}\bar{u}} + \overline{u'\bar{u}} + \overline{u'}\bar{u} + \overline{u'u'}\right)\right\} = \frac{\partial}{\partial x}(\rho \bar{u}\bar{u}) + \frac{\partial}{\partial x}\left(\rho \overline{u'u'}\right)$$

$$\frac{\partial}{\partial x}\overline{(\rho uv)} = \frac{\partial}{\partial y}\left\{\rho\left(\overline{\bar{u}\bar{v}} + \overline{v'}\bar{u} + \overline{u'}\bar{v} + \overline{u'v'}\right)\right\} = \frac{\partial}{\partial y}(\rho \bar{u}\bar{v}) + \frac{\partial}{\partial y}\left(\rho \overline{u'v'}\right)$$

So, after Reynolds averaging, X-momentum equation becomes

$$\begin{aligned} &\frac{\partial}{\partial t}(\rho \bar{u}) + \frac{\partial}{\partial x}(\rho \bar{u}\bar{u}) + \frac{\partial}{\partial y}(\rho \bar{u}\bar{v}) = -\frac{\partial \bar{p}}{\partial x} \\ &+ \frac{\partial}{\partial x}\left(\mu \frac{\partial u}{\partial x} - \rho \overline{u'u'}\right) + \frac{\partial}{\partial y}\left(\mu \frac{\partial u}{\partial y} - \rho \overline{u'v'}\right) \end{aligned} \tag{9.6}$$

Similarly, the Y-momentum becomes

$$\begin{aligned} &\frac{\partial}{\partial t}(\rho \bar{v}) + \frac{\partial}{\partial x}(\rho \bar{u}\bar{v}) + \frac{\partial}{\partial y}(\rho \bar{v}\bar{v}) = -\frac{\partial \bar{p}}{\partial y} \\ &+ \frac{\partial}{\partial x}\left(\mu \frac{\partial v}{\partial x} - \rho \overline{u'v'}\right) + \frac{\partial}{\partial y}\left(\mu \frac{\partial v}{\partial y} - \rho \overline{v'v'}\right) \end{aligned} \tag{9.7}$$

Comparing Esp. (9.3) and (9.4) with Eqs. (9.6) and (9.7) we can see that because of Reynolds averaging, we are getting three extra terms $-\rho\overline{u'u'}$; $-\rho\overline{u'v'}$ *and* $-\rho\overline{v'v'}$.

For three dimension, the momentum transport equation without any body forces in the Reynolds-averaged form is given by

$$\frac{\partial}{\partial t}(\rho\overline{u_i}) + \frac{\partial}{\partial x_j}(\rho\overline{u_i u_j}) = \frac{\partial \overline{p}}{\partial x_i} + \frac{\partial \tau_{ij}}{\partial x_j} + \frac{\partial}{\partial x_j}\left(-\rho\overline{u'_i u'_j}\right) \tag{9.8}$$

$-\rho\overline{u'_i u'_j}$ are unknown and lead to closure problem as number of equations are less than number of unknowns. These terms are also known as *Reynolds (R) stresses*, and the Eq. (9.8) is known as the *Reynolds-averaged Navier-Stokes (RANS) equation.*

To close the system, we need to describe the unknown values as a function of the solution. That is known as *Turbulence Modeling*. Two methods are:

1. Write an algebraic function, resulting in *eddy viscosity models.*
2. Add more differential equations (i.e. transport equations for R), producing *Reynolds Transport Models.* As we introduce new equations, the above problem will recur.

Both options are in use today, but the first massively out-weights the second.

The objectives of turbulence modeling are to close the mean-flow equations by specifying the Reynolds stresses $\left(-\rho\overline{u'_i u'_j}\right)$ and additional transported quantities $\left(-\rho\overline{u'_j \emptyset'}\right)$.

9.4.2 Eddy Viscosity Models Hypothesis (Boussinesq [1877])

We have seen that main idea behind turbulence modeling is to specify the Reynolds stresses $\left(-\rho\overline{u'_i u'_j}\right)$ and additional transported quantities $\left(-\rho\overline{u'_j \emptyset'}\right)$, which are unknown. The eddy viscosity hypothesis is based on the analogy between turbulent transport and molecular transport in laminar flow, in which it is postulated that the Reynolds stress tensor is proportional to the mean strain rate tensor.

Now, the molecular transport is given by

$$\underbrace{\tau_{ij} =}_{\text{Shear Stress}} \quad \underbrace{\mu}_{\text{Laminar Viscosity}} \quad \underbrace{\left(\frac{\partial u_i}{\partial x_j} + \frac{\partial u_j}{\partial x_i}\right) - \frac{2}{3}\mu\frac{\partial u_k}{\partial x_k}\delta_{ij}}_{\text{Mean Shear Strain}}$$

Hence by analogy, turbulent transport can be expressed as

$$\underbrace{\tau_t = -\rho\overline{u'_i u'_j} =}_{\text{Reynolds Stress}} \quad \underbrace{\mu_t}_{\text{Eddy Viscosity}} \quad \underbrace{\left(\frac{\partial \overline{u}_i}{\partial x_j} + \frac{\partial \overline{u}_j}{\partial x_i}\right) - \frac{2}{3}\rho k\delta_{ij}}_{\text{Mean Shear Strain}}$$

$\frac{2}{3}\rho k\delta_{ij}$ ensures the sum of normal Reynolds stress components as 2k.

The RANS equation for incompressible flow becomes

$$\frac{\partial}{\partial t}(\rho \bar{u}_i) + \frac{\partial}{\partial x_j}(\rho \overline{u_i u_j}) = \frac{\partial \bar{p}}{\partial x_i} + \frac{\partial}{\partial x_j}(\mu + \mu_t)\left(\frac{\partial \bar{u}_i}{\partial x_j} + \frac{\partial \bar{u}_j}{\partial x_i}\right) \tag{9.9}$$

It is to be noted that this is a model. Whereas μ is the physical property of the fluid, eddy or turbulent viscosity μ_t is local turbulent of flow property and varies with position. For turbulent simulation modeling of μ_t is required. At high Reynolds number, the value of μ_t is far greater than μ throughout the flow field except in the laminar sub-region.

Eddy-viscosity models are widely used and popular because they are easy to implement. Although these models may not give sufficiently accurate results for complex flows, they may work for design purposes.

9.5 RANS TURBULENCE MODELS

RANS Turbulence models can be classified as

1) **Linear Eddy Viscosity based Models (LEVM)**
   In LEVM, the turbulent stress is proportional to mean strain and eddy viscosity is determined by solving transport equations. LEVM can be further categorized as:
   - **Zero Equation (Algebraic) Model**
     In this model, turbulence length and velocity scales prescribed algebraically.
   - **One Equation Model**
     Here, one transport equation is solved for velocity scale and length scale is prescribed
   - **Two Equation Model**
     In this model, two transport equations are solved - one for Velocity scale and one for Length scale.
2) **Non-Linear Eddy Viscosity based Models (NLEVM)**
   In NLEVM, turbulent stress is non-linear function of mean strain and vorticity. Direct or indirect algebraic expression is used between Reynolds Stress and the Mean Strain Rate tensor.
3) **Reynolds Stress Transport equation-based Models (RSTM)**
   RSTM model solves six transport equations for six components of symmetric Reynolds Stress Tensor.

We shall now discuss some popular linear eddy viscosity models.

9.5.1 ZERO-EQUATION MODELS

These models are based on Prandtl's mixing length theory.

Eddy viscosity is given by:

$$\mu_t = \rho \nu_t \ where\ \nu_t = u l_m \tag{9.10}$$

The *mixing length*, l_m is normally specified by algebraic expressions depending on the type of flow. The velocity scale u is determined from the expression:

$$u = l_m \left| \frac{\partial u}{\partial y} \right|$$

Here the assumption is a lump of fluid from a certain layer is displaced over a distance l_m in the transverse direction, then the difference in velocity of the fluid lump (eddy) will differ from its surrounds by an amount $l_m|\partial u/\partial y|$.

The resulting turbulent shear stress is given by:

$$\tau_t = \mu_t \left(\frac{\partial \bar{u}}{\partial y} \right) = \rho l_m^2 \left(\frac{\partial \bar{u}}{\partial y} \right)^2 \tag{9.11}$$

The *mixing length* l_m is given by
$l_m = \alpha d$, *where* α = Closure Coefficient and d = Characteristic Layer Thickness
Far Wake: $\alpha = 0.180$
Plane jet: $\alpha = 0.098$
Mixing Layer: $\alpha = 0.071$
Round Jet: $\alpha = 0.080$
The model works reasonably well for thin Shear Flows.
Advantages: Simplicity and computational economy.
Disadvantages: Insufficient generality, difficult in prescribing mixing length for complex flows.

9.5.1.1 Structure of the Turbulent Boundary Layer

This section is applicable for a flat-plate boundary layer or fully developed pipe or channel flow. The total shear stress in the y-direction is composed of viscous stress and Reynolds stress and can be expressed as

$$\tau = \mu \frac{\partial \bar{u}}{\partial y} - \rho \overline{u'v'} = \tau_w + \tau_t$$

Near the wall, the viscous stress, τ_w is the most important factor and it becomes negligible in the outer part of a turbulent boundary layer.

For $y < 0.1\delta$, shear stress can be taken as constant and equal to its value at the wall:

$$\tau \approx \tau_w$$

τ_w has dimensions of [*density*] x [*velocity*]^2.

Let us define an important velocity scale, the *friction velocity*, u_τ, by

$$u_\tau = u^* = \sqrt{\frac{\tau_w}{\rho}}$$

Near the wall, the most important parameters are kinematic viscosity ν; and wall shear stress τ_w.

Hence, the characteristic velocity and length scales can be expressed as:

friction velocity: $u_\tau = \sqrt{\tau_w/\rho}$
viscous length scale: $\delta_v = {}^{\nu}/_{u_\tau}$

Now, we can define non-dimensional velocity and distance from the wall as:

$$u^+ = \frac{u}{u_\tau} \text{ and } y^+ = \frac{y}{\delta_v} = \frac{yu_\tau}{\nu}$$

Please note that y^+ is an important factor to determine the importance of viscous and turbulent transport at different distances from the wall.

9.5.2 Key Modifications of Prandtl's Mixing Length Model

9.5.2.1 The Cebaci-Smith Model

The boundary layer is identified in two zones – the inner zone and outer zone.

For the Inner Zone:

$$(\mu_t)_{inner} = \rho l_m^2 |\Omega| \tag{9.12}$$

where,

$$l_m = \kappa y \left[1 - exp\left(\frac{-y^+}{A^+}\right)\right]$$

y = distance normal to the wall
κ = Von Karman constant = 0.41

$$|\Omega| = \text{Vorticity} = \left|\frac{\partial u}{\partial y} - \frac{\partial v}{\partial x}\right|$$

For Outer Zone: $(\mu_t)_{outer} = 0.0168 \rho u_e \delta^*$

$$\delta^* = \textit{displacement thickness}$$

$$u_e = \textit{velocity at the edge of boundary layer}$$

Final eddy viscosity: $\mu_t = \min\left[(\mu_t)_{inner}, (\mu_t)_{outer}\right]$

Advantage: Simple

Disadvantage: requires knowledge of velocity and displacement thickness at the edge of boundary layer which is not easy to calculate for complex flows.

9.5.2.2 The Baldwin-Lomax Model

This model does not require the knowledge of velocity and displacement thickness at the edge of boundary layer. Here, for outer zone,

$$(\mu_t)_{outer} = \rho K C_{cP} F_{wake} F_{Kelb} \tag{9.13}$$

$$F_{wake} \text{ contains mixing length term} = min\left(y_{max}F_{max}, C_{wk}y_{max}U_{diff}^2/F_{max}\right)$$

$$F(y) = y|\Omega|\left[1 - exp\left(-\frac{y^+}{A^+}\right)\right]$$

$$U_{diff} = \left(\sqrt{U^2+V^2}\right)_{max} - \left(\sqrt{U^2+V^2}\right)_{min}$$

$$F_{Kleb}(y) = \left[1 + 5.5\left(\frac{C_{Kleb}y}{y_{max}}\right)^6\right]^{-1}$$

Final eddy viscosity: $\mu_t = \min\left[(\mu_t)_{inner}, (\mu_t)_{outer}\right]$

The value of $(\mu_t)_{inner}$ can be obtained from Cebaci-Smith Model. The constants are given by

$A^+ = 26;\ C_{cP} = 1.6;\ C_{Kleb} = 0.3;\ C_{wk} = 1.0;\ \kappa = 0.41;\ K = 0.68$

9.5.3 The Transport Equation for Turbulent Kinetic Energy (One-Equation Model)

Exact equation for k (=0.5 $\overline{u'v'}$) is derived from the Navier Stokes equation and given by

$$\frac{\partial}{\partial t}(\rho k) + \frac{\partial}{\partial x_j}(\rho \bar{u}_j k) = \frac{\partial}{\partial x_j}\left[\mu\frac{\partial k}{\partial x_j} - \frac{1}{2}\rho\overline{u'_i u'_j u'_k} - \overline{pu'_j}\right] + \tau_{ij}\frac{\partial \bar{u}_i}{\partial x_j} - \rho\mu\frac{\partial u'_i}{\partial x_i}\frac{\partial u'_i}{\partial x_i} \tag{9.14}$$

In the above equation, convection term is

$$\frac{\partial}{\partial x_j}(\rho \bar{u}_j k)$$

Analogy to laminar flow models, diffusion term becomes

$$\frac{\partial}{\partial x_j}\left[\mu\frac{\partial k}{\partial x_j} - \frac{1}{2}\rho\overline{u'_i u'_j u'_k} - \overline{pu'_j}\right] = \frac{\partial}{\partial x_j}\left(\mu + \frac{\mu_t}{\sigma_k}\right)\frac{\partial k}{\partial x_j}$$

Production term:

$$\tau_{ij}\frac{\partial \bar{u}_i}{\partial x_j} = \left\{\mu_t\left(\frac{\partial \bar{u}_i}{\partial x_j} + \frac{\partial \bar{u}_j}{\partial x_i}\right) - \frac{2}{3}\rho k\delta_{ij}\right\}\frac{\partial \bar{u}_i}{\partial x_j}$$

The dissipation term, ε, remains unaltered and is determined from length scale (l) information by

$$\varepsilon \propto \frac{k^{3/2}}{l_m} = C_\mu^{3/2}\frac{k^{3/2}}{l_m}$$

Hence, modelled equation of k becomes:

$$\frac{\partial}{\partial t}(\rho k) + \frac{\partial}{\partial x_j}(\rho \bar{u}_j k) = \frac{\partial}{\partial x_j}\left(\mu + \frac{\mu_t}{\sigma_k}\right)\frac{\partial k}{\partial x_j} + \left\{\mu_t\left(\frac{\partial \bar{u}_i}{\partial x_j} + \frac{\partial \bar{u}_j}{\partial x_i}\right) - \frac{2}{3}\rho k\delta_{ij}\right\}\frac{\partial \bar{u}_i}{\partial x_j} - \rho\varepsilon \quad (9.15)$$

The turbulent viscosity is given as

$$\mu_t = C_\mu k^{1/2} l_m \quad (9.16)$$

One-equation model (k-model) still requires a length scale l as before to define the eddy viscosity.

9.5.4 Two-Equation Models

The main disadvantage of using a one-equation model is the incomplete representation of the two scales required to build eddy viscosity. Several two-equation models have been developed. They attempt to represent both scales independently and are widely used in industry.

Two-equation turbulence models are widely used in the industries. Several two-equation models have been developed by various researchers. In two-equation models, we solve a transport equation for turbulent kinetic energy (k) and a second transport equation for turbulent length scale. Normally, second transport equation is solved for turbulent dissipation (ε) or turbulent specific dissipation (ω) which are related to turbulent length scale l. Two-equation models can be classified as high Reynolds number models and low Reynolds number models. The high Reynolds number models can be applied only outside the inner region of the boundary layer whereas low Reynolds number models can be applied down to the wall. Some of the two-equation models are described here.

9.5.4.1 The Standard k-ε Model

The standard version of the two-equation model involves transport equations for the turbulent kinetic energy (k) and its dissipation rate (ε). The following transport equations are suggested by Jones and Launder [1972].

The transport equation for turbulent kinetic energy (k) is

$$\frac{\partial}{\partial t}(\rho k) + \frac{\partial}{\partial x_j}(\rho u_j k) = \frac{\partial}{\partial x_j}\left[\left(\mu + \frac{\mu_t}{\sigma_k}\right)\frac{\partial k}{\partial x_j}\right] + p_k - \rho\varepsilon \quad (9.17)$$

The transport equation for turbulent dissipation rate (ε):

$$\frac{\partial}{\partial t}(\rho\varepsilon) + \frac{\partial}{\partial x_j}(\rho u_j \varepsilon) = \frac{\partial}{\partial x_j}\left[\left(\mu + \frac{\mu_t}{\sigma_\varepsilon}\right)\frac{\partial \varepsilon}{\partial x_j}\right] + C_{\varepsilon 1}\frac{\varepsilon}{k}p_k - C_{\varepsilon 2}\frac{\rho\varepsilon^2}{k} \quad (9.18)$$

The turbulent viscosity is given by

$$\mu_t = \rho c_\mu \frac{k^2}{\varepsilon} \quad (9.19)$$

p_k in Eq. (9.17) is the production of k which is given by

$$p_k = -\rho\overline{u'_i u'_j}\frac{\partial u_i}{\partial x_j} = \mu_t\left(\frac{\partial u_i}{\partial x_j} + \frac{\partial u_j}{\partial x_i}\right)\frac{\partial u_i}{\partial x_j}$$

The constants are given by $C_\mu = 0.09$, $C_{\varepsilon 1} = 1.44$, $C_{\varepsilon 2} = 1.92$, $\sigma_k = 1.0$, $\sigma_\varepsilon = 1.3$.

This model is good compromise between universality and economy for engineering problems. However, the model needs near wall modification as it is insensitive to freestream boundary conditions. The model suffers from poor performance for flows with sudden change of strain rate (i.e. curvature, swirl, rotation, separated flows).

9.5.4.2 The Wilcox k-ω Model

In the k-ε model, the turbulent length scale l is taken as $l = k^{3/2}/\varepsilon$. As discussed earlier, turbulent length scale l can also be related to turbulent specific dissipation (ω). The k-ω model proposed by Wilcox uses the turbulence frequency ω (in s^{-1}) which is given by $\omega = \varepsilon/k$. Hence, the length scale becomes $l = \sqrt{k}/\omega$ and the eddy viscosity is given by $\mu_t = \rho k/\omega$.

The Reynolds stresses are computed as usual in two-equation models with the Boussinesq expression:

$$\tau_{ij} = -\rho u_i' u_j' = 2\mu_t S_{ij} - \frac{2}{3}\rho k\delta_{ij} = \mu_t\left(\frac{\partial U_i}{\partial x_j} + \frac{\partial U_j}{\partial x_i}\right) - \frac{2}{3}\rho k\delta_{ij}$$

where,

$$S_{ij} = \frac{1}{2}\left(\frac{\partial U_i}{\partial x_j} + \frac{\partial U_j}{\partial x_i}\right)$$

The transport equation for k and ω for turbulent flows at high Reynolds number is as follows:

$$\frac{\partial(\rho k)}{\partial t}+\frac{\partial}{\partial x_j}(\rho u_j k)=\frac{\partial}{\partial x_j}\left[\left(\mu+\frac{\mu_t}{\sigma_k}\right)\frac{\partial k}{\partial x_j}\right]+P_k-\beta^*\rho k\omega \tag{9.20}$$

where

$$p_k=-\overline{\rho u'_i u'_j}\frac{\partial u_i}{\partial x_j}=\mu_t\left(\frac{\partial u_i}{\partial x_j}+\frac{\partial u_j}{\partial x_i}\right)\frac{\partial u_i}{\partial x_j} \tag{9.21}$$

is the rate of production turbulent kinetic energy, and

$$\frac{\partial(\rho\omega)}{\partial t}+\frac{\partial}{\partial x_j}(\rho u_j\omega)=\frac{\partial}{\partial x_j}\left[\left(\mu+\frac{\mu_t}{\sigma_\omega}\right)\frac{\partial\omega}{\partial x_j}\right]+\gamma_1\left(2\rho S_{ij}S_{ij}-\frac{2}{3}\rho\omega\frac{\partial u_{ij}}{\partial x_j}\delta_{ij}\right)-\beta_1\rho\omega^2 \tag{9.22}$$

The model constants are as follows:

$\sigma_k = 2.0$; $\sigma_\omega = 2.0$; $\gamma_1 = 0.533$; $\beta_1 = 0.075$; $\beta^* = 0.09$

This model is valid even in the laminar sublayer and near wall modification not obligatory. However, it suffers from poor performance for flows with sudden change of strain rates. The model over predicts the shear stress in adverse pressure gradient boundary layers and is unreliable for flow with detached free shear layers.

9.5.4.3 The SST k- ω (Menter) Turbulence Model

In practice, the *k*-ε models are generally well behaved in the far field and insensitive to freestream boundary conditions. The *k*-ω models are valid even in the laminar sublayer and behave well in the near-wall region. Taking the strengths and weakness of both models, Menter [1994] suggested a blended model in which he used the *k*-ω model near the wall and transitions and the *k*-ε model away from the wall. The model was enhanced by Menter et al. [2003] and has the form

Kinematic Eddy Viscosity:

$$\mu_t=\frac{a_1\rho k}{max(a_1\omega, SF_2)} \tag{9.23}$$

Turbulent Kinetic Energy:

$$\frac{\partial\rho k}{\partial t}+\frac{\partial}{\partial x_J}(\rho u_j k)=P_k-\rho\beta^* k\omega+\frac{\partial}{\partial x_j}\left[(\mu+\sigma_k\mu_T)\frac{\partial k}{\partial x_j}\right] \tag{9.24}$$

Specific Dissipation Rate:

$$\frac{\partial \rho\omega}{\partial t} + \frac{\partial}{\partial x_j}(\rho u_j \omega) = \rho\alpha S^2 - \rho\beta\omega^2 + \frac{\partial}{\partial x_j}\left[(\mu + \sigma_\omega \mu_T)\frac{\partial k}{\partial x_j}\right] + 2\rho(1 - F_1)\sigma_{\omega 2}\frac{1}{\omega}\frac{\partial k}{\partial x_j}\frac{\partial \omega}{\partial x_j} \tag{9.25}$$

Closure Coefficients and Auxiliary Relations:

$$F_1 = \tan h\left\{\left\{\min\left[max\left(\frac{\sqrt{k}}{\beta^* \omega y}, \frac{500\mu}{\rho y^2 \omega}\right), \frac{4\sigma_{\omega_2} k}{CD_{k\omega} y^2}\right]\right\}^4\right\}$$

$$F_2 = \tan h\left(\left(\max\left(\frac{2\sqrt{k}}{\beta^* \omega y}, \frac{500\mu}{\rho y^2 \omega}\right)\right)^2\right)$$

$$P_k = \min\left(\tau_{ij}\frac{\partial u_i}{\partial x_j}, 10\rho\beta^* k\omega\right)$$

$$CD_{k\omega} = \max\left(2\rho\sigma_{\omega 2}\frac{1}{\omega}\frac{\partial k}{\partial x_i}\frac{\partial \omega}{\partial x_i}, 10^{-10}\right)$$

Each of the constants is blended as

$$\emptyset = \emptyset_1 F + \emptyset_2(1 - F)$$

Where $\emptyset_1$ represents constant 1 and $\emptyset_2$ represents constant 2.

$$\alpha_1 = \frac{5}{9}, \quad \alpha_2 = 0.44, \quad \beta_1 = \frac{3}{40}, \quad \beta = 0.0828, \quad \beta^* = 0.09, \quad a_1 = 0.31$$

$$\sigma_{k_1} = 0.85, \quad \sigma_{k_2} = 1, \quad \sigma_{\omega_1} = 0.5, \quad \sigma_{\omega_2} = 0.856$$

9.5.4.4 Near-Wall Modifications for Two-Equation Models

The presence of wall surface affects the turbulent structure. Near to the wall, the molecular viscosity μ becomes important and it is of the same order of that of turbulent viscosity μ_t. Molecular viscosity diffuses vorticity and damps turbulence and as a result viscous diffusion terms become one of the largest terms to be balanced by the other terms in the Reynolds Stress transport equation.

Standard k- ε model does not automatically allow the eddy viscosity to tend to zero as the wall is approached. This can be done by resolving Boundary Layers in two ways:

Defining Wall Functions

In most engineering applications, we are not interested in finding the detailed turbulence structure near the wall. Instead, we want to find drag, etc. Thus, we assume a velocity profile between the near-wall node and the boundary as shown in Figure 9.2. The near-wall node can be placed in the region 30 < y^+ < 50 (the range 15–150 is just about acceptable). This means that coarse mesh can be used close to solid boundaries.

The standard wall function method based on logarithmic law of wall (Launder and Spalding [1974]; Wilcox [1993]) is

$$\frac{u_P}{u_\tau} = \frac{1}{\kappa} lnEy_P^+$$

where subscript P denotes the near-wall node. Given the value of u_P this could be solved (iteratively) for u_τ and hence the wall stress τ_w. The value of κ = 0.41 and E = 9.0 is

$$k_P = \frac{u_\tau^2}{\rho\sqrt{C_\mu}} \quad and \quad \varepsilon_P = \frac{u_\tau^3}{\kappa y_P}$$

Readers are advised to refer near-wall models available in the literature.

Low-Re Turbulence Models

In many flow situations (e.g. those with large pressure or density gradients), the law of wall is invalid and conservation equations with low Reynolds models may have to be solved. When $\rho\overline{u'v'}$ is not much greater than $\mu^{\partial u}/_{\partial y}$ the turbulent structure is influenced by fluid viscosity and empirical constants may vary with $^{k^2}/_{\nu\varepsilon}$, the turbulent Reynolds number.

The transport equation for turbulent kinetic energy (k):

$$\frac{\partial}{\partial t}\langle\rho k\rangle + \frac{\partial}{\partial x_j}\langle\rho k u_j\rangle = \frac{\partial}{\partial x_j}\left[\left(\mu + \frac{\mu_t}{\sigma_k}\right)\frac{\partial k}{\partial x_j}\right] + \left[\mu_t\left(\frac{\partial u_i}{\partial x_j} + \frac{\partial u_j}{\partial x_i}\right)\frac{\partial u_i}{\partial x_j}\right] - \rho(\varepsilon + D) \tag{9.26}$$

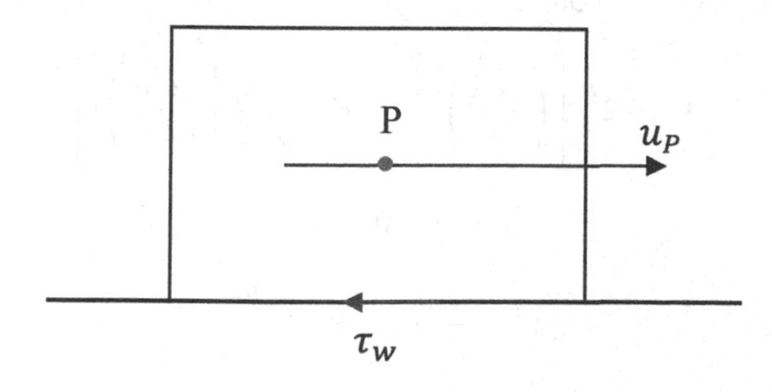

FIGURE 9.2 The Near-Wall Node.

The transport equation for turbulent dissipation rate (ε):

$$\frac{\partial}{\partial t}\langle\rho\varepsilon\rangle + \frac{\partial}{\partial x_j}\langle\rho\varepsilon u_j\rangle = \frac{\partial}{\partial x_j}\left[\left(\mu + \frac{\mu_t}{\sigma_\varepsilon}\right)\frac{\partial\varepsilon}{\partial x_j}\right] + C_{\varepsilon 1}f_1\frac{\varepsilon}{k}\left[\mu_t\left(\frac{\partial u_i}{\partial x_j} + \frac{\partial u_j}{\partial x_i}\right)\frac{\partial u_i}{\partial x_j}\right] - C_{\varepsilon 2}f_2\frac{\varepsilon^2}{k} + \rho E \tag{9.27}$$

Turbulent viscosity is given by

$$\mu_t = C_\mu \rho f_\mu \frac{k^2}{\varepsilon} \tag{9.28}$$

Various low-Re k-ε models are available in the literature: Eqs. (9.26) to (9.28) remain the same, but the coefficients vary. Some popular models are described here:

Chien:

$$C_\mu = 0.09;\quad C_{\varepsilon 1} = 1.35;\quad C_{\varepsilon 2} = 1.8;\quad \sigma_k = 1.0\ ;\ \sigma_\varepsilon = 1.3$$

$$f_\mu = 1 - exp(-0.0115y^+);\ f_1 = 1.0;\ f_2 = 1 - \frac{2}{9}exp\left[-\left(\frac{Re_t}{6}\right)^2\right]$$

$$D = 2\frac{\mu}{\rho}\frac{k}{y_n^2};\qquad E = \frac{2\mu\varepsilon}{\rho y_n^2}exp(-0.5y^+)$$

$$k = \varepsilon = 0\ at\ wall$$

Jones & Launder:

$$C_\mu = 0.09;\quad C_{\varepsilon 1} = 1.44;\quad C_{\varepsilon 2} = 1.92;\quad \sigma_k = 1.0;\quad \sigma_\varepsilon = 1.3$$

$$f_\mu = exp\left(\frac{-2.5}{1 + Re_t/50}\right);\qquad f_1 = 1.0;\qquad f_2 = 1 - 0.3exp\left[-(Re_t)^2\right]$$

$$D = -2\frac{\mu}{\rho}\left[\left(\frac{\partial\sqrt{k}}{\partial x_j}\right)^2\right];\quad E = \frac{2\mu\mu_t}{\rho^2 y_n^2}\left[\left(\frac{\partial^2 u_i}{\partial x_j^2}\right)^2\right]$$

$$k = \varepsilon = 0\ at\ wall$$

$$y^+ = \frac{y_n u_\tau}{\nu};\quad Re_t = \frac{k^2}{\nu\varepsilon}$$

These models require very fine meshes near the wall. The near-wall node can be placed in the region $0.01 < y^+ < 1$. Hence, these models can be computationally very demanding specially for high speed flows.

9.6 REYNOLDS STRESS TRANSPORT (EQUATION-BASED) MODELS (RSTMS)

Also known as differential stress models or second-order closure, this model solves individual equations for Reynolds stresses $\left(-\rho\overline{u'_i u'_j}\right)$ rather than turbulent kinetic energy k.

The equation of motion for instantaneous fluctuating velocity component u_i can be written as

$$\frac{\partial(\rho u_i')}{\partial t}+\frac{\partial}{\partial x_k}\left[\rho\left(u_i'\bar{u}_k+u_k'\bar{u}_i+u_i'u_k'\right)\right]=-\frac{\partial p}{\partial x_i}+\mu\frac{\partial^2 u_i'}{\partial x_k\partial x_k}+\frac{\partial}{\partial x_k}\left(\rho\overline{u_i'u_k'}\right) \quad (9.29)$$

Multiplying the u_i' component of Eq. (9.29) by u_j' and for u_j' component by u_i', adding them and taking time average of the sum we get

$$\frac{\partial}{\partial t}\left(\rho\overline{u'_i u'_j}\right)+\frac{\partial}{\partial x_k}\left(\rho\bar{u}\overline{u'_i u'_j}\right)=-\rho\left[\overline{u'_i u'_k}\frac{\partial\bar{u}_j}{\partial x_k}-\overline{u'_j u'_k}\frac{\partial\bar{u}_i}{\partial x_k}\right]$$
$$-\frac{\partial}{\partial x_k}\left[\rho\overline{u'_i u'_j u'_k}+\overline{pu'_i}\delta_{jk}+\overline{pu'_j}\delta_{ik}-\mu\frac{\partial}{\partial x_k}\left(\overline{u'_i u'_j}\right)\right]+\bar{p}\overline{\left[\frac{\partial u'_i}{\partial x_k}+\frac{\partial u'_j}{\partial x_k}\right]}-2\mu\overline{\frac{\partial u'_i}{\partial x_k}\frac{\partial u'_j}{\partial x_k}} \quad (9.30)$$

The left-hand side of Eq. (9.30) contains the rate of change of fluctuations during its transport with the mean flow. First term of the right-hand side is turbulent production term (P_{ij}) i.e. generation of turbulent stress due to mean velocity gradients i.e. the mean strain. Second term indicates the diffusion of Reynolds Stresses and their redistribution. The diffusion term consists of molecular diffusion, D^{μ}_{ijk} and turbulent diffusion, D^{t}_{ijk}. The Pressure strain term, Π_{ij}, redistributes energy between various turbulent stress components. The last term is turbulent dissipation due to viscosity, ε_{ij}. While the advection, production and molecular diffusion terms are exact, other terms need modeling. Hence, Eq. (9.30) can be written as

$$\frac{\partial}{\partial t}\left(\rho\overline{u'_i u_j}\right)+\frac{\partial}{\partial x_k}\left(\rho\bar{u}_k\overline{u'_i u'_j}\right)=P_{ij}+\frac{\partial}{\partial x_k}\left(D^{\mu}_{ijk}+D^{t}_{ijk}\right)+\Pi_{ij}-\varepsilon_{ij} \quad (9.31)$$

Let us now discuss each term in the right-hand side of Eq. (9.31) one by one.

The production term P_{ij} is exact in nature and is given by

$$P_{ij}=-\rho\left[\overline{u'_i u'_k}\frac{\partial\bar{u}_j}{\partial x_k}-\overline{u'_j u'_k}\frac{\partial\bar{u}_i}{\partial x_k}\right]$$

Similarly, the molecular diffusion term D^{μ}_{ijk} is exact but generally negligible except very near to wall and is given by

$$D^{\mu}_{ijk} = \mu \frac{\partial}{\partial x_k}\left(\overline{u'_i u'_j}\right)$$

The turbulent diffusion term D^{t}_{ijk} needs modeling. There are various models available. The simplest form is gradient diffusion model as used for k-equation in eddy viscosity model

$$D^{t}_{ijk} = \mu_t \frac{\partial}{\partial x_k}\left(\overline{u'_i u'_j}\right)$$

Another form is the generalized gradient diffusion hypothesis of Daly and Harlow [1970]

$$D^{t}_{ijk} = C_s \frac{k}{\varepsilon}\overline{u'_k u'_l}\frac{\partial\left(\overline{u'_i u'_j}\right)}{\partial x_l}$$

There are various models for pressure strain rate suggested by various researchers. The generalized model can be expressed in the following form

$$\Pi_{ij} = \Big[-C_1 b_{ij} + C'_1\varepsilon\left(b_{ik}b_{kj} - \frac{1}{3}b_{mn}b_{mn}\delta_{ij}\right) + C_2 k S_{ij}$$
$$+ C_3 k\left(b_{ik}S_{jk} + b_{jk}S_{ik} - \frac{2}{3}b_{mn}S_{mn}\delta_{ij}\right) + C_4 k\left(b_{ik}W_{jk} + b_{jk}W_{ik}\right)\Big] \quad (9.32)$$

where
Anisotropic tensor

$$b_{ij} = \frac{\overline{u'_i u'_j}}{2k} - \frac{1}{3}\delta_{ij}$$

Strain Tensor

$$S_{ij} = \frac{1}{2}\left(\frac{\partial u_i}{\partial x_j} + \frac{\partial u_j}{\partial x_i}\right)$$

and rotational tensor

$$W_{ij} = \frac{1}{2}\left(\frac{\partial u_i}{\partial x_j} - \frac{\partial u_j}{\partial x_i}\right)$$

At the smallest scales, the turbulent kinetic energy is dissipated into heat due to viscous stresses. At high Reynolds numbers, the smallest eddies can be assumed to be isotropic. Hence, the dissipation term becomes

$$\varepsilon_{ij} = \frac{2}{3}\varepsilon\delta_{ij}$$

where ε is the total rate of energy dissipation.

Hence, to close the system of equations, we need to solve the equation for dissipation, which is given as

$$\frac{\partial}{\partial t}(\rho\varepsilon) + \frac{\partial}{\partial x_k}(\rho u_k \varepsilon) = \frac{\partial}{\partial x_k}\left[\mu \frac{\partial \varepsilon}{\partial x_k}\right] + \frac{\partial}{\partial x_i}\left(D^t_{\varepsilon j}\right) + C_{\varepsilon 1}\frac{\varepsilon}{k}p_k - C_{\varepsilon 2}\frac{\rho\varepsilon^2}{k} \quad (9.33)$$

where

$$p_\varepsilon = -\rho\overline{u'_i u'_j}\frac{\partial u_i}{\partial x_j}$$

$$D^t_{\varepsilon j} = C_\varepsilon \frac{k}{\varepsilon}\overline{u'_k u'_l}\frac{\partial \varepsilon}{\partial x_j}$$

The constant values in these equations for the Launder-Reece-Rodi (LRR) [1975], Gibson-Launder (GL) [1976] and Speziale-Sarker-Gatski (SSG) [1991] RSTM models are given in Table 9.1.

Hence, for RSTM models, seven transport equations (six Reynolds stress components and one turbulent dissipation equation) are to be solved at each time step for obtaining the Reynolds stresses. Note that near the wall, the above coefficients must be modified. Normally, many models use wall functions. These models are computationally expensive and have numerical stability problem.

To reduce computational time, Algebraic Stress Model (ASM) models were developed. In these models, a system of algebraic equations is solved along with two transport equations, for turbulent kinetic energy (k) and turbulent dissipation (ε). However, the results obtained with ASM models depend on the choice of the model used to provide k and ε.

9.7 LARGE EDDY SIMULATION

In Reynolds averaging we assume eddies are isotropic and homogeneous although anisotropic behavior is dealt in NLEVMs. Large-scale turbulence

TABLE 9.1

Constant Values for RSTM Models

Model	C_1	C'_1	C_2	C_3	C_4	C_s	$C_{\varepsilon 1}$	$C_{\varepsilon 2}$	C_ε
LRR	1.5	0.0	0.4	0.6	0.0	0.25	1.44	1.92	0.15
GL	3.6	0.0	0.8	1.2	1.2	0.11	1.44	1.92	0.11
SSG	3.4+1.8 P/ε	4.2	$0.8 - 1.3\left(b_{ij}b_{ij}\right)^{\frac{1}{2}}$	1.25	0.4	0.11	1.44	1.83	0.11

depends on the mean flow, geometry and boundary conditions, making it case-dependent and difficult to model. However, small-scale eddies are nearly isotropic; they are flow-independent and modelled. This is achieved by filtering N-S equations to some scale where vortices smaller than the filter scale are resolved and larger-scale turbulence and coherent structures are simulated. Consequently, the LES approach is computationally more demanding than RANS. The computational effort required by LES is less than DNS by a factor of 10, while RANS models have a computing time of only about 5% of the LES.

To reduce the computational cost, detached eddy simulation was developed, in which RANS is used in the near-wall region to mitigate the mesh-resolution requirement. This model can be thought as a replacement of LES for high Reynolds number flows for external aerodynamics applications.

It should be noted that turbulence modeling is a subject by itself. Various models have been developed for different applications, and universal applications are still in development. Interested readers may refer to Tennekes and Lumley [1972], Wilcox [1993], Pope [2000], Rodi [2017] and current publications on turbulence modeling.

9.8 SUMMARY

- Turbulent flow is disorderly, non-linear, three-dimensional and non-repetitive in nature.
- Large-scale vortices or eddies are created by the flow and are anisotropic. Through the process of vortex stretching, vortices are broken up into smaller vortices. This moves the energy from larger eddies to smaller eddies, thereby causing energy redistribution by pressure fluctuation. At the smallest scales, the turbulent kinetic energy is dissipated into heat due to viscosity.
- The energy transfer process is irreversible and dissipative. This causes increased transfer of momentum, energy, species, etc.
- Turbulence modeling is therefore about manipulating N-S equations in such a manner that we can simulate turbulence interaction in a simpler form.
- In RANS, the instantaneous flow parameter is expressed by superimposing the fluctuating component over the mean component.
- Turbulent fluctuations are responsible for the transport of momentum and other quantities. By Reynolds averaging, we obtain an additional term called Reynolds stresses (e.g. $\tau_t = -\overline{\rho u'v'}$) in mean momentum equations.
- We need to describe the unknown values (i.e. Reynolds stresses) as a function of the solution. That is known as turbulence modeling.
- Two important parameters identified to characterize the turbulence effects on the mean flow are the length scale and the velocity scale of turbulence.
- The k-ε model is the most widely used two-equation model for engineering application problems. However, it is inaccurate for flows with

adverse pressure gradient and does not allow integration of the conservation equations through the viscous sublayer where low Reynolds number corrections are usually recommended.

- The k- ω model is claimed to be accurate for flow with variable pressure gradient. One weakness is that it is sensitive to freestream boundary conditions for free-shear flows.
- Any two-equation model based on the concept of isotropic eddy viscosity fails in flow situations where anisotropy in the normal Reynolds stress components is significant due to sudden changes in mean strain rate.
- There is a trend in moving from the RANS modeling to LES on a case-by-case basis and depending on problems with current models and available computer resources.
- DNS remains out of reach for all engineering use but provides a very good insight to turbulent structure.

QUESTIONS

1) What is turbulence?
2) Does the N-S equation represent turbulent flows? Give reasons.
3) Define Reynolds stress.
4) Explain the origin of the Reynolds stresses in the RANS equations
5) What do you understand by turbulence modeling, and why is it needed?
6) Discuss why turbulence must be modelled.
7) Explain what is meant by the closure problem.
8) Write the desirable features of a turbulence model.
9) Explain what is meant by the Boussinesq approximation.
10) What do you understand about eddy-viscosity models?
11) Write the advantages and disadvantages of the algebraic model for turbulence.
12) Name at least two turbulence models used widely in the industry.
13) Discuss the various turbulence models available and the advantages and disadvantages of each of model.
14) Derive the RANS equation.
15) Discuss the eddy-viscosity concept.
16) The stress tensor in the N-S equation is given by

$$\tau_{ij} = \mu\left[\frac{\partial u_i}{\partial x_j} + \frac{\partial u_j}{\partial x_i} - \frac{2}{3}\delta_{ij}\frac{\partial u_k}{\partial x_k}\right]$$

Write down the expressions for τ_{xx}, τ_{yy}, τ_{xy} and τ_{yx}.

17) Briefly discuss about the zero-equation turbulence models. What are the advantages and disadvantages of using these models.
18) Show that sufficiently close to the wall, the mean velocity is linear and write down an expression for mean velocity $\bar{u}$ in terms of τ_w, μ and distance from wall y.

19) The turbulence stress at large distance from the wall can expressed as

$$-\rho\overline{u'v'} = \mu_t \frac{\partial \bar{u}}{\partial y}$$

where $\mu_t = \rho\, u_o\, l_m$; $l_m = \kappa y$ and $u_o = l_m {}^{\partial \bar{u}}/_{\partial y}$. Show that this leads to a velocity profile

$$\frac{\bar{u}}{u_\tau} = \frac{1}{\kappa}\left(E\frac{yu_\tau}{\upsilon}\right)$$ where E is a constant of integration.

20) Discuss the differences between Reynolds stress modeling and eddy-viscosity based modeling.
21) Explain what is meant by wall functions, why they are used and when it is appropriate to use them.
22) Discuss how the near-wall treatment can be improved (i.e. when the wall-function approach is not appropriate).

10 Grid Generation

10.1 INTRODUCTION

So far, we have dealt with conversion of PDE to difference equations (DE). PDEs are discretized to get DEs at discrete points. The resulting algebraic equations are solved to get flow parameter values at these discrete points. Therefore, a set of grid points within the domain of interest must be specified.

Grid generation plays an important part of CFD and consists of cells or elements on which the flow parameters are to be solved. Grids should represent the geometry properly including boundaries where boundary conditions are applied. The grid plays a major role in the convergence rate, solution accuracy and computational efforts required. Grid quality is very important for obtaining a good solution.

One should remember that grid generation and solving PDEs are two independent methods. A numerical (flow) solver can, in principle, be developed independently of the grid; a grid generator then gives the metrics (weights) and the one-to-one correspondence between the spatial grid and computational grid.

Grid generation methods is a separate subject altogether. In this chapter, we shall concentrate on structured and unstructured grid generation methods and the qualities of grid generation needed. Interested readers may refer to books written by Muralidhar and Sundararajan [2008], Cebeci et al. [2005], Ferziger and Peric [2002] and Thompson et al. [1985] for further study.

10.2 GEOMETRY

The starting point for grid generation is defining the geometry. Before analyzing a problem, we need to specify the geometry (i.e. the shape of the body where analysis is to be carried out). The geometry can be very simple or very complex in nature. Simple geometries can be created in the pre-processor of commercial software like FLUENT, STARCCM, etc. used for analysis. For very complex geometries, dedicated CAD software like SOLIDWORKS, CATIA and PRO-E is often used and then exported to other software for grid generation. Sometimes, dedicated grid generation software like GRIDGEN and ICEM-CFD is used for generating quality grids. Terms used in meshing are explained in Figure 10.1.

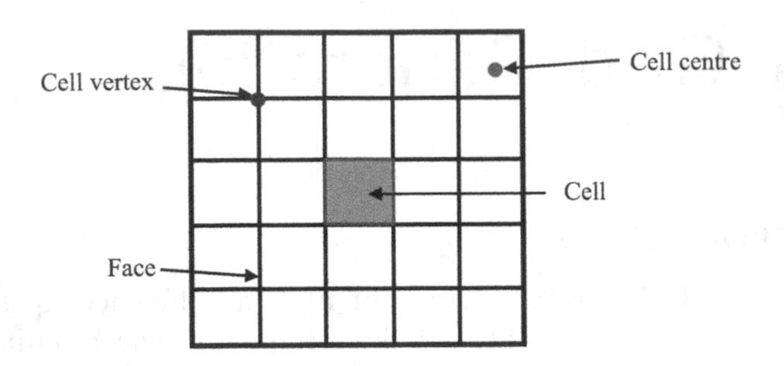

FIGURE 10.1 Definition of Various Terms.

10.3 GRID STRUCTURE

The main objective of the grid generator is to divide the flow domain into discrete control volumes or cells. By this process, we obtain the information of the locations of cell *vertices* and their connectivity information. The *nodes* are the locations where we store the flow and geometry variables and can be cell-centered or at the cell-vertex storage. However, their locations (i.e. cell-centered or cell-vertex storage) depend on the type of the solver used. When we employ a staggered velocity grid, the complexity is further increased. The locations of the nodes are explained in Figure 10.2.

The grid that comprises control volumes may be structured or unstructured. The shapes of control volumes depend on the types of grids used in the solver. The structured grid uses quadrilaterals in two-dimension and hexahedra in three-dimension flows. In the unstructured grid, triangles (two dimension) or tetrahedra (three dimension) are used. However, newer codes are available that use an arbitrary polyhedral. This is shown in Figure 10.3.

Whatever types of grid are employed, it is necessary to specify the cells that are adjacent to each other and through which face they are connected. For structured grids, this is straightforward and can be stored in a simple fashion. However, for unstructured grids, the connectivity information is very complicated to store and requires lots of computer memory.

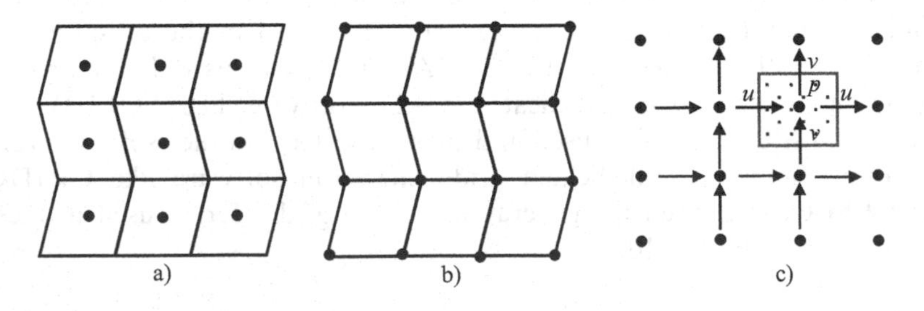

FIGURE 10.2 Type of Storage Locations: a) Cell-Centered storage; b) Cell-Vertex storage; c) Staggered Grid.

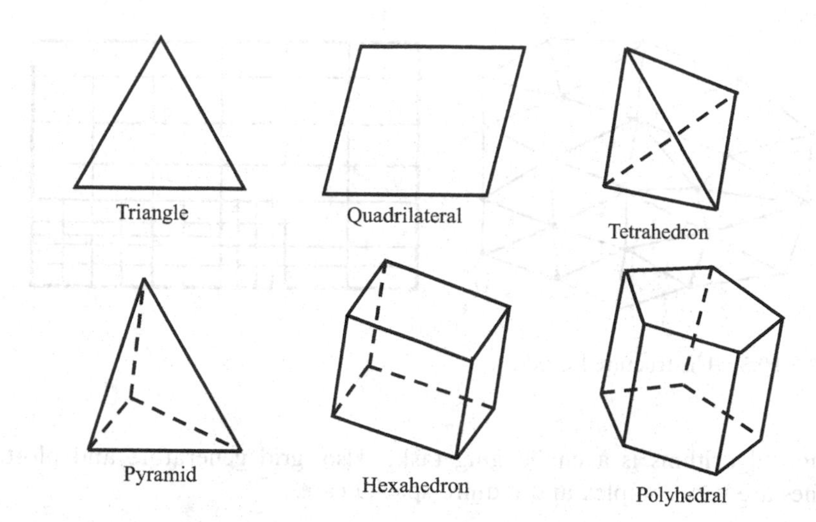

FIGURE 10.3 Types of Grid Used.

10.4 CLASSIFICATION OF GRID TYPES

Grids can be *Cartesian or curvilinear.* Grid lines are always parallel to the coordinate axes in a Cartesian grid. In a curvilinear grid, grid lines are curved, conforming the boundaries. The grids may be *orthogonal* and *non-orthogonal.* In orthogonal grids (for example, Cartesian or polar meshes) all grid lines are 90° of each other. Some flows, like the flow in a circular pipe, can be treated as axisymmetric. In these cases, flow equations are normally expressed in terms of polar coordinates (r, θ), rather than Cartesian coordinates (x, y).

Structured grids (Figure 10.4) are those whose control volumes have a fixed number of adjacent grids and can be indexed by (i, j, k). Multi-block structured meshes can be used for mapping the complex geometry arising in many practical flow situations.

Unstructured grids (Figure 10.5) have the flexibility of mapping any arbitrary geometry. However, connectivity data structures for control volumes and

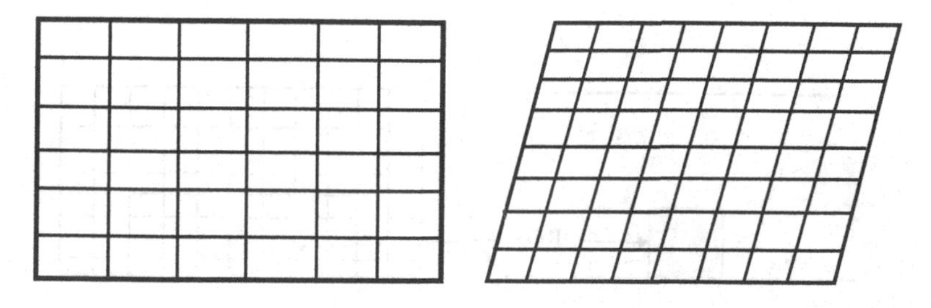

FIGURE 10.4 Structured Grids.

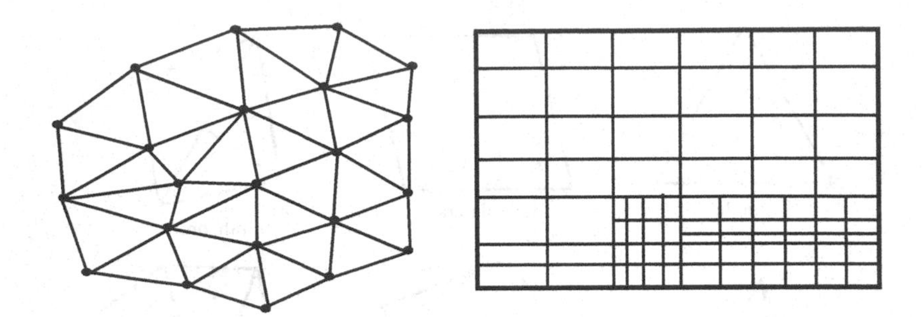

FIGURE 10.5 Unstructured Grids.

solution algorithms is a challenging task. Also, grid generators and plotting routines are very complex and require special care.

10.5 GENERATING STRUCTURED GRIDS FITTING COMPLEX BOUNDARIES

10.5.1 Blocking Out Cells

In most practical applications, the flow is very complex, and the use of a Cartesian grid is limited. However, reasonably smooth geometry (e.g. flow over a bluff body), can be computed using Cartesian mesh by blocking out cells as shown in Figure 10.6. In this case, the velocity and other flow variables in blocked cells are forced to zero by modifying the source terms for these cells.

We have the discretized equation as

$$a_P\emptyset_P - \sum a_{nb}\emptyset_{nb} = b_P + S_P\emptyset_P$$

$$Put \;\; b_P \rightarrow 0 \quad and \;\; S_P \rightarrow -\gamma$$

where γ is a very large number (e.g. $\gamma = 10^{30}$). By this way, we can force $\emptyset_P$ values to zero in these cells. However, still we need to do computations for

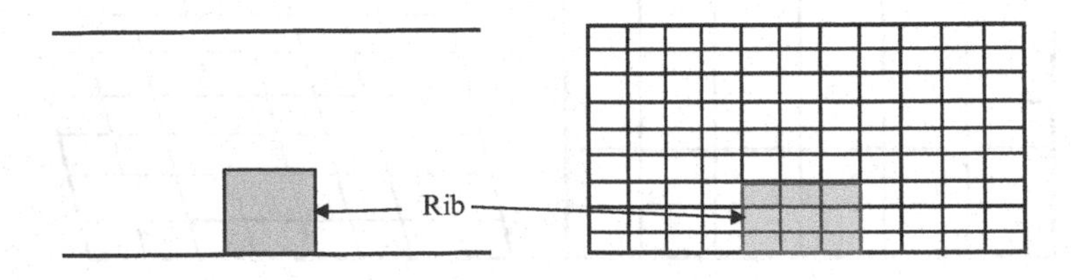

FIGURE 10.6 Grid over Rib with Blocked-out Cells.

these cells and store the values of $\emptyset_P$ thereby doing unnecessary work. This can be avoided by using multi-block grids.

10.5.2 Multi-Block Structured Grids

Instead of blocking out cells, the whole domain can be divided into several smaller sub-domains. In each sub-domain, structured grids can be generated so that grid lines at each adjacent sub-domain match each other at the interface. These structured grids are known as multi-block structured grids and are shown in Figure 10.7.

Normally the blocks are generated in such a way that they do not overlap each other. However, some solvers allow overlapping of the blocks (e.g. chimera grid). For data transfer between the blocks, interpolation is needed at the boundaries of each block. Overlapping grids have more flexibility to map complex geometries.

Figure 10.8 shows the use of the multi-block method in generating structured grids around complex boundaries. It is better to have a gradual change in grid direction, especially near the solid boundaries, for better accuracy. Also, it is better to minimize the non-orthogonality of the grid.

10.5.3 Body-Fitted Grids

Body-fitted grids are used to analyse flow over a curved surface like aerofoil. To improve the accuracy in turbulent-flow calculations, grid clustering is

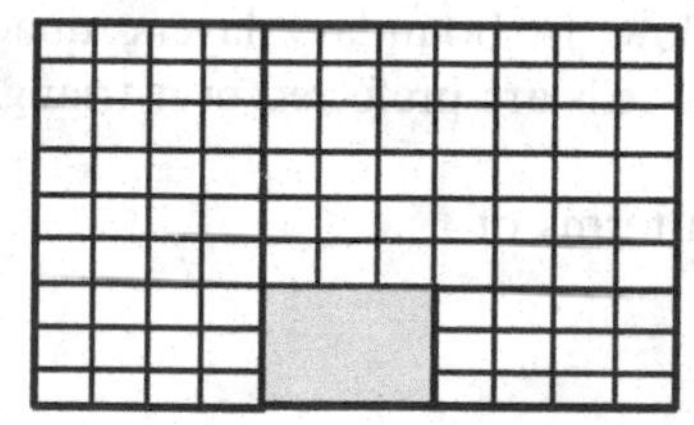

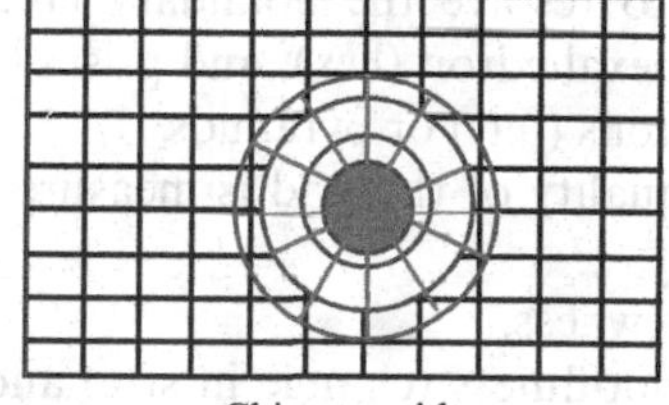

FIGURE 10.7 Multi-Block Structured Grids.

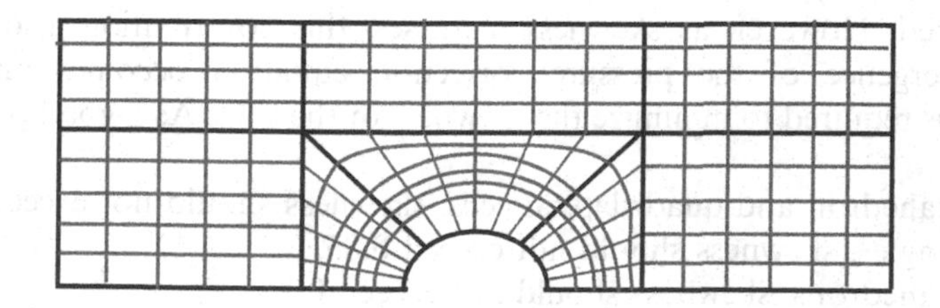

FIGURE 10.8 The Multi-Block Structured Grid around Complex Geometry.

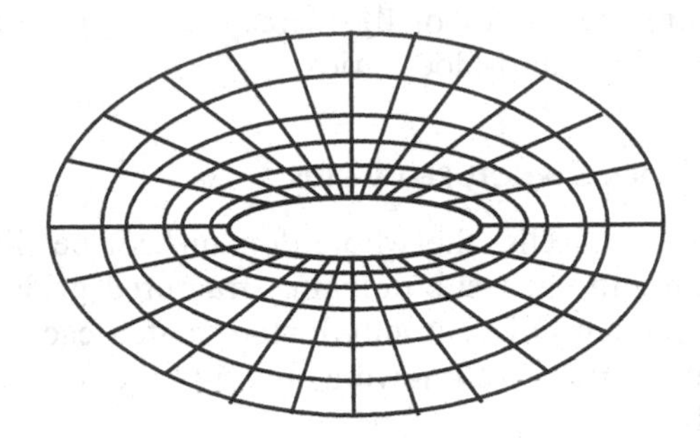

FIGURE 10.9 A Body-Fitted Grid over an Aerodynamic Body.

needed near solid boundaries. By refining the grids near the surface by using body-fitted grids, we can achieve considerable saving of computer resources, as seen in Figure 10.9. Body-fitted grids can be orthogonal or non-orthogonal.

10.6 MESH QUALITY

As earlier discussed, the grid has a significant impact on the rate of convergence, solution accuracy and computational efforts required. Grid quality is very important for obtaining a good solution. For example, for the same number of meshes, hexahedral meshes normally give better solutions. The mesh density near the sudden change in geometry should be high enough to capture all relevant flow features. Also, the mesh adjacent to the solid surface needs to be fine enough to resolve the boundary layer flow. In boundary layers, quadrilateral (quad), hexahedron (hex), and prism layer cells are preferred over triangles (tris), tetrahedrons (tets) or pyramids.

The quality of the grid is measured in terms of

- Skewness,
- Smoothness (change in size) and
- Aspect ratio.

For a quadrilateral grid, skewness refers to the non-orthogonality of the grid. Because of non-orthogonality, cross-derivative terms come into the picture. If the skewness is small, the contribution of the cross-derivative terms is small and can be neglected. However, as skewness increases, this contribution also increases, and convergence of the pressure correction equation becomes problematic. Hence, it is required to minimize the skewness of the cells. As a good practice, for

- Hexahedron and quadrilateral cells, skewness should not exceed 0.85.
- Triangles, skewness should not exceed 0.85.
- Tetrahedrons, skewness should not exceed 0.9.

Smoothness refers to the change in size of the adjacent cells. Normally, the change should be gradual and as a best practice, the maximum change in grid spacing should be less than 20%.

Aspect ratio is the ratio of width to height of the control volume. Ideally, the ratio should be equal to 1 (e.g. equilateral triangle or square). It is very difficult to control the aspect ratio to its ideal value. However, cell aspect ratio for multi-dimensional flow should be near 1.

Normally, maintaining skewness, smoothness and aspect ratio of all the cells to their ideal values is very difficult. However, by re-meshing, one can try to keep the variations in those parameters as low as possible. If we increase the number of cells, we get higher accuracy, but it will result in increasing memory size and CPU time. To reduce the number of grid-points, we can use a non-uniform grid to cluster cells only where they are needed or use solution-adaptive grids.

10.7 ADAPTIVE GRID

We know that the spacing of the grid points determines the accuracy of the solution. The spacing also determines the computation effort required for solving a problem. For well-behaved problems, a grid of uniform mesh gives satisfactory results. However, there are classes of problems in which flow is very complex in certain regions, and it is more difficult to know those regions beforehand. If we use a uniform grid, the grid spacing should be very fine so that the local flow features can be captured, and errors estimated in these difficult regions are within acceptable limits. This approach is computationally costly because of the increase in the number of grids. For many problems, it is very difficult to refine grids where required that will give acceptable results because the flow features are not known *a-priori.* In the adaptive mesh refinement technique, we start with a base coarse grid. As the solution proceeds, we identify the regions requiring more resolution by some parameter characterising the solution, say the local truncation error or gradient. We superimpose finer and finer sub-grids only on these regions until a given level of refinement is reached or the local truncation error has dropped below the desired level. This is illustrated in Figure 10.10.

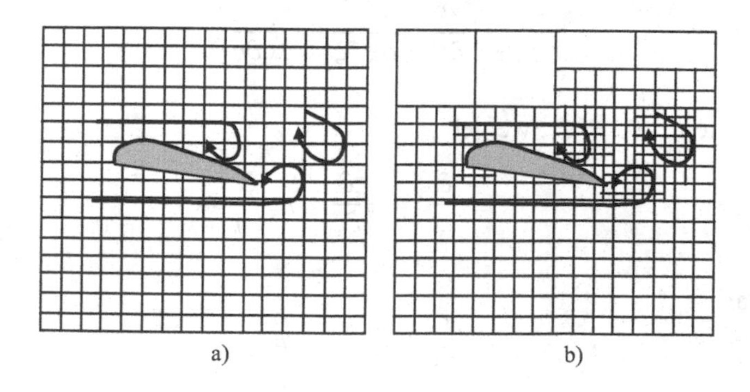

FIGURE 10.10 Adaptive Grids: a) Uniform Fixed Grid; b) Adaptive Grid.

The advantages of adaptive grid refinement are:

- A comparatively smaller number of overall grid points as the grids are refined where necessary,
- Increased computational savings,
- Increased storage savings and
- Complete control of grid resolution.

10.8 GRID-GENERATION TECHNIQUES

For a simple rectangular domain, grid generation is very simple. However, most physical domains are very complex, and imposing rectangular grids on such domains requires interpolation of the boundary conditions at the boundary cells. One should remember that boundary conditions influence the solution of the PDE. Again, using non-uniform grids near the boundary creates further complications with finite difference methods as discretization with non-uniform step size is required. Hence, a transformation of complex physical domain to rectangular computational domain is necessary.

There are three major classes of grid generation: i) algebraic methods, ii) differential equation methods and iii) conformal mapping methods. We shall discuss briefly the first two methods applied to the structured grid.

10.8.1 Coordinate Transformation

Let us consider two-dimensional spaces where physical space is represented by Cartesian coordinate x and y whereas the computational space is represented by ξ and η coordinates. Let the relation between them is given by

$$\xi = \xi(x, y) \tag{10.1}$$

$$\eta = \eta(x, y) \tag{10.2}$$

Applying the chain rule for partial differentiation, we get

$$\frac{\partial}{\partial x} = \frac{\partial \xi}{\partial x}\frac{\partial}{\partial \xi} + \frac{\partial \eta}{\partial x}\frac{\partial}{\partial \eta} \tag{10.3}$$

Let us denote $\partial\xi/\partial x = \xi_x$, we get

$$\frac{\partial}{\partial x} = \xi_x \frac{\partial}{\partial \xi} + \eta_x \frac{\partial}{\partial \eta} \tag{10.4}$$

and similarly,

$$\frac{\partial}{\partial y} = \xi_y \frac{\partial}{\partial \xi} + \eta_y \frac{\partial}{\partial \eta} \tag{10.5}$$

Let us consider the continuity equation

$$\frac{\partial u}{\partial x}+\frac{\partial v}{\partial y}=0 \tag{10.6a}$$

This can be transformed from physical space to computational space using Eqs. (10.4) and (10.5) resulting in

$$\xi_x\frac{\partial u}{\partial \xi}+\eta_x\frac{\partial u}{\partial \eta}+\xi_y\frac{\partial v}{\partial \xi}+\eta_y\frac{\partial v}{\partial \eta}=0 \tag{10.6b}$$

This is the equation which will be solved in the computational domain. However, we need to know the derivatives ξ_x, ξ_y, η_x and η_y. From Eqs. (10.1) and (10.2), we get

$$d\xi = \xi_x dx + \xi_y dy$$

$$d\eta = \eta_x dx + \eta_y dy$$

or

$$\begin{bmatrix} d\xi \\ d\eta \end{bmatrix} = \begin{bmatrix} \xi_x & \xi_y \\ \eta_x & \eta_y \end{bmatrix}\begin{bmatrix} dx \\ dy \end{bmatrix} \tag{10.7}$$

Again

$$x = x(\xi,\eta)$$

$$y = y(\xi,\eta)$$

So, we get

$$dx = x_\xi d\xi + x_\eta d\eta$$

$$dy = y_\xi d\xi + y_\eta d\eta$$

or

$$\begin{bmatrix} dx \\ dy \end{bmatrix} = \begin{bmatrix} x_\xi & x_\eta \\ y_\xi & y_\eta \end{bmatrix}\begin{bmatrix} d\xi \\ d\eta \end{bmatrix} \tag{10.8}$$

Comparing Eqs. (10.7) and (10.8), we can write

$$\begin{bmatrix} \xi_x & \xi_y \\ \eta_x & \eta_y \end{bmatrix} = \begin{bmatrix} x_\xi & x_\eta \\ y_\xi & y_\eta \end{bmatrix}^{-1}$$

From which,

$$\xi_x = Jy_\eta \ ; \quad \xi_y = -Jx_\eta \ ; \quad \eta_x = -Jy_\xi \ ; \quad \eta_y = Jx_\xi$$

and

$$J = \frac{1}{x_\xi y_\eta - y_\xi x_\eta}$$

J is known as the Jacobian of transformation, and it is interpreted as the ratios of areas in two dimensions and volume in three dimensions in the physical space to the computational space.

10.8.2 Grid Generation

A grid system should have the following features:

- One-to-one mapping of grid lines,
- Grid lines of the same family not crossing each other,
- Smooth grid-point distribution,
- Orthogonal or near-orthogonal grid lines and
- Clustered grid points.

It is possible that not all features will be achieved while generating the grid.

10.8.2.1 Algebraic Grid Generation

This is the simplest method of generating a grid. An algebraic equation is used to relate the physical domain to the computational domain. Interior grid points are obtained by interpolation from the specified corresponding boundary grid points, as illustrated in Figure 10.11.

The relations of this non-rectangular physical domain to the rectangular domain are

$$\xi = x$$

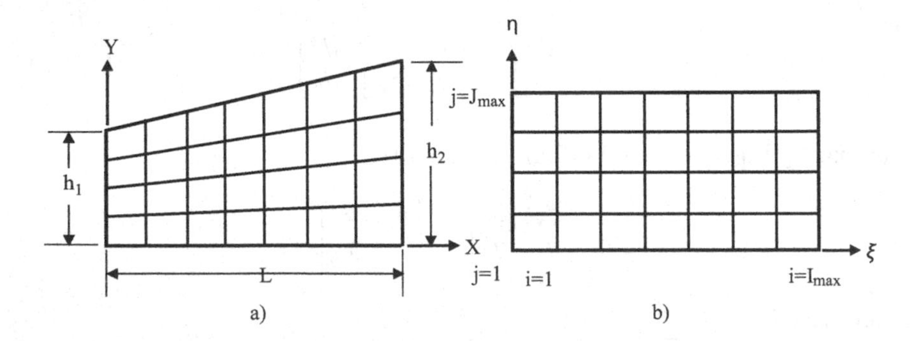

FIGURE 10.11 Method of Transforming Physical Domain to Computational Domain: a) Physical Domain; b) Computational Domain.

$$\eta = \frac{y}{y_t}$$

where,

$$y_t = h_1 + \frac{h_2 - h_1}{L}x$$

Hence,

$$\xi = x$$

$$\eta = \frac{y}{h_1 + \frac{h_2 - h_1}{L}x}$$

The above equations can be rearranged as

$$x = \xi$$

$$y = \left(h_1 + \frac{h_2 - h_1}{L}x\right)\eta$$

Let the total number of grid points are I_{max} in ξ direction and J_{max} in η direction. If we generate grids of equal spacing, then

$$\Delta\xi = \frac{L}{I_{max} - 1}$$

$$\Delta\eta = \frac{1.0}{J_{max} - 1}$$

By this method, we shall be knowing the values of ξ and η at each grid points. Note that while ξ varies from 0 to L, η varies from 0 to 1 due to normalization.

As discussed earlier, we need to know the matrices and the Jacobian of transformation for solving the transformed PDE. This can be found out as follows:

$$\xi_x = 1 \textit{ and } \xi_y = 0$$

$$\eta_x = -\frac{\frac{h_2 - h_1}{L}y}{\left(h_1 + \frac{h_2 - h_1}{L}x\right)^2} = -\frac{h_2 - h_1}{L}\frac{\eta}{\left(h_1 + \frac{h_2 - h_1}{L}\xi\right)}$$

$$\eta_y = \frac{1}{h_1 + \frac{h_2 - h_1}{L}x} = \frac{1}{h_1 + \frac{h_2 - h_1}{L}\xi}$$

$$J = \frac{1}{x_\xi y_\eta - y_\xi x_\eta}$$

Initially we must compute x_ξ, y_ξ, x_η *and* y_η from which the Jacobian is calculated. This expression is numerically computed by finite difference method. For example, central difference approximation for interior points can be obtained from

$$x_\xi = \frac{x_{i+1,j} - x_{i-1,j}}{2\Delta\xi}$$

For boundary (at i=1), this may be obtained from

$$x_\xi = \frac{-3x_{1,j} + 4x_{2,j} - x_{3,j}}{2\Delta\xi}$$

For flow problems, where large gradients are concentrated at a specific region, we need to cluster the grids at that region to capture the physical phenomena with a smaller number of total grid points. For example, near the solid surface in a viscous flow, a large gradient exists. Instead of using uniform grids with fine spacing, we can use non-uniform grids where the grids are clustered near the solid boundary, as shown in Figure 10.12.

Consider the transformation given by

$$\xi = x$$

$$\eta = 1 - \frac{ln[A(y)]}{lnB}$$

where

$$A(y) = \frac{\beta + (1 - {}^{y}/_{h})}{\beta - (1 - {}^{y}/_{h})} \quad and \quad B = \frac{\beta + 1}{\beta - 1}$$

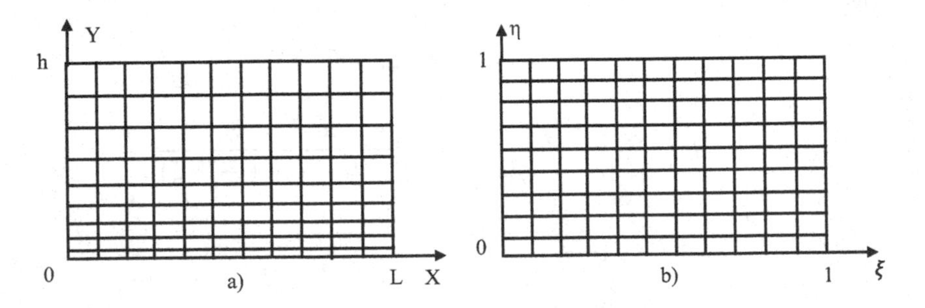

FIGURE 10.12 One-Dimensional Clustering Transformation: a) Physical Domain; b) Computational Domain.

The parameter β (1 < β < ∞) is clustering parameter. As the value of β approaches 1, more grid points are clustered near the surface y = 0. The inverse transformation is given by

$$x = \xi$$

$$y = h\frac{(\beta + 1) - (\beta - 1)B^{1-\eta}}{1 + B^{1-\eta}}$$

Similarly, the clustering can be carried out both at y = 0, y= h or at interior of the physical domain by using suitable transformation functions.

The advantages of algebraic grid generation methods are that they

- Are computationally very fast,
- Can be evaluated analytically, thus avoiding numerical error and
- Allow for grid clustering.

The disadvantages are that

- Control of grid smoothness and skewness is difficult and
- Discontinuities in the boundary may propagate into the interior regions.

10.8.2.2 Differential-Equation Based Techniques

Some of the disadvantages posed by algebraic grid generation can be eliminated by PDE grid generators. This method of structured grid generation is based on solving PDEs. These techniques may be categorized as elliptic grid generator, hyperbolic grid generator or parabolic grid generator, depending on the type of PDE used.

Elliptic PDE-Based Methods

These methods are particularly useful for confined physical domains, where all the physical boundaries are specified. The elliptic PDEs for grid generation describe the variation of the body fitting coordinates (ξ, η) in the interior of the physical domain, with prescribed values or slopes at the boundary. In the computational domain, however, the physical coordinates (x, y) are treated as the unknown variables on the grid formed by the ξ = constant and η = constant lines. They are determined by numerically solving the transformed grid generation equations.

Consider the transformation functions, which are solutions of an elliptic Dirichlet boundary value problem. The mathematical problem is given by equation below with fixed ξ, η on the boundaries. P and Q are source functions, which can be used to grid point controlling.

$$\xi_{xx} + \xi_{yy} = P(\xi, \eta)$$

$$\eta_{xx} + \eta_{yy} = Q(\xi, \eta)$$

The condition: $P = Q = 0$, results in uniform distribution of the points. The system then becomes the Laplacian.

Normally computations to generate the grid mapping are actually carried out in the computational domain (ξ, η) itself, as we don't want to solve the elliptic problem in the complex physical domain.

Using the general rule, the elliptic problem is transformed into:

$$ax_{\xi\xi} - 2bx_{\xi\eta} + cx_{\eta\eta} + J^2(Px_\xi + Qx_\eta) = 0$$

$$ay_{\xi\xi} - 2by_{\xi\eta} + cy_{\eta\eta} + J^2(Py_\xi + Qy_\eta) = 0$$

$$a = x_\eta^2 + y_\eta^2 \ ; \ b = x_\xi x_\eta + y_\xi y_\eta \ ; \ c = x_\xi^2 + y_\xi^2 \ ; \ J = x_\xi y_\eta - y_\xi x_\eta$$

These equations may be solved by any iterative scheme provided P and Q are specified. The functions P and Q are specified depending on the specific need.

The advantages are

- Smooth grid distribution and
- Options for clustering and surface orthogonality.

The disadvantages are that

- Computation time is large,
- The specification of forcing functions P and Q is difficult and
- A numerical solution is required.

10.9 UNSTRUCTURED GRID GENERATION

We have discussed the generation of structured grids. Normally, the discretization process is carried out using local indices. For building the system of equation, global indices over the computational domain are used during the solution process. In structured grid systems, each cell is identified with an ordered set of indices (*i, j, k*), and each index varies over a fixed range. Hence, bookkeeping of grid information and its connectivity is easier for structured grids. However, for complex geometries, grid generation with structured grids requires more computational efforts than the flow computations.

The use of unstructured grids can reduce the computational efforts for grid generation considerably as it can be used to fit the complex geometries very easily. Unstructured meshes normally use triangles in two dimensions and tetrahedrons in three dimensions, thus providing the flexibility to fill the region defined by boundary elements (i.e. edges in two dimensions and triangles in three dimensions). Unstructured meshes are better for concentrating cells in specific regions for better resolution and accuracy. For local mesh refining or adaptive meshes, the use of unstructured grids is very useful. Automatic grid generation is easier with the use of unstructured grids, and they are gaining popularity. Hence, most commercial computer codes have adapted grid generation by unstructured grids as a default. However, in an unstructured grid system, the cells are numbered sequentially and cannot be identified only by local indices, as can be seen in Figure 10.13.

Detailed connectivity information regarding neighboring cells, faces and nodes is to be mapped with respect to the global index of the grid. Therefore, bookkeeping is more difficult and needs an experienced programmer. Also, solver efficiency is not as good as in the structured grid system because of random cell location and connectivity.

10.9.1 Connectivity Information

As discussed earlier, we need a pointer system to obtain information regarding grid topology. In the finite volume method, all fluxes enter/exit the control volume through the faces; hence, we need a pointer that gives information on starting and end points of a face and the cells adjacent to it.

For example, in Figure 10.14 we can see that the face k_1 starts at point n_1 and ends at n_2. Cell P_0 is at the left and P_1 at the right side of the face k_1. This information can be stored as follows:

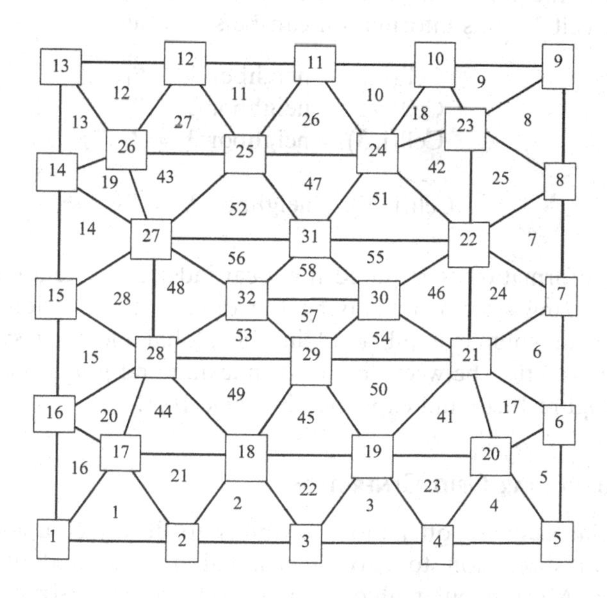

FIGURE 10.13 A Typical Unstructured Mesh with Global Indexing.

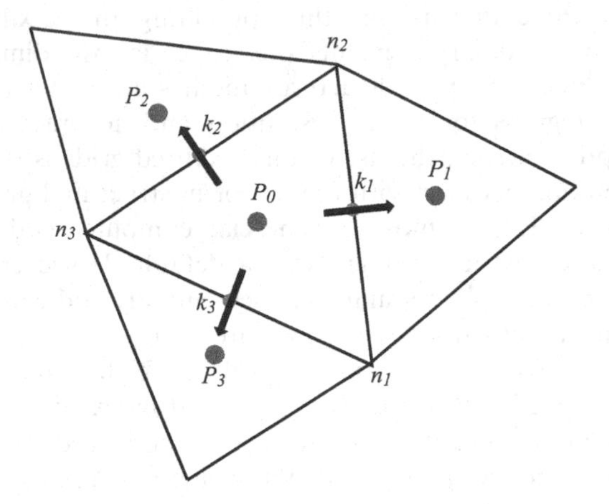

FIGURE 10.14 Typical Control Volume with Local Indexing.

$$\begin{aligned} \text{Face(i, 1)} &= n_1 \\ \text{Face(i, 2)} &= P_0 \\ \text{Face(i, 3)} &= P_1 \\ \text{Face(i, 4)} &= n_2 \end{aligned},$$

where i is the index of the face, which is k_1 in this case.

During the discretization process, we need a pointer giving details of neighboring cells for the control volume. The number of neighboring cells may be arbitrary. For cell P_0, this information can be stored as

$$\begin{aligned} \text{Cell(i, 1)} &= \text{neighbor } 1 = P_1 \\ \text{Cell(i, 2)} &= \text{neighbor } 2 = P_2 \\ \text{Cell(i, 3)} &= \text{neighbor } 3 = P_3 \\ &\ldots\ldots \\ \text{Cell}(\text{i}, N) &= \text{neighbor } N = P_N \end{aligned}.$$

Now, this information is required for local indexing system connected with the global indexing system for solving the discretized equation. Let us consider the control volume number 49 in the global indexing system (Figure 10.13). The connection between the local indexing and global indexing system is shown in Figure 10.15 and explained in Table 10.1.

10.9.2 Triangular Grid Generation

Since triangular elements offer most flexibility to fit any complex domain, we shall limit our discussion to two-dimensional unstructured grid generation with triangles. Most popular algorithms of generating unstructured grids by

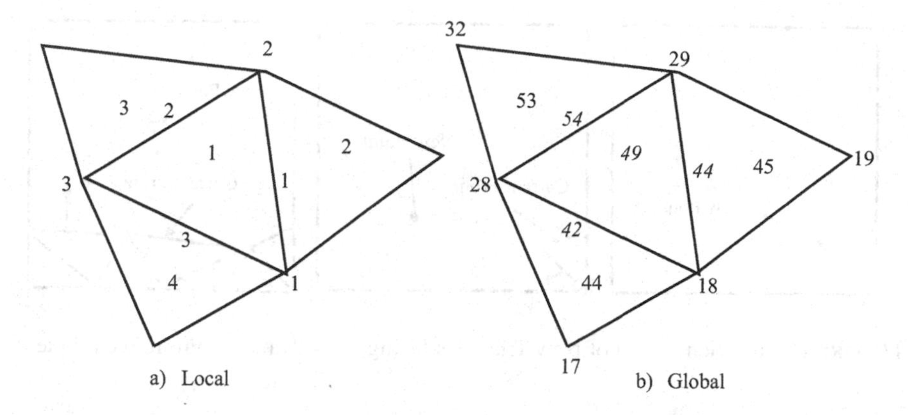

FIGURE 10.15 Connectivity Details between Local and Global Indexing Systems.

TABLE 10.1
Connectivity Details between Local and Global Indexing Systems

	Local indexing	Global indexing
Cell No.	1	49
Face	1, 2, 3	44, 54, 42
Neighbor	2, 3, 4	45, 53, 44
Nodes	1, 2, 3	18, 29, 28

the advancing-front technique and the Delaunay-based method will be discussed here.

10.9.2.1 The Advancing-Front Technique

In this method, we proceed from the boundary of the computational domain by adding elements in the interior space until the domain is filled completely, as shown in Figure 10.16.

The algorithm is as follows:

- Start with boundary meshing, thus forming a set of edges. These edges form the initial/current front.
- Select a particular edge of this front. Form a new triangle by joining the two ends of the current edge with a newly created point or existing point on the front.
- Remove the current edge from the front as the triangle is formed and assign the two edges of the new triangle to the front.

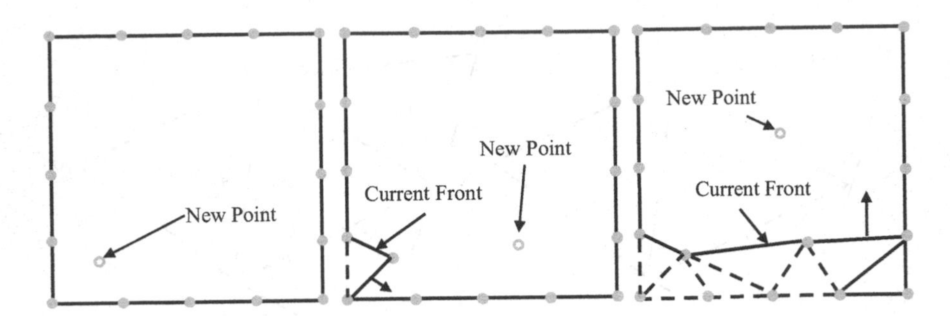

FIGURE 10.16 Generation of New Triangles Using the Advancing-Front Technique

- While generating a new triangle, position a new point that results in an optimum triangle.
- Reject any front that results in intersecting with the existing front.
- Continue this process until no front is left and the whole domain is covered.

This algorithm is simple and relatively easy to implement. It allows us to triangularize the concave domain without any problem. However, an efficient search algorithm is required for checking the intersection of the fronts.

10.9.2.2 The Delaunay-Based Method

Given a set of points in the region, many possibilities exist for triangulations with these points. The Delaunay triangulation method arranges random sets of nodes (points) in the region in a uniform pattern or in accordance with the density we specify, thus providing optimized triangles.

Bowyer-Watson (Bowyer [1981]; Watson [1981]) proposed an *empty circle property* test for generating Delaunay triangles. This property states that no other point (vertex) can be contained within the circumcircle of any triangle. It can be seen in Figure 10.17a, point D lies within the circumcircle formed by vertices of triangle ABC. Hence, it violates the Delaunay property. In Figure 10.17b, point D lies outside the circumcircle formed by vertices of triangle ABC; hence, it is a valid Delaunay triangle.

The Bower-Watson algorithm can be summarized as follows:

- Generate a set of required grid points on the boundary and interior locations in the domain.
- Create initial triangulations enclosing the whole domain. This can be a large single triangle or a rectangle.
- Introduce a new point P from the list in the existing triangulations.
- Find all triangles whose circumcircles contain P. Delete these triangles.
- Find all external faces of the cavity resulting from the deletion of these triangles.

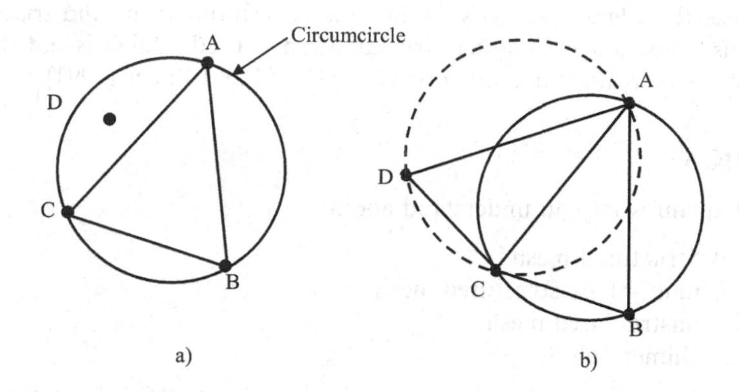

FIGURE 10.17 Delaunay Criteria: a) Violates Delaunay Criteria; b) Valid Delaunay Triangle.

- Form new triangles by joining external faces to the new point P.
- Add the new elements and points to the list to update the data structure.
- Proceed until all points from the list are inserted.

The above procedure is illustrated in Figure 10.18.

Delaunay triangulation method algorithms are simpler and more efficient than the advancing-front technique. Extension to three dimensions is also straightforward. However, this method is applicable for convex domains. In

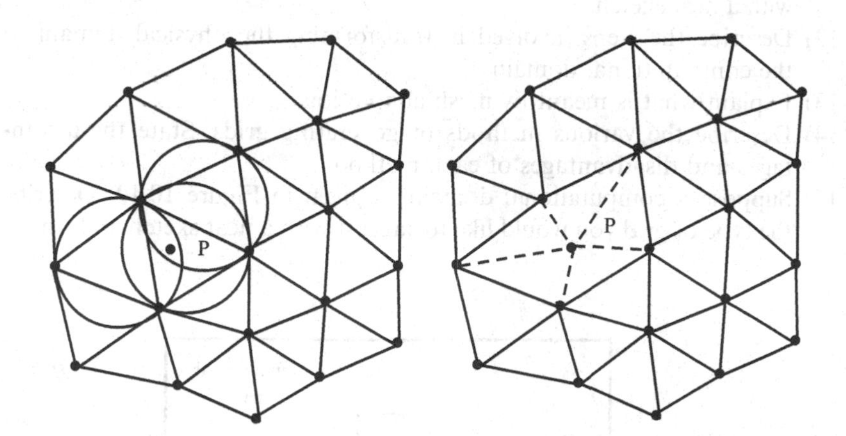

FIGURE 10.18 The Procedure of the Bowyer-Watson Algorithm for Constructing the Delaunay Triangle.

some cases the edges may cross the boundary, resulting in invalid triangles. To avoid this, constrained Delaunay triangulation is used, which is not discussed here. Interested readers may refer to Chew [1989] and Baker [1991].

QUESTIONS

1) Explain what you understand about
   i) Structured mesh,
   ii) multi-block structured mesh,
   iii) unstructured mesh,
   iv) chimera mesh.
2) With respect to structured meshes when do you use following meshes?
   i) Cartesian
   ii) curvilinear
   iii) orthogonal
3) What do you understand about structured grid?
4) What do you understand about unstructured grid?
5) State the desirable features of grid generation.
6) State two methods of generating grids.
7) What are the basic requirements of grid generation?
8) What do you understand about Cartesian grids?
9) What do you understand about a body-fitted grid, and why is it needed?
10) What do you understand about *orthogonal* and *non-orthogonal* grids?
11) What do you understand about a multi-block structured grid? Explain with a neat sketch.
12) Describe the steps involved in transforming the physical domain to the computational domain.
13) Explain what is meant by mesh adaptation.
14) Describe the various methods of generating grids. State the advantages and disadvantages of each method.
15) Suppose a computational domain as given in Figure 10.19. Describe the type of grid you would like to suggest with a neat sketch and why.

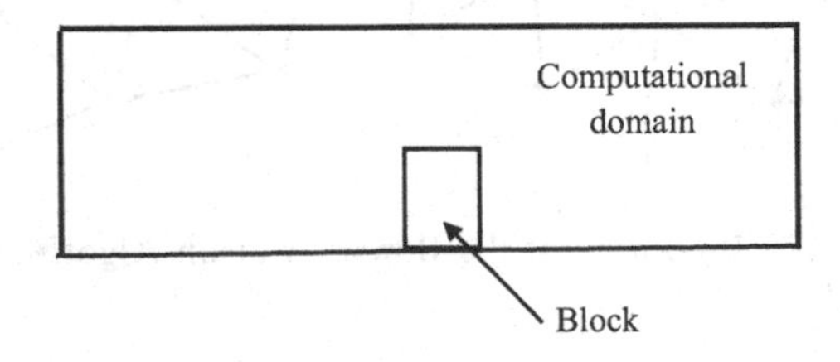

FIGURE 10.19 Figure for Problem 15.

16) Describe the steps involved in the algebraic grid-generation method. State the advantages and disadvantages of this method.
17) Describe the elliptic grid generation technique. State the advantages and disadvantages of using this method.
18) What are the advantages and disadvantages of unstructured grids?
19) Describe the procedure for triangulation by a) the advancing-front technique and b) the Delaunay-based method.

11 Best Practice Guidelines in CFD

11.1 INTRODUCTION

Over the last three decades, Computational fluid dynamics (CFD) has matured into a reliable tool for analyzing fluid-flow problems. General-purpose computer software such as Fluent, CFX, Star-CD, FLOW-3D and Phoenics is available, some very specialized programs are developed to simulate combustion in engines, electronic cooling systems, etc. With the availability of robust commercial software and high-speed computing, CFD has become a very popular and powerful tool for fluid-flow analysis of components in the design stage. CFD has already been successfully applied to various fields of engineering and even medicine. The range of applications is wide and encompasses many different fluid phenomena.

CFD does not provide an exact solution to the problem because of various approximations involved. However, an experienced CFD practitioner can provide reliable results. In the hands of an inexperienced user, colorful graphs may be produced, but the results may be far from physical reality. Problems arise due to errors and uncertainties involved in CFD simulation. An error is a recognizable deficiency that can be identified and removed with proper care and effort, but uncertainties are related to lack of knowledge. For example, commercial CFD codes provide many default settings that may not be appropriate for a specific problem. Also, commercial codes provide many turbulence models, and not all are suitable for all cases. The selection of a proper turbulence model for a specific case is essential for obtaining meaningful results. One should critically study the results of simulation and carry out sensitivity analysis, verification, validation and calibration to check the robustness and reliability of the simulation. In this chapter, some guidelines are discussed for avoiding the most common mistakes. Interested readers may refer to Franke et al. [2007], the AIAA guide [1998], Roache [1998] and Oberkampf and Trucano [2002].

11.2 SOURCES OF ERROR

Industrial flow is often complex and involves intricate geometries as in the case of heat exchangers, turbomachinery, aircraft, electrical and electronic components, meteorology, biomedical engineering and nuclear reactors. In practical situations, the flow is generally three-dimensional and turbulent. Hence, simulation demands a CFD analyst to perform various activities like

- problem definition,
- selection of solution strategy,

- development of a computational model and
- analysis and interpretation of the results.

During simulation, each of these steps may lead to errors and uncertainties due to human error or inadequacies in the modeling methods and model equations. ERCOFTAC BPG identified following sources of error and uncertainty:

1) Model errors and uncertainties
2) Discretization or numerical errors
3) Iteration or convergence errors
4) Round-off errors
5) Application uncertainties
6) User errors
7) Code errors

11.2.1 Model Errors and Uncertainties

Many times, exact governing equations are not solved. Instead, they are replaced by physical models that may not represent the real flow situation. For example, DNS represents the exact N-S equation for turbulent flow. Since simulating complex flows with DNS is very expensive, we use the RANS approach, which involves approximations. In fact, there is no universal turbulence model. Approximations introduce errors and uncertainties into simulation results.

11.2.2 Discretization or Numerical Errors

Governing equations are normally expressed in terms of partial differential equations (PDEs), which are continuously varying functions. We normally approximate PDEs at a finite number of points, and this process is called *discretization*. In this process, we get a set of equations known as *difference equations (DEs)*. We solve these discretized equations numerically. This leads to discretization errors because of numerical approximations involved in spatial as well as temporal variation of flow variables.

11.2.3 Iteration or Convergence Errors

Discretization of PDEs leads to a set of algebraic equations consisting of nodes. The system of algebraic equations thus obtained is solved by an iterative method to obtain values at the nodes. Normally, the iteration continues until convergence is achieved (i.e. the residuals are within the tolerance limit). Many times, the user terminates the iterative process and does not allow the solution to reach convergence. This leads to errors in the results.

11.2.4 Round-off Errors

Normally, computer calculations are carried out to a finite number of decimal places or significant digits. Single precision numbers are stored in 32 bits and double precision in 64 bits. Errors caused by approximation at a finite number of decimal places are called *round-off errors.*

11.2.5 Application Uncertainties

The selection of the computational domain and choice of boundaries plays a major role in the simulation process as it influences the results of simulation. Specifying complete and correct boundary conditions, especially in-flow and out-flow boundary conditions, are essential. All factors influence uncertainty in results and may lead to error. Also, to reduce computational cost, complex geometries are often simplified and intricate details of the geometry are omitted. In addition, there is uncertainty in deciding whether flow is steady or unsteady This can introduce large errors in simulation.

11.2.6 User Errors

This error is introduced due to mistakes and oversights of the user. With increasing experience, this type of error can be reduced but may not be fully eliminated.

11.2.7 Code Errors

Code errors happen during development of the CFD code and are the responsibility of code developer. During code verification, errors made during programming can be eliminated. Other errors may be originated while porting the computer code to different platforms like hardware, operating systems and compilers. Such errors are difficult to find because of the complexity of the CFD software.

11.3 BEST PRACTICES GUIDELINES

Here we shall discuss guidelines for reducing errors and uncertainties in simulation results.

11.3.1 Geometry and Grid Design

11.3.1.1 Geometry Generation

- Try to simplify the geometry to reduce computational cost. However, make sure that CAD geometry is not over-simplified, as critical details may be lost.
- It is not necessary to provide geometrical details less than the smallest computational cell. However, sometimes it is necessary to provide details like surface roughness, which influences flow features.

- Check whether geometrical data transferred from the CAD to the CFD system is proper.
- Check that the geometry in the CAD system is in the correct coordinate system with proper units required by the computer code.

11.3.1.2 Grid Design

- Use body-fitted grids and ensure that the surface grid conforms to the CAD geometry.
- Avoid highly skewed cells so that the angles between the grid lines are orthogonal or near orthogonal. Angles with less than 40 degrees or more than 140 degrees may lead to numerical instabilities or poor results.
- Maintain a proper grid aspect ratio. Normally, the aspect ratio should be below 5 but can be up to 10 inside the boundary layer.
- Ensure the change in grid size is gradual and, as a best practice, the maximum change in grid spacing should be less than 20%.
- Use hex grids if possible.
- Use hybrid grids if pure hex grids are not possible. For example, use prism layer grids in the boundary layer and tetrahedral cells in the core flow region.
- Use grid refinement where local details are required.
- Carry out grid independence tests to select the number of grids.

11.3.2 Discretization Schemes

11.3.2.1 Spatial Discretization Errors

- Avoid the use of first-order upwind schemes, as they lead to numerical diffusion.
- Higher-order schemes, at least second order, are recommended for all transported variables.
- Initially, using a first-order upwind scheme may be necessary for robustness. However, switch over to higher-order schemes as convergence approaches.
- Estimate the error by mesh refining if possible.
- Compare the results with different discretization schemes.

11.3.2.2 Time Discretization Errors

- Take a small time step so that it can be increased to check the influence of time steps on the results.
- Time steps can be decided based on CFL criteria. However, they should be small enough to capture transient phenomena.
- Ensure that the solution converges at each time step.
- For temporal accuracy, use a second-order accurate discretization scheme like Crank–Nicolson.

11.3.3 Convergence

Poor convergence leads to numerical error. Normally, residuals are monitored to check convergence, but monitoring residuals alone may not be enough to assure convergence.

- Use different convergence criteria for different variables. Some variables like concentration may need a lower residual than other variables.
- Ensure global balances for mass, momentum and energy.
- Monitor the solution at specific important points.
- Check residuals along with the rate of change of the residuals with increasing iteration numbers.
- Test for steady state by switching to a transient solver.
- Monitor the residual at least at a single point in the region of interest to check convergence.

There are many ways to enhance convergence. These are

- Use more robust numerical schemes. Use first-order upwind initially, changing to a higher order during final iterations.
- Initially use lower under-relaxation or a CFL number specially for high Reynolds number flows for ensuring stability. Normally, pressure correction requires less under-relaxation than velocity corrections. Increase it afterwards during final iterations.
- Examine the local residual. A convergence problem may be restricted to a small region.
- Use grid adaptation or refine the mesh in areas of large gradient to reduce computational cost.
- Using a fine grid throughout may reduce the convergence rate.
- Solve steady-state problems transiently.
- For high-speed compressible flows use the coupled solver, although a coupled solver requires 1.5–2 times more memory than a segregated solver.

11.3.4 Modeling Uncertainty

11.3.4.1 Solution Algorithms

- Check whether a proper solution procedure is selected, keeping in mind the physical properties of the flow.
- For the solution algorithm, use controlling convergence parameters like relaxation parameters as suggested by the CFD-code vendor or developer.
- Do not change all the parameters at once for accelerating convergence rate. This will make it difficult to analyse which change influences the convergence most.

If divergence still persists,

- Check carefully boundary conditions, grid, discretization and convergence errors.
- Check whether the flow is steady or unsteady.
- Check the time step used for unsteady simulation.
- If a steady solution is carried out and there is doubt about the nature of flow, then carry out an unsteady simulation with the existing steady flow field as the initial condition.
- Examine the time development of the physical quantities in the locations of interest to identify whether the flow is steady.

11.3.4.2 Turbulence Modeling

DNS is an exact formulation of the Navier–Stokes equations. All other turbulence models are model approximations.

- The k-ε model is inaccurate for flows with adverse pressure gradients, and the model does not allow integration of the conservation equations through the viscous sublayer where low Reynolds number corrections are usually recommended.
- The k- ω model is claimed to be accurate for flow with variable pressure gradients. One weakness of this model is that it is sensitive to the freestream boundary conditions for free-shear flows.
- For highly swirling flows use RSTM or non-linear eddy viscosity models.
- For mixing flows with strong buoyancy effects or high streamline curvature, use RSTM.
- For flow separation from surfaces under the action of adverse pressure gradients, use either Baldwin and Lomax [1978] one-equation model or shear stress transport (SST) version of Menter's k-ω based, near-wall resolved model.
- Transition from laminar-to-turbulent flow is very difficult to simulate. The use of low Reynolds k-ε or the transitional SST version of Menter's k-ω based near-wall resolved model may be an option.
- Be aware of the limitations of a specific model. Test different turbulence models to check the results
- The trend is moving from the RANS modeling to LES on a case-by-case basis and depending on problems with current models and available computer resources.

11.3.4.3 Near-Wall Modeling

- Ensure using a proper turbulence model capable of resolving the flow structure near the wall.
- Meshing should be done in such a way that for RANS models, y^+ values adjacent to wall are greater than 30. The upper value of y^+ should not be greater than 100.

- For low-*Re* turbulence models, the first grid point should be $y^+ \approx 1$. There should be a small stretching factor so that 10 mesh points are within $y^+ = 20$.
- Check the manual of the commercial CFD programs for guidance in specifying actual values.
- Standard wall functions are not suitable with flow separation at the wall.
- Roughness at the wall is important for predicting friction and drag. Take proper care in placing near-wall meshing.
- Check that the correct form of the wall function is used to account for wall roughness.

11.3.5 Round-off Errors

In high Reynolds number flows, we may need small cells near the wall for proper boundary-layer resolution. Round-off errors can be significant in this situation. The 32-bit (i.e. single-precision) simulation can lead to poor convergence in such cases. Use a double-precision version (i.e. 64-bit) representation of real numbers to avoid round-off errors.

11.3.6 User Errors

User errors may result from the following factors:

- Lack of attention to geometrical details, carelessness and mistakes
- Lack of familiarity with a CFD code; lack of experience with commercial software
- Over-simplification of geometry or governing equations without proper understanding of a given problem
- Poor geometry and grid generation
- Use of incorrect boundary conditions
- Use of incorrect or inadequate solver parameters like time step and under-relaxation parameter
- Acceptance of non-converged solutions and post-processing errors

The user should give careful thought to the requirements and objectives of the simulation.

11.3.6.1 Boundary Conditions

Inlet Conditions:

- Place inlet boundaries in a position such that it has weak influence on the downstream flow. The boundary conditions are easily identified and can be precisely specified.

- The exact inlet conditions for the turbulence properties are usually not known exactly. For external flows $\mu_t/\mu \approx 10$ or turbulent intensity, $\approx 1\%$ can be assumed.

Perform a sensitivity analysis on

- inlet flow direction and magnitude,
- inlet velocity profile,
- variation of physical parameters and
- variation of turbulence properties at inlet.

Outlet Conditions:

- The outlet boundary position should be such that it has a weak influence on the upstream flow.
- Normally, static pressure is specified as an outlet boundary condition at the outlet plane.
- For multiple outlets, use either pressure boundary conditions or mass flow specifications depending on the known quantities.
- If pressure is specified at the inlet, then provide mass flow conditions at the outlet.
- A good practice is to set the convective derivative normal to the boundary face equal to zero at the outlet plane and extrapolate the transported quantities streamwise.
- For flow with strong pressure gradients, special non-reflecting boundary conditions may be required.

Solid Walls:

- Check whether proper boundary conditions are imposed on solid walls.
- If roughness on the wall affects the flow parameters, correct specification of roughness within the wall function is necessary to reduce the uncertainty level.

Steady Flow, Symmetry, and Periodicity:

- In symmetry planes, gradients normal to the plane are zero.
- Periodic boundary conditions are determined from the flow field.
- If the symmetry or periodicity planes cross the inlet or outlet boundaries, proper care is necessary in specifying consistent inlet or outlet variables.
- Check carefully whether the geometry is symmetric.
- Check whether flow is asymmetric, turbulent or unsteady by estimating the Reynolds number of the inflow.

- After obtaining a steady solution, switch to the transient mode and check that the solution remains stable.
- If there is an oscillation of the residuals, switch to the transient mode.
- In case of doubt, carry out the simulation in unsteady mode and without a symmetry assumption at the boundaries.

11.4 ANALYSIS OF RESULTS, SENSITIVITY STUDIES AND UNCERTAINTIES

11.4.1 Analysis of Results

- Check overall mass balance.
- Check whether values of temperatures, velocities, forces and pressures are realistic.
- Check whether the distribution of fluid variables, such as temperature, velocity and pressure, are physically meaningful.
- Perform some simple hand calculations in a smaller number of grids to check the orders of magnitudes of variables.
- Ensure that a proper solution algorithm is used and approximations made are realistic.
- Ensure that proper input conditions are specified, as the accuracy of the solution depends on that.
- Compare the result with similar problems or simplified versions of the same problem.

11.4.2 Sensitivity Analysis

This analysis is done to examine the effect of a specific choice on the results. It is advisable to start sensitivity analysis from a converged solution.

- Compare the results by changing the number of grid points.
- Examine the effect of boundary conditions on the results.
- Run the problem using a different computer code and compare the results.
- Investigate the effects of different turbulence models.

11.4.3 Uncertainties

- Recognize the important features of the geometry and incorporate them into the model to avoid over-simplification.
- Recognize various errors like convergence error, discretization error, turbulence model and errors in geometrical representation in the model. Take care of round-off errors.
- Recognize the importance of the scale factor. While a solution is much easier at model scale, scaling up the results may be difficult.

11.5 VERIFICATION, VALIDATION AND CALIBRATION

The definitions of verification, validation and calibration as per the guidelines given in the AIAA guide [1998] and Roache [1998] are

- Verification: Procedure to ensure that the program solves the equations correctly
- Validation: Procedure to test the extent to which the model accurately represents reality
- Calibration: Procedure to assess the ability of a CFD code to predict global quantities of interest for specific geometries of engineering design interest.

Verification and validation procedures assess accuracy and build confidence and credibility in computational simulations. As per Roache [1998], verification is the domain of mathematics and validation is the domain of physics.

11.5.1 VERIFICATION

To establish confidence in the computational model, Verification is done in two stages:

- Code verification – to check whether the mathematical model and solution algorithms are working correctly
- Calculation verification – to check whether the discrete solution of the mathematical model is accurate.

11.5.1.1 Code Verification

Code verification is done by software developers by using software quality assurance techniques and testing each version of the software before release. Users of software also take part in code verification without having access to the source code.

Code verification methods compare code outputs with analytical solutions, Ordinary Differential Equation (ODE) benchmark solution and PDE benchmark solution. Analytical solution provides an exact solution. However, the numbers and variety of exact solutions are limited.

11.5.1.2 Calculation Verification

Calculation verification estimates the errors in the numerical solution due to discretization. Errors involved during the discretization of PDE to DE are related to truncation error. Discretization at the boundary with lower orders of accuracy than interior points can dominate the numerical accuracy of the simulation.

The three aspects of calculation verification are

- checking and ensuring the correctness of input data,

- estimating numerical errors due to discretization, meshing, insufficient iterative convergence and round-off errors and
- checking the output data and correctness of post-processing.

For every simulation, calculation verification must be performed, and it must be sufficiently different from previous solutions.

11.5.2 Validation

Verification processes do not address questions of the adequacy of the selected models for representing the reality of interest. This is the domain of validation and ensures whether enough physics are incorporated in providing reliable answers for the problem statement. Validation procedures assure that the mathematical model accurately relates to experimental measurements.

However, experimental data may not be always correct because of experimental uncertainty and unknown bias error. Hence, there should be a close interaction between the experimentalists and computationalists for designing and executing validation experiments. The design should be able to capture the relevant physics and all initial and boundary conditions. Finally, the computational and experimental results should be obtained independently to avoid a tendency to produce favorable results.

Appendix 1

Area and Volume Calculation

A1.1 FOR TWO DIMENSIONS

Referring to Figure A.1.1, areas and volumes associated with the control volume around point P can be calculated as follows:

$$A^1_{Px} = \text{projection of area } A^1_P \text{ along X-axis} = (Y_n - Y_s)$$
$$A^1_{Py} = \text{projection of area } A^1_P \text{ along Y-axis} = -(X_n - X_s)$$
$$A^2_{Px} = \text{projection of area } A^2_P \text{ along X-axis} = -(Y_e - Y_w)$$
$$A^2_{Py} = \text{projection of area } A^2_P \text{ along Y-axis} = (X_e - X_w)$$

where A^1_P and A^2_P are the area vectors at face P as shown in Figure A.1.1.

$$\begin{aligned} V_p &= \text{volume of the control volume associated with P} \\ &= (X_e - X_w)(Y_n - Y_s) - (X_n - X_s)(Y_e - Y_w) \end{aligned}$$

The areas and volume associated with face e are given by

$$A^1_{ex} = \text{projection of area } A^1_e \text{ along X-axis} = (Y_{ne} - Y_{se})$$
$$A^1_{ey} = \text{projection of area } A^1_e \text{ along Y-axis} = -(X_{ne} - X_{se})$$

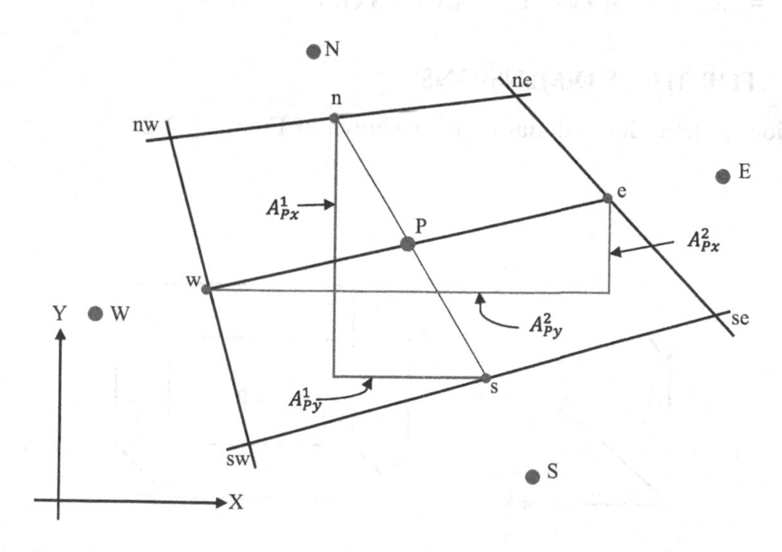

FIGURE A.1.1 Area and Volume Calculation for Faces at point P.

A_{ex}^2 = projection of area A_e^2 along X-axis $= -(Y_E - Y_P)$
A_{ey}^2 = projection of area A_e^2 along Y-axis $= (X_E - X_P)$
V_e = volume of the control volume associated with e
$= (X_E - X_P)(Y_{ne} - Y_{se}) - (X_{ne} - X_{se})(Y_E - Y_P)$

The areas and volume associated with face w are given by

A_{wx}^1 = projection of area A_w^1 along X-axis $= (Y_{nw} - Y_{sw})$
A_{wy}^1 = projection of area A_w^1 along Y-axis $= -(X_{nw} - X_{sw})$
A_{wx}^2 = projection of area A_w^2 along X-axis $= -(Y_P - Y_W)$
A_{wy}^2 = projection of area A_w^2 along Y-axis $= (X_P - X_W)$
V_w = volume of the control volume associated with w
$= (X_P - X_W)(Y_{nw} - Y_{sw}) - (X_{nw} - X_{sw})(Y_P - Y_W)$

The areas and volume associated with face n are given by

A_{nx}^1 = projection of area A_n^1 along X-axis $= (Y_N - Y_P)$
A_{ny}^1 = projection of area A_n^1 along Y-axis $= -(X_N - X_P)$
A_{nx}^2 = projection of area A_n^2 along X-axis $= -(Y_{ne} - Y_{nw})$
A_{ny}^2 = projection of area A_n^2 along Y-axis $= (X_{ne} - X_{nw})$
V_n = volume of the control volume associated with e
$= (X_{ne} - X_{nw})(Y_N - Y_P) - (X_N - X_P)(Y_{ne} - Y_{nw})$

The areas and volume associated with face s are given by

A_{sx}^1 = projection of area A_s^1 along X-axis $= (Y_P - Y_S)$
A_{sy}^1 = projection of area A_s^1 along Y-axis $= -(X_P - X_S)$
A_{sx}^2 = projection of area A_s^2 along X-axis $= -(Y_{se} - Y_{sw})$
A_{sy}^2 = projection of area A_s^2 along Y-axis $= (X_{se} - X_{sw})$
= volume of the control volume associated with w
$= (X_{se} - X_{sw})(Y_P - Y_S) - (X_P - X_S)(Y_{se} - Y_{sw})$

A1.2 FOR THREE DIMENSIONS

Consider a three-dimensional body as shown in Figure A.1.2.

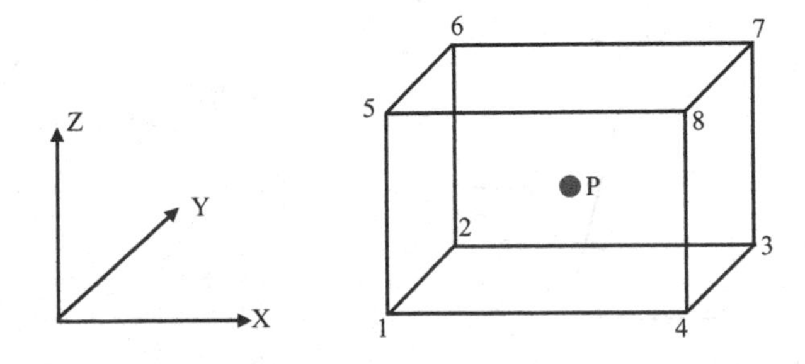

FIGURE A.1.2 Area and Volume Calculation for Three-Dimensional Body.

Area Calculation

a) West Face (Face 1265)

$$X_{56} = \frac{X_5 + X_6}{2};\ Y_{56} = \frac{Y_5 + Y_6}{2};\ Z_{56} = \frac{Z_5 + Z_6}{2}$$

$$X_{12} = \frac{X_1 + X_2}{2};\ Y_{12} = \frac{Y_1 + Y_2}{2};\ Z_{12} = \frac{Y_1 + Y_2}{2}$$

$$X_{15} = \frac{X_1 + X_5}{2};\ Y_{15} = \frac{Y_1 + Y_5}{2};\ Z_{15} = \frac{Z_1 + Z_5}{2}$$

$$X_{26} = \frac{X_2 + X_6}{2};\ Y_{26} = \frac{Y_2 + Y_6}{2};\ Z_{26} = \frac{Z_2 + Z_6}{2}$$

$$x_1 = X_{56} - X_{12};\ x_2 = X_{15} - X_{26}$$

$$y_1 = Y_{56} - Y_{12};\ y_2 = Y_{15} - Y_{26}$$

$$z_1 = z_{56} - z_{12};\ z_2 = Z_{15} - Z_{26}$$

$$A^1_{wx} = y_1 z_2 - z_1 y_2;\ A^1_{wy} = z_1 x_2 - x_1 z_2;\ A^1_{wz} = x_1 y_2 - y_1 x_2$$

b) East Face (Face 4378)

$$X_{87} = \frac{X_8 + X_7}{2};\ Y_{87} = \frac{Y_8 + Y_7}{2};\ Z_{87} = \frac{Z_8 + Z_7}{2}$$

$$X_{43} = \frac{X_4 + X_3}{2};\ Y_{43} = \frac{Y_4 + Y_3}{2};\ Z_{43} = \frac{Y_4 + Y_3}{2}$$

$$X_{48} = \frac{X_4 + X_8}{2};\ Y_{48} = \frac{Y_4 + Y_8}{2};\ Z_{48} = \frac{Z_4 + Z_8}{2}$$

$$X_{37} = \frac{X_2 + X_6}{2};\ Y_{26} = \frac{Y_2 + Y_6}{2};\ Z_{26} = \frac{Z_2 + Z_6}{2}$$

$$x_1 = X_{87} - X_{43};\ x_2 = X_{48} - X_{37}$$

$$y_1 = Y_{87} - Y_{43};\ y_2 = Y_{48} - Y_{37}$$

$$z_1 = z_{87} - z_{43};\ z_2 = Z_{48} - Z_{37}$$

$$A^1_{ex} = y_1 z_2 - z_1 y_2;\ A^1_{ey} = z_1 x_2 - x_1 z_2;\ A^1_{ez} = x_1 y_2 - y_1 x_2$$

c) *South Face (Face 1485)*

$$X_{48} = \frac{X_4 + X_8}{2}; \; Y_{48} = \frac{Y_4 + Y_8}{2}; \; Z_{48} = \frac{Z_4 + Z_8}{2}$$

$$X_{14} = \frac{X_1 + X_4}{2}; \; Y_{14} = \frac{Y_1 + Y_4}{2}; \; Z_{14} = \frac{Z_1 + Z_4}{2}$$

$$X_{58} = \frac{X_5 + X_8}{2}; \; Y_{58} = \frac{Y_5 + Y_8}{2}; \; Z_{58} = \frac{Z_5 + Z_8}{2}$$

$$x_1 = X_{48} - X_{15}; \; x_2 = X_{14} - X_{58}$$

$$y_1 = Y_{48} - Y_{15}; \; y_2 = Y_{14} - Y_{58}$$

$$z_1 = z_{48} - z_{15}; \; z_2 = Z_{14} - Z_{58}$$

$$A^2_{sx} = y_1 z_2 - z_1 y_2; \; A^2_{sy} = z_1 x_2 - x_1 z_2; \; A^2_{sz} = x_1 y_2 - y_1 x_2$$

d) *North Face (Face 2376)*

$$X_{37} = \frac{X_3 + X_7}{2}; \; Y_{37} = \frac{Y_3 + Y_7}{2}; \; Z_{37} = \frac{Z_3 + Z_7}{2}$$

$$X_{23} = \frac{X_2 + X_3}{2}; \; Y_{23} = \frac{Y_2 + Y_3}{2}; \; Z_{23} = \frac{Z_2 + Z_3}{2}$$

$$X_{76} = \frac{X_7 + X_6}{2}; \; Y_{76} = \frac{Y_7 + Y_6}{2}; \; Z_{76} = \frac{Z_6 + Z_7}{2}$$

$$x_1 = X_{37} - X_{26}; \; x_2 = X_{23} - X_{76}$$

$$y_1 = Y_{37} - Y_{26}; \; y_2 = Y_{23} - Y_{76}$$

$$z_1 = z_{37} - z_{26}; \; z_2 = Z_{23} - Z_{76}$$

$$A^2_{nx} = y_1 z_2 - z_1 y_2; \; A^2_{ny} = z_1 x_2 - x_1 z_2; \; A^2_{nz} = x_1 y_2 - y_1 x_2$$

e) *Bottom Face (Face 1432)*

$$X_{23} = \frac{X_5 + X_6}{2}; \; Y_{56} = \frac{Y_5 + Y_6}{2}; \; Z_{56} = \frac{Z_5 + Z_6}{2}$$

$$x_1 = X_{23} - X_{14}; \; x_2 = X_{12} - X_{43}$$

$$y_1 = Y_{23} - Y_{14}; \; y_2 = Y_{12} - Y_{43}$$

$$z_1 = z_{23} - z_{14}; \; z_2 = Z_{12} - Z_{43}$$

$$A_{bx}^3 = y_1z_2 - z_1y_2;\ A_{by}^3 = z_1x_2 - x_1z_2;\ A_{bz}^3 = x_1y_2 - y_1x_2$$

f) Top Face (Face 5876)

$$x_1 = X_{76} - X_{58};\ x_2 = X_{65} - X_{87}$$

$$y_1 = Y_{76} - Y_{58};\ y_2 = Y_{65} - Y_{87}$$

$$z_1 = z_{76} - z_{58};\ z_2 = Z_{65} - Z_{87}$$

$$A_{tx}^3 = y_1z_2 - z_1y_2;\ A_{ty}^3 = z_1x_2 - x_1z_2;\ A_{tz}^3 = x_1y_2 - y_1x_2$$

$$xc1 = \frac{X_4 + X_3 + X_7 + X_8}{4};\ yc1 = \frac{Y_4 + Y_3 + Y_7 + Y_8}{4};zc1 = \frac{Z_4 + Z_3 + Z_7 + Z_8}{4}$$

$$xc2 = \frac{X_2 + X_3 + X_7 + X_6}{4};\ yc2 = \frac{Y_2 + Y_3 + Y_7 + Y_6}{4};zc2 = \frac{Z_2 + Z_3 + Z_7 + Z_6}{4}$$

$$xc3 = \frac{X_1 + X_2 + X_6 + X_5}{4};\ yc3 = \frac{Y_1 + Y_2 + Y_6 + Y_5}{4};zc3 = \frac{Z_1 + Z_2 + Z_6 + Z_5}{4}$$

$$xc4 = \frac{X_1 + X_4 + X_8 + X_5}{4};\ yc4 = \frac{Y_1 + Y_4 + Y_8 + Y_5}{4};zc4 = \frac{Z_1 + Z_4 + Z_8 + Z_5}{4}$$

$$xc5 = \frac{X_5 + X_8 + X_7 + X_6}{4};\ yc5 = \frac{Y_5 + Y_8 + Y_7 + Y_6}{4};zc5 = \frac{Z_5 + Z_8 + Z_7 + Z_6}{4}$$

$$xc6 = \frac{X_1 + X_4 + X_3 + X_2}{4};\ yc6 = \frac{Y_1 + Y_4 + Y_3 + Y_2}{4};zc6 = \frac{Z_1 + Z_4 + Z_3 + Z_2}{4}$$

a) Area A_P in X-direction (Refer to Figure A.1.3)

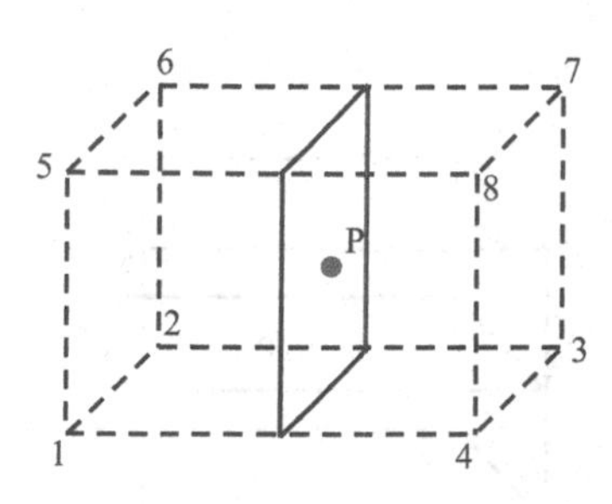

FIGURE A.1.3 Area A_P in X Direction.

$$x_1 = xc5 - xc6; \; x_2 = xc4 - xc2$$

$$y_1 = yc5 - yc6; \; y_2 = yc4 - yc2$$

$$z_1 = zc5 - zc6; \; z_2 = zc4 - zc2$$

$$A^1_{Px} = y_1z_2 - z_1y_2; A^1_{Py} = z_1x_2 - x_1z_2; A^1_{Pz} = x_1y_2 - y_1x_2$$

b) *Area A_P in Y-direction* (Refer to Figure A.1.4)

$$x_1 = xc1 - xc3; \; x_2 = xc6 - xc5$$

$$y_1 = yc1 - yc3; \; y_2 = yc6 - yc5$$

$$z_1 = zc1 - zc3; \; z_2 = zc6 - zc5c$$

$$A^2_{Px} = y_1z_2 - z_1y_2; \; A^2_{Py} = z_1x_2 - x_1z_2; \; A^2_{Pz} = x_1y_2 - y_1x_2$$

c) *Area A_P in Z-direction* (Refer Figure A.1.5)

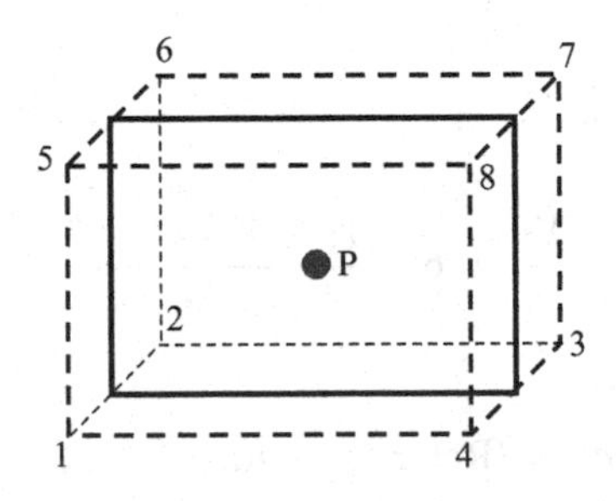

FIGURE A.1.4 Area A_P in Y Direction.

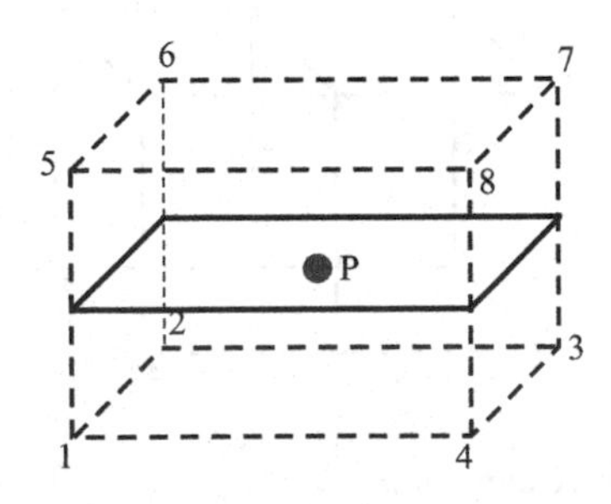

FIGURE A.1.5 Area A_P in Z Direction

$$x_1 = xc2 - xc4; \; x_2 = xc3 - xc1$$

$$y_1 = yc2 - yc4; \; y_2 = yc3 - yc1$$

$$z_1 = zc2 - zc4; z_2 = zc3 - zc1$$

$$A^3_{Px} = y_1 z_2 - z_1 y_2; \; A^3_{Py} = z_1 x_2 - x_1 z_2; \; A^3_{Pz} = x_1 y_2 - y_1 x_2$$

Volume Calculation

$$x11 = (A^3_{tx} + A^2_{nx})(X_{76} - X_{12}) + (A^1_{ex} - A^2_{sx})(X_{48} - X_{12})$$

$$y11 = \left(A^3_{ty} + A^2_{ny}\right)(Y_{76} - Y_{12}) + \left(A^1_{ey} - A^2_{sy}\right)(Y_{48} - Y_{12})$$

$$z11 = (A^3_{tz} + A^2_{nz})(Z_{76} - Z_{12}) + (A^1_{ez} - A^2_{sz})(Z_{48} - Z_{12})$$

$$V_P = \frac{x11 + y11 + z11}{3}$$

Appendix 2

Transformation of Governing Equations to Generalized Curvilinear Coordinates

A2.1 INTRODUCTION

As discussed in Chapter 7, for complex geometry, discretization of Navier-Stokes Equations can be carried out in physical space by the finite volume method. However, finite differences are mainly employed on Cartesian geometries. So, the irregular boundary of a complex geometry is approximated in a stepwise manner in the finite difference method. These boundary cells can introduce considerable errors in the implementation of a boundary condition for such a scheme. This can be avoided by mapping more complex geometries into a regular shape (Thompson et al. [1982]). Here, by suitable transformations, the computational mesh is always made rectangular. This method allows for grids that coincide with the boundary of the domain in the computational space, thereby simplifying the application of boundary conditions. Here we shall discuss the coordinate transformation of two-dimensional Navier-Stokes equations from physical space to computational space.

A2.2 GOVERNING EQUATIONS

Incompressible Navier-Stokes equations without body forces for two dimensions can be written as

Continuity equation:

$$\frac{\partial u}{\partial x}+\frac{\partial v}{\partial y}=0 \tag{A.2.1}$$

X-momentum equation:

$$\frac{\partial}{\partial t}(\rho u)+\frac{\partial}{\partial x}(\rho uu)+\frac{\partial}{\partial y}(\rho uv)=-\frac{\partial p}{\partial x}+\frac{\partial}{\partial x}\left(\mu\frac{\partial u}{\partial x}\right)+\frac{\partial}{\partial y}\left(\mu\frac{\partial u}{\partial y}\right) \tag{A.2.2}$$

Y-Momentum equation:

$$\frac{\partial}{\partial t}(\rho v)+\frac{\partial}{\partial x}(\rho uv)+\frac{\partial}{\partial y}(\rho vv)=-\frac{\partial p}{\partial y}+\frac{\partial}{\partial x}\left(\mu\frac{\partial v}{\partial x}\right)+\frac{\partial}{\partial y}\left(\mu\frac{\partial v}{\partial y}\right) \tag{A.2.3}$$

A2.3 TRANSFORMATION FROM PHYSICAL SPACE TO COMPUTATIONAL SPACE

Let us consider two-dimensional spaces, where physical space is represented by Cartesian coordinates x and y, and computational space is represented by ξ and η coordinates as shown in Figure A.2.1.
The relation between them is given by

$$\begin{aligned} \tau &= \tau(t) = t \\ \xi &= \xi\,(x, y, t) \\ \eta &= \eta\,(x, y, t) \end{aligned} \tag{A.2.4}$$

Applying the chain rule for partial differentiation, we get

$$\begin{aligned} \frac{\partial}{\partial t} &= \frac{\partial}{\partial \tau} + \frac{\partial \xi}{\partial t}\frac{\partial}{\partial \xi} + \frac{\partial \eta}{\partial t}\frac{\partial}{\partial \eta} \\ \frac{\partial}{\partial x} &= \frac{\partial \xi}{\partial x}\frac{\partial}{\partial \xi} + \frac{\partial \eta}{\partial x}\frac{\partial}{\partial \eta} \\ \frac{\partial}{\partial y} &= \frac{\partial \xi}{\partial y}\frac{\partial}{\partial \xi} + \frac{\partial \eta}{\partial y}\frac{\partial}{\partial \eta} \end{aligned} \tag{A.2.5}$$

Let us denote $\partial\xi/\partial t = \xi_t$; $\partial\xi/\partial x = \xi_x$; *etc.* we get for a fixed grid

$$\begin{aligned} \frac{\partial}{\partial t} &= \frac{\partial}{\partial \tau} + \xi_t\frac{\partial}{\partial \xi} + \eta_t\frac{\partial}{\partial \eta} = \frac{\partial}{\partial \tau} \\ \frac{\partial}{\partial x} &= \xi_x\frac{\partial}{\partial \xi} + \eta_x\frac{\partial}{\partial \eta} \end{aligned} \tag{A.2.6}$$

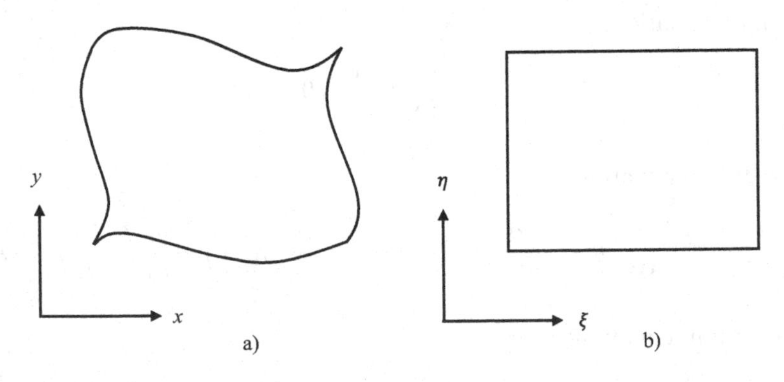

FIGURE A.2.1 Correspondence between Physical Plane and Computational plane: a) Physical Plane; b) Computational Plane.

$$\frac{\partial}{\partial y} = \xi_y \frac{\partial}{\partial \xi} + \eta_y \frac{\partial}{\partial \eta}.$$

However, we need to know the derivatives ξ_x, ξ_y and η_x, η_y. From Eq. (A.2.4) by inverse transformation we get

$$t = t(\tau) = \tau$$

$$x = x(\xi, \eta, \tau)$$

$$y = y(\xi, \eta, \tau).$$

Hence,
$$dt = t_\tau d\tau + t_\xi d\xi + t_y d\eta.$$

For a fixed grid

$$\frac{\partial t}{\partial \tau} = 1 \text{and} \frac{\partial t}{\partial \xi} = \frac{\partial t}{\partial \eta} = 0.$$

Hence,

$$dt = d\tau$$

$$dx = x_\tau d\tau + x_\xi d\xi + x_\eta d\eta$$

$$dy = y_\tau d\tau + y_\xi d\xi + y_\eta d\eta$$

or in matrix form

$$\begin{bmatrix} dt \\ dx \\ dy \end{bmatrix} = \begin{bmatrix} 1 & 0 & 0 \\ x_\tau & x_\xi & x_\eta \\ y_\tau & y_\xi & y_\eta \end{bmatrix} \begin{bmatrix} d\tau \\ d\xi \\ d\eta \end{bmatrix} \tag{A.2.7}$$

Similarly, transformation of Eq. (A.2.4) gives

$$\begin{bmatrix} d\tau \\ d\xi \\ d\eta \end{bmatrix} = \begin{bmatrix} 1 & 0 & 0 \\ \xi_t & \xi_x & \xi_y \\ \eta_t & \eta_x & \eta_y \end{bmatrix} \begin{bmatrix} dt \\ dx \\ dy \end{bmatrix} \tag{A.2.8}$$

Comparing Eqs. (A.2.7) and (A.2.8) we can write

$$\begin{bmatrix} 1 & 0 & 0 \\ \xi_t & \xi_x & \xi_y \\ \eta_t & \eta_x & \eta_y \end{bmatrix} = \begin{bmatrix} 1 & 0 & 0 \\ x_\tau & x_\xi & x_\eta \\ y_\tau & y_\xi & y_\eta \end{bmatrix}^{-1},$$

From which,
$$\xi_x = Jy_\eta;\ \xi_y = -Jx_\eta;\ \eta_x = -Jy_\xi;\ \eta_y = Jx_\xi$$

$$\xi_t = -(\tau_t x_\tau \xi_x + \tau_t y_\tau \xi_y) = 0; \; \eta_t = -\left(\tau_t x_\tau \eta_x + \tau_t y_\tau \eta_y\right) = 0 \tag{A.2.9}$$

and

$$J = \frac{\partial(\xi, \eta)}{\partial(x, y)} = \frac{1}{x_\xi y_\eta - y_\xi x_\eta}$$

J is known as the Jacobian of transformation and it is interpreted as the ratios of areas in two-dimension and volume in three-dimension in the physical space to the computational space.

For three-dimension J is given by

$$J = \frac{\partial(\xi, \eta, \zeta)}{\partial(x, y, z)} = \frac{1}{x_\xi(y_\eta z_\zeta - z_\eta y_\zeta) - x_\eta(y_\xi z_\zeta - z_\zeta y_\xi) + x_\zeta(y_\xi z_\eta - z_\xi y_\eta)}.$$

Let us consider the continuity equation

$$\frac{\partial u}{\partial x} + \frac{\partial v}{\partial y} = 0.$$

This can be transformed from physical space to computational space using Eq. (A.2.6) resulting in

$$\xi_x \frac{\partial u}{\partial \xi} + \eta_x \frac{\partial u}{\partial \eta} + \; \xi_y \frac{\partial v}{\partial \xi} + \eta_y \frac{\partial v}{\partial \eta} = 0$$

or

$$\xi_x u_\xi + \eta_x u_\eta + \; \xi_y v_\xi + \eta_y v_\eta = 0$$

$$J(y_\eta u_\xi - y_\xi u_\eta - x_\eta v_\xi + x_\xi v_\eta) = 0$$

$$J\Big[(y_\eta u)_\xi - u y_{\xi\eta} - (y_\xi u)_\eta + u y_{\xi\eta} - (x_\eta v)_\xi + v x_{\xi\eta} + (x_\xi v)_\eta - v x_{\xi\eta}\Big] = 0 \tag{A.2.10}$$

$$J\left[\frac{\partial}{\partial \xi}(u y_\eta - v x_\eta) + \frac{\partial}{\partial \eta}(v x_\xi - u y_\xi)\right]$$

Let us define

$$U = \frac{1}{J}(u\xi_x + v\xi_y) = u y_\eta - v x_\eta \text{ and } V = \frac{1}{J}\left(u\eta_x + v\eta_y\right) = v x_\xi - u y_\xi, \tag{A.2.11}$$

where U and V are known as contravariant velocities.

Eq. (A.2.10) becomes

$$\frac{\partial U}{\partial \xi} + \frac{\partial V}{\partial \eta} = 0 \tag{A.2.12}$$

Advection Term:

$$\frac{\partial}{\partial x}(\rho uu) + \frac{\partial}{\partial y}(\rho uv) = \xi_x \frac{\partial}{\partial \xi}(\rho uu) + \eta_x \frac{\partial}{\partial \eta}(\rho uu) + \xi_y \frac{\partial}{\partial \xi}(\rho uv) + \eta_y \frac{\partial}{\partial \eta}(\rho uv)$$
$$= J\left[y_\eta \frac{\partial}{\partial \xi}(\rho uu) - y_\xi \frac{\partial}{\partial \eta}(\rho uu) - x_\eta \frac{\partial}{\partial \xi}(\rho uv) + x_\xi \frac{\partial}{\partial \eta}(\rho uv)\right]$$
$$= J\left[\frac{\partial}{\partial \xi}\{\rho u(uy_\eta - vx_\eta)\} + \frac{\partial}{\partial \eta}\{\rho u(vx_\xi - uy_\xi)\}\right] = J\left[\frac{\partial}{\partial \xi}(\rho uU) + \frac{\partial}{\partial \eta}(\rho uV)\right]$$

Pressure term:

$$\frac{\partial p}{\partial x} = \xi_x \frac{\partial p}{\partial \xi} + \eta_x \frac{\partial p}{\partial \eta} = J(y_\eta p_\xi - y_\xi p_\eta)$$

Now

$$\frac{\partial}{\partial x}\left(\frac{\partial u}{\partial x}\right) = J\left[\left(\frac{\partial u}{\partial x}\right)_\xi y_\eta - \left(\frac{\partial u}{\partial x}\right)_\eta y_\xi\right]$$
$$= J\left[\left(J(u_\xi y_\eta - u_\eta y_\xi)\right)_\xi y_\eta - \left(J(u_\xi y_\eta - u_\eta y_\xi)\right)_\eta y_\xi\right]$$

$$\frac{\partial}{\partial y}\left(\frac{\partial u}{\partial y}\right) = J\left[\left(\frac{\partial u}{\partial y}\right)_\eta x_\xi - \left(\frac{\partial u}{\partial y}\right)_\xi x_\eta\right]$$
$$= J\left[\left(J(u_\eta x_\xi - u_\xi x_\eta)\right)_\eta x_\xi - \left(J(u_\eta x_\xi - u_\xi x_\eta)\right)_\xi x_\eta\right]$$

Hence, for constant property, the diffusion term becomes

$$\mu\left(\frac{\partial^2 u}{\partial x^2} + \frac{\partial^2 u}{\partial y^2}\right) = \mu J\left[\frac{\partial}{\partial \xi}\left(J\left(u_\xi y_\eta^2 - u_\eta y_\xi y_\eta - u_\eta x_\xi x_\eta + u_\xi x_\eta^2\right)\right)\right.$$
$$\left. + \frac{\partial}{\partial \eta}\left(J\left(u_\eta x_\xi^2 - u_\xi x_\xi x_\eta - u_\xi y_\xi y_\eta + u_\eta y_\xi^2\right)\right)\right]$$
$$= \mu J\left[\frac{\partial}{\partial \xi}\left(J(\alpha u_\xi - \beta u_\eta)\right) + \frac{\partial}{\partial \eta}\left(J(\gamma u_\eta - \beta u_\xi)\right)\right],$$

where

$$\alpha = x_\eta^2 + y_\eta^2;\ \beta = x_\xi x_\eta + y_\xi y_\eta;\ \gamma = x_\xi^2 + y_\xi^2.$$

Hence, the incompressible Navier-Stokes equation in the curvilinear coordinate system becomes

Continuity:

$$\frac{\partial U}{\partial \xi} + \frac{\partial V}{\partial \eta} = 0 \tag{A.2.13}$$

Momentum equations:

$$\begin{aligned}\frac{\partial}{\partial t}(\rho u) + J\left[\frac{\partial}{\partial \xi}(\rho u U) + \frac{\partial}{\partial \eta}(\rho u V)\right] &= -J\left(y_\eta \frac{\partial p}{\partial \xi} - y_\xi \frac{\partial p}{\partial \eta}\right) \\ &+\mu J\left[\frac{\partial}{\partial \xi}(J(\alpha u_\xi - \beta u_\eta)) + \frac{\partial}{\partial \eta}(J(\gamma u_\eta - \beta u_\xi))\right]\end{aligned} \tag{A.2.14}$$

$$\begin{aligned}\frac{\partial}{\partial t}(\rho v) + J\left[\frac{\partial}{\partial \xi}(\rho U v) + \frac{\partial}{\partial \eta}(\rho V v)\right] &= -J\left(x_\xi \frac{\partial p}{\partial \eta} - y_\eta \frac{\partial p}{\partial \xi}\right) \\ &+\mu J\left[\frac{\partial}{\partial \xi}(J(\alpha v_\xi - \beta v_\eta)) + \frac{\partial}{\partial \eta}(J(\gamma v_\eta - \beta v_\xi))\right]\end{aligned} \tag{A.2.15}$$

Eqs. A.2.13 to A.2.15 represent governing equations in computational space. It can be observed that by transformation, complex geometries can be mapped into a simple rectangular domain. However, the governing equations become more complex in nature.

Appendix 3

Review of Vector Calculus

A3.1 INTRODUCTION

This appendix provides a short introduction of linear algebra, which includes vectors, matrices and tensor operations. This will help the readers revise basic concepts in linear algebra, which are important in solving conservation equations numerically using computational fluid dynamics. However, for obtaining better insight, readers are directed to books dealing with linear algebra.

A3.2 VECTORS AND VECTOR OPERATIONS

In fluid dynamics, we deal mostly with velocity vector **v**. In Cartesian coordinates, it consists of three velocity components u, v, w in x, y and z directions, respectively. It can be represented as

$$\mathbf{v} = u\mathbf{i} + v\mathbf{j} + w\mathbf{k}$$

where **i**, **j**, **k** are unit vectors in x, y, z directions as shown in Figure A.3.1.

The velocity vector **v** can also be represented in matrix form as shown below

$$\mathbf{v} = \begin{bmatrix} u \\ v \\ w \end{bmatrix}$$

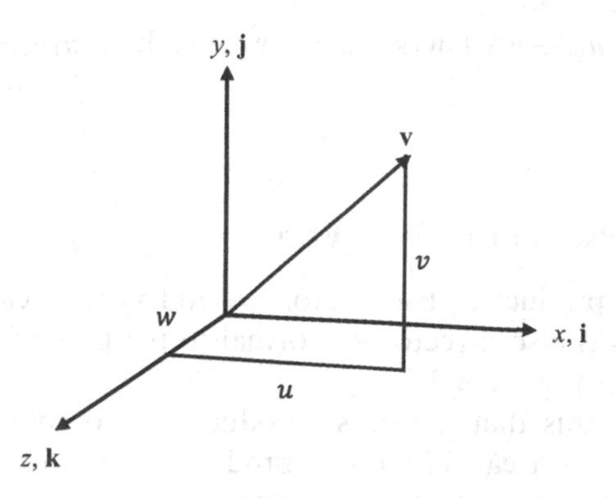

FIGURE A.3.1 Components of Velocity Vector v in Cartesian Coordinate System.

The magnitude of a vector is given as

$$\mathbf{v} = \sqrt{u^2 + v^2 + w^2}.$$

Summation of two vectors $\mathbf{v_1}$ and $\mathbf{v_2}$ is given as

$$\begin{aligned}\mathbf{v}_1 + \mathbf{v}_2 &= (u_1\mathbf{i} + v_1\mathbf{j} + w_1\mathbf{k}) + (u_2\mathbf{i} + v_2\mathbf{j} + w_2\mathbf{k}) \\ &= (u_1 + u_2)\mathbf{i} + (v_1 + v_2)\mathbf{j} + (w_1 + w_2)\mathbf{k}\end{aligned}$$

or

$$\mathbf{v}_1 + \mathbf{v}_2 = \begin{bmatrix} u_1 \\ v_1 \\ w_1 \end{bmatrix} + \begin{bmatrix} u_2 \\ v_2 \\ w_2 \end{bmatrix} = \begin{bmatrix} u_1 + u_2 \\ v_1 + v_2 \\ w_1 + w_2 \end{bmatrix}.$$

The multiplication of a scalar s with a vector $\mathbf{v}$ is given by

$$s\mathbf{v} = s(u\mathbf{i} + v\mathbf{j} + w\mathbf{k}) = su\mathbf{i} + sv\mathbf{j} + sw\mathbf{k}.$$

However, the multiplication of two vectors $\mathbf{v_1}$ and $\mathbf{v_2}$ is not so straightforward. This leads to dot product and cross product.

A3.2.1 The Dot Product of Two Vectors

The dot product of two vectors $\mathbf{v_1}$ and $\mathbf{v_2}$ is given as

$$\mathbf{v}_1.\mathbf{v}_2 = \|\mathbf{v}_1\|\|\mathbf{v}_2\|cos(\mathbf{v}_1, \mathbf{v}_2)$$

where $cos(\mathbf{v}_1, \mathbf{v}_2)$ denotes the cosine of the angle between vectors $\mathbf{v_1}$ and $\mathbf{v_2}$.

By definition

$$\mathbf{i.i} = \mathbf{j.j} = \mathbf{k.k} = 1 \quad and \quad \mathbf{i.j} = \mathbf{i.k} = \mathbf{j.i} = \mathbf{j.k} = \mathbf{k.i} = \mathbf{k.j} = 0$$

Hence,

$$\mathbf{v}_1.\mathbf{v}_2 = (u_1\mathbf{i} + v_1\mathbf{j} + w_1\mathbf{k}).(u_2\mathbf{i} + v_2\mathbf{j} + w_2\mathbf{k}) = u_1u_2 + v_1v_2 + w_1w_2$$

is a scalar.

A3.2.2 Cross Product of Two Vectors

Although the dot product of two vectors $\mathbf{v_1}$ and $\mathbf{v_2}$ is a scalar, the cross product of two vectors is also a vector $\mathbf{v_3}$ normal to the plane formed by vectors $\mathbf{v_1}$ and $\mathbf{v_2}$ as shown in Figure A.3.2.

It follows from this that the cross product of two collinear vectors is zero as they represent no area. The cross product of two unit vectors results in a unit vector perpendicular to two unit vectors.

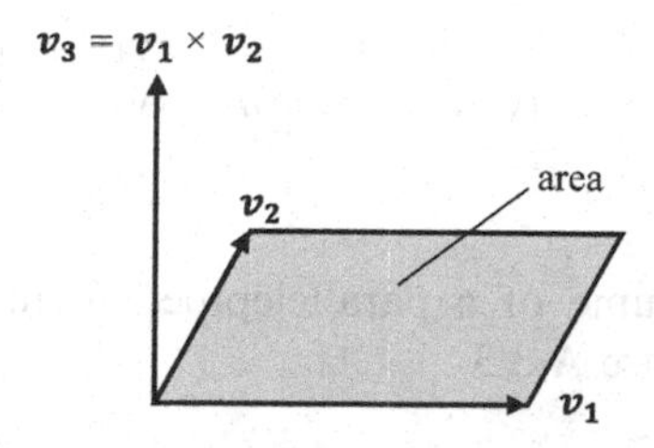

FIGURE A.3.2 Cross Product of Two Vectors.

Hence, the cross product of two collinear vectors is zero, and the cross product of two orthogonal unit vectors is a unit vector perpendicular to the two unit vectors.

By definition

$$\begin{aligned}
\mathbf{i}\times\mathbf{i} &= \mathbf{j}\times\mathbf{j} = \mathbf{k}\times\mathbf{k} = 0\\
\mathbf{i}\times\mathbf{j} &= \mathbf{k} = -\mathbf{j}\times\mathbf{i}\\
\mathbf{j}\times\mathbf{k} &= \mathbf{i} = -\mathbf{k}\times\mathbf{j}\\
\mathbf{k}\times\mathbf{i} &= \mathbf{j} = -\mathbf{i}\times\mathbf{k}
\end{aligned}$$

Using the above notation, the cross product of two vectors are given by

$$\begin{aligned}
\mathbf{v}_1\times\mathbf{v}_2 &= (u_1\mathbf{i}+v_1\mathbf{j}+w_1\mathbf{k})\times(u_2\mathbf{i}+v_2\mathbf{j}+w_2\mathbf{k})\\
&= u_1u_2\mathbf{i}\times\mathbf{i}+u_1v_2\mathbf{i}\times\mathbf{j}+u_1w_2\mathbf{i}\times\mathbf{k}\\
&\quad+v_1u_2\mathbf{j}\times\mathbf{i}+v_1v_2\mathbf{j}\times\mathbf{j}+v_1w_2\mathbf{j}\times\mathbf{k}\\
&\quad+w_1u_2\mathbf{k}\times\mathbf{i}+w_1v_2\mathbf{k}\times\mathbf{j}+w_1w_2\mathbf{k}\times\mathbf{k}\\
&= u_1u_2 0+u_1v_2\mathbf{k}+u_1w_2(-\mathbf{j})\\
&\quad+v_1u_2(-\mathbf{k})+v_1v_2 0+v_1w_2\mathbf{i}\\
&\quad+w_1u_2\mathbf{j}+w_1v_2\mathbf{i}+w_1w_2 0
\end{aligned}$$

or $$\mathbf{v}_1\times\mathbf{v}_2 = (v_1w_2-w_1v_2)\mathbf{i}+(w_1u_2-u_1w_2)\mathbf{j}+(u_1v_2-v_1u_2)\mathbf{k}.$$

This can be written as

$$\mathbf{v}_1\times\mathbf{v}_2 = \begin{vmatrix}\mathbf{i} & \mathbf{j} & \mathbf{k}\\ u_1 & v_1 & w_1\\ u_2 & v_2 & w_2\end{vmatrix} = \begin{bmatrix}(v_1w_2-w_1v_2)\\(w_1u_2-u_1w_2)\\(u_1v_2-v_1u_2)\end{bmatrix}$$

A3.2.3 Scalar Triple Product

The combined product of three vectors is represented as

$$\mathbf{v}_1.(\mathbf{v}_2 \times \mathbf{v}_3) = \begin{vmatrix} u_1 & v_1 & w_1 \\ u_2 & v_2 & w_2 \\ u_3 & v_3 & w_3 \end{vmatrix}$$

This represents the volume of a parallelepiped formed by the vectors $\mathbf{v_1}$, $\mathbf{v_2}$ and $\mathbf{v_3}$ as shown in Figure A.3.3.

A3.2.4 Operation with "Del" Operator

The "del" or "nabla" operator, which frequently arises in fluid dynamics, is defined as

$$\nabla = \frac{\partial}{\partial x}\mathbf{i} + \frac{\partial}{\partial y}\mathbf{j} + \frac{\partial}{\partial z}\mathbf{k}$$

The "del" operator applied to a scalar s results in gradient of s

$$\nabla \mathrm{s} = \frac{\partial s}{\partial x}\mathbf{i} + \frac{\partial s}{\partial y}\mathbf{j} + \frac{\partial s}{\partial z}\mathbf{k}$$

It represents a vector field in which value of s changes with both magnitude and direction.

Dot product of the del operator with a vector $\mathbf{v}$ results in divergence of the vector, which is a scalar quantity and is given by

$$\nabla.\mathbf{v} = \frac{\partial u}{\partial x} + \frac{\partial v}{\partial y} + \frac{\partial w}{\partial z}$$

It is a measure of how much the vector field goes into or out the region. You may recall, this represents the mass balance equation for incompressible flow where $\mathbf{v}$ is the velocity vector.

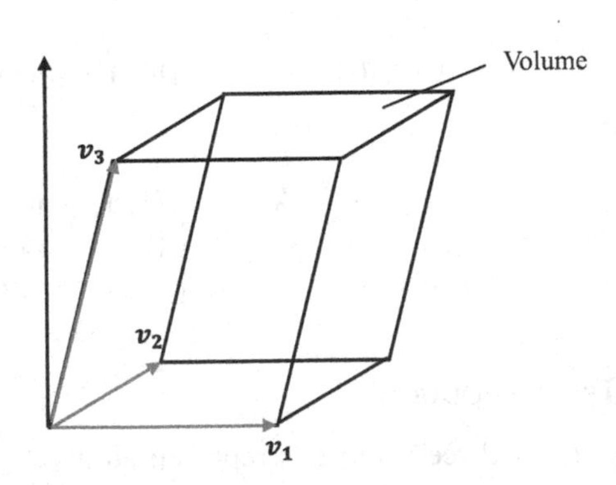

FIGURE A.3.3 Representation of Scalar Triple Product.

Divergence of the gradient is denoted by Laplacian of s and is a scalar quantity.

$$\nabla.\nabla s = \nabla^2 s = \frac{\partial^2 s}{\partial x^2} + \frac{\partial^2 s}{\partial y^2} + \frac{\partial^2 s}{\partial z^2}$$

Laplacian of a vector $\mathbf{v}$, also a vector, is given as

$$\nabla^2 \mathbf{v} = (\nabla^2 u)\mathbf{i} + (\nabla^2 v)\mathbf{j} + (\nabla^2 w)\mathbf{k}$$

Another important operation is given by

$$\nabla \times \mathbf{v} = \left(\frac{\partial}{\partial x}\mathbf{i} + \frac{\partial}{\partial y}\mathbf{j} + \frac{\partial}{\partial z}\mathbf{k}\right) \times (u\mathbf{i} + v\mathbf{j} + w\mathbf{k}) = \begin{vmatrix} \mathbf{i} & \mathbf{j} & \mathbf{k} \\ \frac{\partial}{\partial x} & \frac{\partial}{\partial y} & \frac{\partial}{\partial z} \\ u & v & w \end{vmatrix}$$

$$= \left(\frac{\partial w}{\partial y} - \frac{\partial v}{\partial z}\right)\mathbf{i} + \left(\frac{\partial u}{\partial z} - \frac{\partial w}{\partial x}\right)\mathbf{j} + \left(\frac{\partial v}{\partial y} - \frac{\partial u}{\partial x}\right)\mathbf{k}$$

The divergence of a vector $\mathbf{v}$ with its gradient can be expressed as

$$[(\mathbf{v}.\nabla)\mathbf{v}] = (u\mathbf{i} + v\mathbf{j} + w\mathbf{k}).\left(\frac{\partial}{\partial x}\mathbf{i} + \frac{\partial}{\partial y}\mathbf{j} + \frac{\partial}{\partial z}\mathbf{k}\right)(u\mathbf{i} + v\mathbf{j} + w\mathbf{k})$$

$$= \left(u\frac{\partial}{\partial x} + v\frac{\partial}{\partial y} + w\frac{\partial}{\partial z}\right)(u\mathbf{i} + v\mathbf{j} + w\mathbf{k})$$

$$= \left(u\frac{\partial u}{\partial x} + v\frac{\partial u}{\partial y} + w\frac{\partial u}{\partial z}\right)\mathbf{i} + \left(u\frac{\partial v}{\partial x} + v\frac{\partial v}{\partial y} + w\frac{\partial v}{\partial z}\right)\mathbf{j}$$

$$+ \left(u\frac{\partial w}{\partial x} + v\frac{\partial w}{\partial y} + w\frac{\partial w}{\partial z}\right)\mathbf{k}$$

A3.3 MATRICES AND MATRIX OPERATIONS

During finite difference/finite volume discretization, we obtain a system of algebraic equations, which can be written in matrix form as

$$A\emptyset = b$$

where **A** is the coefficient matrix, $\emptyset$ is unknown vector and **b** is the known vector. Here we shall discuss about the properties of matrices and their operations.

A matrix of order m × n is a rectangular array of quantities in m rows and n columns and can be written as

$$\mathbf{A} = a_{ij} = \begin{bmatrix} a_{11} & a_{12} & a_{13} & \cdots\cdots & a_{1n} \\ a_{21} & a_{22} & a_{23} & \cdots\cdots & a_{2n} \\ a_{31} & a_{32} & a_{33} & \cdots\cdots & a_{3n} \\ & & \vdots & & \\ a_{m1} & a_{m2} & a_{m3} & \cdots\cdots & a_{mn} \end{bmatrix}$$

($j \rightarrow$ *Column*, n; $i \downarrow$ *Row*, m)

If m = 1 (n > 1), it is a row matrix, and if n = 1 while m > 1, it is a column matrix. Row and column matrices are also known as vectors

A column vector **v** can be expressed as

$$\mathbf{v} = \begin{bmatrix} v_1 \\ v_2 \\ v_3 \\ \vdots \\ v_n \end{bmatrix}$$

Transpose of a matrix **A** of order m × n is a matrix $\mathbf{A}^{\mathrm{T}}$ of order n × m in which the row of column **A** is column of matrix $\mathbf{A}^{\mathrm{T}}$ and column of **A** is row of $\mathbf{A}^{\mathrm{T}}$ i.e. by interchanging rows and columns. For example, transpose of matrix **A** shown above is given as

$$\mathbf{A}^{\mathrm{T}} = a_{ji} = \begin{bmatrix} a_{11} & a_{21} & a_{31} & \cdots\cdots & a_{m1} \\ a_{12} & a_{22} & a_{32} & \cdots\cdots & a_{m2} \\ a_{13} & a_{23} & a_{33} & \cdots\cdots & a_{m3} \\ & & \vdots & & \\ a_{1n} & a_{2n} & a_{3n} & \cdots\cdots & a_{mn} \end{bmatrix}$$

For example, if

$$\mathbf{A} = \begin{bmatrix} a_{11} & a_{12} & a_{13} \\ a_{21} & a_{22} & a_{23} \\ a_{31} & a_{32} & a_{33} \end{bmatrix} = \begin{bmatrix} 3 & 5 & 2 \\ -1 & 6 & 1 \\ 5 & -3 & -3 \end{bmatrix}$$

then, transpose of it is

$$\mathbf{A}^{\mathrm{T}} = \begin{bmatrix} a_{11} & a_{21} & a_{31} \\ a_{12} & a_{22} & a_{32} \\ a_{13} & a_{23} & a_{33} \end{bmatrix} = \begin{bmatrix} 3 & -1 & 5 \\ 5 & 6 & -3 \\ 2 & 1 & -3 \end{bmatrix}$$

Two matrices are of equal order, if their corresponding elements are same.

Addition and subtraction of two matrices **A** and **B**:

$$\mathbf{A} = \begin{bmatrix} 4 & 4 & 6 \\ 1 & 4 & 2 \\ 2 & -2 & 6 \end{bmatrix}; \; \mathbf{B} = \begin{bmatrix} 2 & 1 & 5 \\ 6 & 4 & 2 \\ 7 & 2 & 1 \end{bmatrix}$$

$$\mathbf{C} = \mathbf{A} + \mathbf{B} = \begin{bmatrix} 4+2 & 4+1 & 6+5 \\ 1+6 & 4+4 & 2+2 \\ 2+7 & -2+2 & 6+1 \end{bmatrix} = \begin{bmatrix} 6 & 5 & 11 \\ 7 & 8 & 4 \\ 9 & 0 & 7 \end{bmatrix}$$

$$\mathbf{C} = \mathbf{A} - \mathbf{B} = \begin{bmatrix} 4-2 & 4-1 & 6-5 \\ 1-6 & 4-4 & 2-2 \\ 2-7 & -2-2 & 6-1 \end{bmatrix} = \begin{bmatrix} 2 & 3 & 1 \\ -5 & 0 & 0 \\ -5 & -4 & 5 \end{bmatrix}$$

If a matrix **A** is multiplied by a scalar s, then all elements of matrix **A** are multiplied by s. For example, if

$$\mathbf{A} = \begin{bmatrix} 4 & 4 & 6 \\ 1 & 4 & 2 \\ 2 & -2 & 6 \end{bmatrix}$$

then

$$s\mathbf{A} = \begin{bmatrix} 4s & 4s & 6s \\ s & 4s & 2s \\ 2s & -2s & 6s \end{bmatrix}$$

For multiplying matrix, **A,** with matrix **B**, the number of columns of matrix **A** must be equal to the number of rows of matrix **B**. For example, if the order of matrix **A** is m × p, then for multiplication the order of matrix **B** must be p × n, and the resulting matrix **C** = **AB** will be of order m × n. The elements of **C**, c_{ij} are obtained as

$$c_{ij} = \sum_{k}^{p} a_{ik} b_{kj}$$

For example,

$$\mathbf{A} = \begin{bmatrix} 2 & 3 \\ 6 & -8 \\ 5 & 2 \end{bmatrix}$$

$$\mathbf{B} = \begin{bmatrix} 2 & 4 & 6 \\ 0 & -4 & 8 \end{bmatrix}$$

Then

$$\mathbf{C} = \mathbf{A.B} = \begin{bmatrix} c_{11} & c_{12} & c_{13} \\ c_{21} & c_{22} & c_{23} \\ c_{31} & c_{32} & c_{33} \end{bmatrix}$$

Now,

$$\begin{aligned}
c_{11} &= 2 \times 2 + 3 \times 0 = 4; \; c_{12} = 2 \times 4 + 3 \times (-4) = -4; \\
c_{13} &= 2 \times 6 + 3 \times 8 = 36 \\
c_{21} &= 6 \times 2 + (-8) \times 0 = 12; \; c_{22} = 6 \times 4 + (-8) \times (-4) = 56; \\
c_{23} &= 6 \times 6 + (-8) \times 8 = -28 \\
c_{31} &= 5 \times 2 + 2 \times 0 = 10; \; c_{32} = 5 \times 4 + 2 \times (-4) = 12; \\
c_{33} &= 5 \times 6 + 2 \times 8 = 46
\end{aligned}$$

So

$$\mathbf{C} = \begin{bmatrix} 4 & -4 & 36 \\ 12 & 56 & -28 \\ 10 & 12 & 46 \end{bmatrix}$$

Please note that **A** + **B** = **B** + **A**. However, **A.B** ≠ **B.A**.

A3.3.1 Square Matrices

A matrix is called a square matrix if number of rows is equal to number columns, i.e. the order of a square matrix is n × n. It can be expressed as

$$\mathbf{A} = a_{ij} = \begin{bmatrix} a_{11} \; a_{12} \; a_{13} \cdots\cdots & a_{1n} \\ a_{21} \; a_{22} \; a_{23} \cdots\cdots & a_{2n} \\ a_{31} \; a_{32} \; a_{33} \cdots\cdots & a_{3n} \\ \vdots & \\ a_{n1} \; a_{n2} \; a_{n3} \cdots\cdots & a_{nn} \end{bmatrix}$$

The elements a_{ii} i.e. a_{11}, a_{22}, …… a_{nn} are the main diagonals of the matrix. A square matrix **A** is called symmetric if $a_{ij} = a_{ji}$ and antisymmetric or skew symmetric if $a_{ij} = -a_{ji}$. Hence, for a symmetric matrix, $\mathbf{A}^T = \mathbf{A}$. Example of symmetric matrix is

$$\mathbf{A} = \begin{bmatrix} 4 & 1 & 2 \\ 1 & 4 & -2 \\ 2 & -2 & 6 \end{bmatrix}$$

whereas the example of antisymmetric or skew symmetric matrix is

$$\mathbf{A} = \begin{bmatrix} 4 & 1 & -2 \\ -1 & 4 & 2 \\ 2 & -2 & 6 \end{bmatrix}$$

In diagonal square matrix all elements off the main diagonals are zero and the main diagonal elements are arbitrary. Example

$$\mathbf{A} = \begin{bmatrix} 4 & 0 & 0 \\ 0 & 4 & 0 \\ 0 & 0 & 6 \end{bmatrix}$$

A diagonal square matrix is called an identity or unit matrix if all main diagonal elements are equal to 1 and is denoted as **I**.

$$\mathbf{I} = \begin{bmatrix} 1 & 0 & 0 \\ 0 & 1 & 0 \\ 0 & 0 & 1 \end{bmatrix}$$

If in a matrix, all elements below the main diagonal are zero then it is called Upper triangular matrix. Mathematically it can be expressed as

$$\mathbf{U} = \begin{cases} u_{ij} \ \textit{if} \ i \leq j \\ 0 \ \textit{if} \ i > j \end{cases}$$

Similarly, if all elements above the main diagonal are zero then it is called Lower triangular matrix. Mathematically it can be expressed as

$$\mathbf{L} = \begin{cases} l_{ij} \ \textit{if} \ ij \\ 0 \ \textit{if} \ i < j \end{cases}$$

Examples of upper triangular and lower triangular matrices are shown below:

$$\mathbf{U} = \begin{bmatrix} 4 & 1 & 2 \\ 0 & 4 & -2 \\ 0 & 0 & 6 \end{bmatrix} \ \textit{and} \ \mathbf{L} = \begin{bmatrix} 4 & 0 & 0 \\ 1 & 4 & 0 \\ 2 & -2 & 6 \end{bmatrix}$$

A3.3.2 The Determinant of a Square Matrix

Determinant of a square matrix is a number denoted by det (**A**) or |**A**| which can be obtained from the elements of matrix **A**. The determinant of a 2×2 matrix is straight forward. However, for higher order matrices the calculation procedure is much more involved. Let us discuss here the method to find the determinant of a matrix.

Let **A** is a matrix of order 2 and is given by

$$\mathbf{A} = \begin{bmatrix} a_{11} & a_{12} \\ a_{21} & a_{22} \end{bmatrix}$$

The determinant is det(**A**) *or* $\begin{vmatrix} a_{11} & a_{12} \\ a_{21} & a_{22} \end{vmatrix} = a_{11}a_{22} - a_{21}a_{12}$

Example A3.1 Find the determinant of

$$\mathbf{A} = \begin{bmatrix} 1 & 3 \\ 5 & 6 \end{bmatrix}$$

Answer:

$$\det(A) = \begin{vmatrix} 1 & 3 \\ 5 & 6 \end{vmatrix} = 1 \times 6 - 5 \times 3 = 6 - 15 = -9$$

We see that finding the determinant of a matrix of order 2 is straightforward. However, for higher-order matrix, the process is more involved. Before going into details of finding the determinant of a higher-order matrix, let us first define the minor and co-factor of a matrix.

Let **A** be a matrix of $n \times n$. The *i,j* minor of elements **A,** a_{ij} is denoted by $(mi)_{ij}$ and can be find out by finding the determinant of the matrix formed by deleting i^{th} row and j^{th} column of matrix **A**. The cofactor of element is given by the relation

$$(co)_{ij} = (-1)^{i+j}(mi)_{ij}$$

Let

$$\mathbf{A} = \begin{bmatrix} a_{11} & a_{12} & a_{13} \\ a_{21} & a_{22} & a_{23} \\ a_{31} & a_{32} & a_{33} \end{bmatrix}$$

Then

$$(mi)_{11} = \begin{vmatrix} a_{22} & a_{23} \\ a_{32} & a_{33} \end{vmatrix} = a_{22} \times a_{33} - a_{32} \times a_{23}$$

and

$$(co)_{11} = (-1)^{1+1}(mi)_{11} = \begin{vmatrix} a_{22} & a_{23} \\ a_{32} & a_{33} \end{vmatrix} = a_{22} \times a_{33} - a_{32} \times a_{23}$$

Similarly,

$$(mi)_{12} = \begin{vmatrix} a_{21} & a_{23} \\ a_{31} & a_{33} \end{vmatrix} = a_{21} \times a_{33} - a_{31} \times a_{23}$$

$$(co)_{12} = (-1)^{1+2}(mi)_{12} = -\begin{vmatrix} a_{21} & a_{23} \\ a_{31} & a_{33} \end{vmatrix} = -(a_{21} \times a_{33} - a_{31} \times a_{23})$$

$$(mi)_{13} = \begin{vmatrix} a_{21} & a_{22} \\ a_{31} & a_{32} \end{vmatrix} = a_{21} \times a_{32} - a_{31} \times a_{22}$$

$$(co)_{13} = (-1)^{1+3}(mi)_{13} = \begin{vmatrix} a_{21} & a_{22} \\ a_{31} & a_{32} \end{vmatrix} = (a_{21} \times a_{32} - a_{31} \times a_{22})$$

$$(mi)_{21} = \begin{vmatrix} a_{12} & a_{13} \\ a_{32} & a_{33} \end{vmatrix} = a_{12} \times a_{33} - a_{32} \times a_{13}$$

$$(co)_{21} = (-1)^{2+1}(mi)_{21} = -\begin{vmatrix} a_{12} & a_{13} \\ a_{32} & a_{33} \end{vmatrix} = -(a_{12} \times a_{33} - a_{32} \times a_{13})$$

and so on.

The determinant of matrix **A** is given by

$$\det(\mathbf{A}) = a_{11}(co)_{11} + a_{12}(co)_{12} + a_{13}(co)_{13}.$$

If the matrix **A** is $n \times n$, then the determinant is given by

$$\det(\mathbf{A}) = a_{11}(co)_{11} + a_{12}(co)_{12} + a_{13}(co)_{13} + \ldots + a_{1n}(co)_{1n}$$

Example A3.2 Find the determinant of matrix

$$\mathbf{A} = \begin{bmatrix} 1 & 2 & 3 \\ 4 & 5 & 6 \\ 7 & 8 & 9 \end{bmatrix}$$

Answer:

The determinant of matrix **A** is given by

$$\det(\mathbf{A}) = a_{11}(co)_{11} + a_{12}(co)_{12} + a_{13}(co)_{13}$$

Now

$$(mi)_{11} = \begin{vmatrix} 5 & 6 \\ 8 & 9 \end{vmatrix} = 5 \times 9 - 8 \times 6 = 45 - 48 = -3$$

$$(co)_{11} = (-1)^{1+1}(mi)_{11} = -3$$

$$(mi)_{12} = \begin{vmatrix} 4 & 6 \\ 7 & 9 \end{vmatrix} = 4 \times 9 - 7 \times 6 = 36 - 42 = -6$$

$$(co)_{12} = (-1)^{1+2}(mi)_{12} = -(-6) = 6$$

$$(mi)_{13} = \begin{vmatrix} 4 & 5 \\ 7 & 8 \end{vmatrix} = 4 \times 8 - 7 \times 5 = 32 - 35 = -3$$

$$(co)_{13} = (-1)^{1+3}(mi)_{13} = -3$$

So

$$\det(\mathbf{A}) = 1 \times (-3) + 2 \times 6 + 3 \times (-3) = -3 + 12 - 9 = 0$$

Example A3.3 Find the determinant of matrix

$$\mathbf{A} = \begin{bmatrix} 1 & -2 & 1 \\ 5 & 5 & 4 \\ 4 & 0 & 0 \end{bmatrix}$$

Answer:
The determinant of matrix **A** is given by

$$\det(\mathbf{A}) = a_{11}(co)_{11} + a_{12}(co)_{12} + a_{13}(co)_{13}$$

$$(co)_{11} = (-1)^{1+1}(5 \times 0 - 0 \times 4) = 0$$

$$(co)_{12} = (-1)^{1+2}(5 \times 0 - 4 \times 4) = 16$$

$$(co)_{13} = (-1)^{1+3}(5 \times 0 - 4 \times 5) = -20$$

So

$$\det(\mathbf{A}) = 1 \times 0 + (-2) \times (16) + 1 \times (-20) = 0 - 32 - 20 = -52.$$

Example A3.4 Find the determinant of matrix

$$\mathbf{A} = \begin{bmatrix} 2 & -1 & 4 & 4 \\ 3 & -3 & 3 & 2 \\ 0 & 4 & -5 & 1 \\ -2 & -5 & -2 & -5 \end{bmatrix}$$

Answer:
The determinant of matrix **A** is given by

$$\det(\mathbf{A}) = a_{11}(co)_{11} + a_{12}(co)_{12} + a_{13}(co)_{13} + a_{14}(co)_{14}$$

Let us find the co-factors first.

$$(co)_{11} = (-1)^{1+1}\begin{vmatrix} -3 & 3 & 2 \\ 4 & -5 & 1 \\ -5 & -2 & -5 \end{vmatrix}$$

Here we must calculate the determinant of 3×3 matrix.

Hence,

$$\begin{aligned}(co)_{11} &= -3 \times (-1)^{1+1}.\{(-5 \times -5) - (-2 \times 1)\} + 3 \times (-1)^{1+2}.\{(4 \times -5) \\ &\quad -(-5 \times 1)\} + 2 \times (-1)^{1+3}.\{(4 \times -2) - (-5 \times -5)\} \\ &= -3 \times (25 + 2) - 3 \times (-20 + 5) + 2 \times (-8 - 25) \\ &= -3 \times 27 + 3 \times 15 - 2 \times 33 \\ &= -102\end{aligned}$$

$$(co)_{12} = (-1)^{1+2}\begin{vmatrix} 3 & 3 & 2 \\ 0 & -5 & 1 \\ -2 & -2 & -5 \end{vmatrix}$$

$$\begin{aligned}(co)_{12} &= -\Big[3 \times (-1)^{1+1}.\{(-5 \times -5) - (-2 \times 1)\} + 3 \times (-1)^{1+2}.\{(0 \times -5) \\ &\quad -(-2 \times 1)\} + 2 \times (-1)^{1+3}.\{(0 \times -2) - (-2 \times -5)\}\Big] \\ &= -[3 \times (25 + 2) - 3 \times (0 + 2) + 2 \times (0 - 10)] \\ &= -[3 \times 27 - 3 \times 2 - 2 \times 10] \\ &= -55\end{aligned}$$

$$(co)_{13} = (-1)^{1+3}\begin{vmatrix} 3 & -3 & 2 \\ 0 & 4 & 1 \\ -2 & -5 & -5 \end{vmatrix}$$

$$\begin{aligned}(co)_{13} &= \Big[3 \times (-1)^{1+1}.\{(4 \times -5) - (-5 \times 1)\} + (-3) \times (-1)^{1+2}.\{(0 \times -5) \\ &\quad -(-2 \times 1)\} + 2 \times (-1)^{1+3}.\{(0 \times -5) - (-2 \times 4)\}\Big] \\ &= [3 \times (-20 + 5) + 3 \times (0 + 2) + 2 \times (0 + 8)] \\ &= [3 \times (-15) + 3 \times 2 + 2 \times 8] \\ &= -23\end{aligned}$$

$$(co)_{14} = (-1)^{1+4} \begin{vmatrix} 3 & -3 & 3 \\ 0 & 4 & -5 \\ -2 & -5 & -2 \end{vmatrix}$$

$$\begin{aligned}(co)_{14} &= -\Big[3 \times (-1)^{1+1}.\{(4 \times -2) - (-5 \times -5)\} + (-3) \times (-1)^{1+2}.\{(0 \times -2) \\ &\quad -(-2 \times -5)\} + 3 \times (-1)^{1+3}.\{(0 \times -5) - (-2 \times 4)\}\Big] \\ &= -[3 \times (-8 - 25) + 3 \times (0 - 10) + 3 \times (0 + 8)] \\ &= -[3 \times (-33) - 3 \times 10 + 3 \times 8] \\ &= -(-105) \\ &= 105\end{aligned}$$

Hence, $\quad \det(\mathbf{A}) = 2 \times (-102) - 1 \times (-55) + 4 \times (-23) + 4 \times 105 = 179$

Clearly, we can see that as the order of the matrix increases, the effort in obtaining the determinant also increases.

A3.3.3 The Inverse of a Square Matrix

Already we have seen that during finite difference/finite volume discretization we obtain a system of algebraic equations, which can be written in matrix form as

$$\mathbf{A}\emptyset = \boldsymbol{b}$$

where **A** is n × n coefficient matrix, $\emptyset$ is unknown vector and **b** is the known n × 1 column vector. Now, the values of $\emptyset$ can be found out from

$$\emptyset = \mathbf{A}^{-1}\boldsymbol{b}$$

We can get unique solution of $\emptyset$ if matrix **A** is invertible.

Let us now see how to find the inverse of a Matrix **A.** First, we shall find the inverse for a 2×2 matrix. Let

$$\mathbf{A} = \begin{bmatrix} a & b \\ c & d \end{bmatrix}$$

Then, the inverse is given by

$$\mathbf{A}^{-1} = \frac{1}{\det(\mathbf{A})}\begin{bmatrix} d & -b \\ -c & a \end{bmatrix} = \frac{1}{ad - cb}\begin{bmatrix} d & -b \\ -c & a \end{bmatrix}$$

i.e. change the position of a and d and put (−) sign before b and c. Then divide the resultant matrix by det(**A**). Please note, we cannot obtain $\mathbf{A}^{-1}$ if the value of det(**A**) is zero as division by zero is not possible.

Example A3.5 Find the inverse of the following matrix

$$\mathbf{A} = \begin{bmatrix} 4 & 6 \\ 3 & 7 \end{bmatrix}$$

Answer:

$$\mathbf{A}^{-1} = \frac{1}{\det(\mathbf{A})}\begin{bmatrix} 7 & -6 \\ -3 & 4 \end{bmatrix} = \frac{1}{4\times 7 - 3\times 6}\begin{bmatrix} 7 & -6 \\ -3 & 4 \end{bmatrix} = \frac{1}{10}\begin{bmatrix} 7 & -6 \\ -3 & 4 \end{bmatrix}$$
$$= \begin{bmatrix} 0.7 & -0.6 \\ -0.3 & 0.4 \end{bmatrix}$$

Now let us consider of finding inverse of n × n matrix. To calculate the inverse of a higher order matrix, the procedure are as follows:

Step 1: Calculate matrix of minors of matrix **A**.
Step 2: Convert it to matrix of cofactors.
Step 3: Take the transpose of matrix of cofactors.
Step 4: Multiply the resultant matrix by $^{1}/_{\det(\mathbf{A})}$ to obtain $\mathbf{A}^{-1}$.

Example A3.6 Find the determinant of a matrix given in Example A3.2.

$$\mathbf{A} = \begin{bmatrix} 1 & 2 & 3 \\ 4 & 5 & 6 \\ 7 & 8 & 9 \end{bmatrix}$$

Answer:
Step 1: The matrix of minors is given by

$$\begin{bmatrix} 5\times 9 - 8\times 6 & 4\times 9 - 7\times 6 & 4\times 8 - 7\times 5 \\ 2\times 9 - 8\times 3 & 1\times 9 - 7\times 3 & 1\times 8 - 7\times 2 \\ 2\times 6 - 5\times 3 & 1\times 6 - 4\times 3 & 1\times 5 - 4\times 2 \end{bmatrix} = \begin{bmatrix} -3 & -6 & -3 \\ -6 & -12 & -6 \\ -3 & -6 & -3 \end{bmatrix}$$

Step 2: The matrix of cofactors can be obtained by adding sign of $(-1)^{i+j}$ before the minors. The sign convention is as follows

$$\begin{bmatrix} + & - & + \\ - & + & - \\ + & - & + \end{bmatrix}$$

Hence, the matrix of cofactors becomes

$$\begin{bmatrix} -3 & 6 & -3 \\ 6 & -12 & 6 \\ -3 & 6 & -3 \end{bmatrix}$$

Step 3: Transpose of matrix of cofactor is

$$\begin{bmatrix} -3 & 6 & -3 \\ 6 & -12 & 6 \\ -3 & 6 & -3 \end{bmatrix}$$

Step 4: Determinant of matrix A is

$$\det(\mathbf{A}) = 1 \times (-3) + 2 \times 6 + 3 \times (-3) = -3 + 12 - 9 = 0$$

Step 5: Inverse of matrix **A**

$$\mathbf{A}^{-1} = \frac{1}{det(\mathbf{A})} \begin{bmatrix} -3 & 6 & -3 \\ 6 & -12 & 6 \\ -3 & 6 & -3 \end{bmatrix}$$

Since the value of $\det(\mathbf{A}) = 0$, the inverse of matrix A does not exist.

Example A3.7 Find the determinant of matrix given in example A3.3

$$\mathbf{A} = \begin{bmatrix} 1 & -2 & 1 \\ 5 & 5 & 4 \\ 4 & 0 & 0 \end{bmatrix}$$

Answer:
Step 1: The matrix of minors is given by

$$\begin{bmatrix} 5 \times 0 - 0 \times 4 & 5 \times 0 - 4 \times 4 & 5 \times 0 - 4 \times 5 \\ -2 \times 0 - 0 \times 1 & 1 \times 0 - 4 \times 1 & 1 \times 0 - 4 \times (-2) \\ -2 \times 4 - 5 \times 1 & 1 \times 4 - 5 \times 1 & 1 \times 5 - 5 \times (-2) \end{bmatrix} = \begin{bmatrix} 0 & -16 & -20 \\ 0 & -4 & 8 \\ -13 & -1 & 15 \end{bmatrix}$$

Step 2: Hence, the matrix of cofactors becomes

$$\begin{bmatrix} 0 & 16 & -20 \\ 0 & -4 & -8 \\ -13 & 1 & 15 \end{bmatrix}$$

Step 3: Transpose of matrix of cofactors is

$$\begin{bmatrix} 0 & 0 & -13 \\ 16 & -4 & -1 \\ -20 & -8 & 15 \end{bmatrix}$$

Step 4: Determinant of matrix A is

$$\det(\mathbf{A}) = 1 \times 0 + (-2) \times (16) + 1 \times (-20) = 0 - 32 - 20 = -52$$

Step 5: Inverse of matrix **A** is

$$\mathbf{A}^{-1} = \frac{1}{-52}\begin{bmatrix} 0 & 0 & -13 \\ 16 & -4 & -1 \\ -20 & -8 & 15 \end{bmatrix} = \begin{bmatrix} 0 & 0 & \frac{-13}{-52} \\ \frac{16}{-52} & \frac{-4}{-52} & \frac{-1}{-52} \\ \frac{-20}{-52} & \frac{-8}{-52} & \frac{15}{-52} \end{bmatrix} = \begin{bmatrix} 0 & 0 & 0.25 \\ -0.307 & 0.077 & 0.019 \\ 0.385 & 0.154 & -0.288 \end{bmatrix}$$

Note:

$$\mathbf{A}.\mathbf{A}^{-1} = \mathbf{I}$$

This you can check by multiplying matrix **A** with its inverse.

A.3.3.4 Eigenvectors and Eigenvalues

Let **A** be a n × n matrix, **x** be a non-zero n × 1 column vector and λ is a scalar such that

$$\mathbf{Ax} = \lambda\mathbf{x}$$

Then **x** is known as *eigenvector* and λ as *eigenvalue* of matrix **A**. *Eigenvector* of matrix **A** provide us the characteristic of **A**. The eigenvalues of matrix **A** can be found out in the following way

$$\mathbf{Ax} = \lambda\mathbf{x} \ or \ \mathbf{Ax} - \lambda\mathbf{x} = 0 \ or \ (\mathbf{A} - \lambda\mathbf{I})\mathbf{x} = 0$$

Since **x** is non-zero, it follows that

$$\det(\mathbf{A} - \lambda\mathbf{I}) = 0$$

If the matrix **A** is given by

$$\mathbf{A} = \begin{bmatrix} a_{11} & a_{12} & a_{13} & \cdots\cdots & a_{1n} \\ a_{21} & a_{22} & a_{23} & \cdots\cdots & a_{2n} \\ a_{31} & a_{32} & a_{33} & \cdots\cdots & a_{3n} \\ & & & \vdots & \\ a_{n1} & a_{n2} & a_{n3} & \cdots\cdots & a_{nn} \end{bmatrix}$$

Then the eigenvalues can be obtained from

$$det\begin{bmatrix} a_{11}-\lambda & a_{12} & a_{13}\ldots\ldots\ldots a_{1n} \\ a_{21} & a_{22}-\lambda & a_{23}\ldots\ldots\ldots a_{2n} \\ a_{31} & a_{32} & a_{33}-\lambda\ldots\ldots\ldots a_{3n} \\ & & \vdots \\ a_{n1} & a_{n2} & a_{n3}\ldots\ldots\ldots a_{nn}-\lambda \end{bmatrix} = 0$$

Example A3.8 Find the eigenvalues of the following matrix

$$\mathbf{A} = \begin{bmatrix} -3 & 0 \\ 5 & 1 \end{bmatrix}$$

Answer:

$$\det(\mathbf{A}-\lambda\mathbf{I}) = \begin{vmatrix} -3-\lambda & 0 \\ 5 & 1-\lambda \end{vmatrix}$$

Hence,

$$(-3-\lambda)\times(1-\lambda) = 0 \; or \; \lambda^2+2\lambda-3=0$$

Solving we get, $\lambda = 1, -3$, i.e. $\lambda_1 = 1$ *and* $\lambda_2 = -3$.

Example A3.9 Find the eigen values of the following matrix

$$A = \begin{bmatrix} 5 & 0 & 0 \\ 1 & 1 & 0 \\ -1 & 5 & -2 \end{bmatrix}$$

Answer:

$$\det(\mathbf{A}-\lambda\mathbf{I}) = \begin{vmatrix} 5-\lambda & 0 & 0 \\ 1 & 1-\lambda & 0 \\ -1 & 5 & -2-\lambda \end{vmatrix}$$
$$= (5-\lambda)\begin{vmatrix} 1-\lambda & 0 \\ 5 & -2-\lambda \end{vmatrix} - 0\begin{vmatrix} 1 & 0 \\ -1 & -2-\lambda \end{vmatrix} + 0\begin{vmatrix} 1 & 1-\lambda \\ -1 & 5 \end{vmatrix}$$
$$= (5-\lambda)(1-\lambda)(-2-\lambda)$$

Hence, $(5-\lambda)(1-\lambda)(-2-\lambda) = 0$; which leads to $\lambda_1 = 5$; $\lambda_2 = 1$ *and* $\lambda_3 = -2$.

Thus, we can see that for a triangular matrix, the diagonal elements are the eigenvalues of the matrix.

A.3.3.4 MATRIX OPERATIONS

$$\mathbf{A} + (\mathbf{B} + \mathbf{C}) = (\mathbf{A} + \mathbf{B}) + \mathbf{C}$$

$$\mathbf{A} + \mathbf{B} = \mathbf{B} + \mathbf{A}$$

$$\mathbf{A}(\mathbf{BC}) = (\mathbf{AB})\mathbf{C}$$

$$\mathbf{AI} = \mathbf{IA} = \mathbf{A}$$

$$\mathbf{A}(\mathbf{B} + \mathbf{C}) = \mathbf{AB} + \mathbf{AC}$$

$$(\mathbf{A} + \mathbf{B})\mathbf{C} = \mathbf{AC} + \mathbf{BC}$$

$$(\mathbf{A} + \mathbf{B})^{\mathrm{T}} = \mathbf{A}^{\mathrm{T}} + \mathbf{B}^{\mathrm{T}}$$

$$(\mathbf{AB})^{\mathrm{T}} = \mathbf{B}^{\mathrm{T}}\mathbf{A}^{\mathrm{T}}$$

$$(\mathbf{AB})^{-1} = \mathbf{B}^{-1}\mathbf{A}^{-1}$$

A3.4 TENSORS AND TENSOR OPERATIONS

Tensors can be thought of extensions of scalar and vector. A scalar is a physical quantity represented by one piece of information i.e. magnitude and having one component. A vector is a physical quantity which provides two pieces of information i.e. magnitude and direction and having three components. A tensor can be thought of a physical quantity which provides three pieces of information like magnitude, direction and the plane in which it is acting. Hence, a scalar can be thought of a tensor of rank zero i.e. only 1 component. A vector is a tensor of rank one leading to three components. A tensor can be of rank two or three. However, in fluid dynamics we often come across stress tensor which has 3^2 i.e. 9 components. Our discussion will be restricted to tensor of rank two.

Let x, y and z be the directions of the Cartesian coordinate system. Then the stress tensor $\boldsymbol{\tau}$ is given by

$$\tau = \begin{bmatrix} \tau_{xx} & \tau_{xy} & \tau_{xz} \\ \tau_{yx} & \tau_{yy} & \tau_{yz} \\ \tau_{zx} & \tau_{zy} & \tau_{zz} \end{bmatrix}$$

where first subscript indicates the plane on which the component is acting while second subscript indicates the direction. The stress tensor is always symmetric i.e. $\tau_{ij} = \tau_{ji}$.

Let $\mathbf{v}_1$ and $\mathbf{v}_2$ be two vectors. Consider product of $\mathbf{v}_1\mathbf{v}_2$, which is neither a *dot* product nor a *cross* product but is known as *dyad* product.

$$
\begin{aligned}
\{\mathbf{v}_1\mathbf{v}_2\} &= (u_1\mathbf{i}+v_1\mathbf{j}+w_1\mathbf{k})(u_2\mathbf{i}+v_2\mathbf{j}+w_2\mathbf{k}) \\
&= \mathbf{ii}u_1u_2+\mathbf{ij}u_1v_2+\mathbf{ik}u_1w_2+\mathbf{ji}v_1u_2+\mathbf{jj}v_1v_2+\mathbf{jk}v_1w_2+\mathbf{ki}w_1u_2 \\
&\quad +\mathbf{kj}w_1v_2+\mathbf{kk}w_1w_2
\end{aligned}
$$

or

$$
\{\mathbf{v}_1\mathbf{v}_2\} = \begin{bmatrix} u_1u_2 & u_1v_2 & u_1w_2 \\ v_1u_2 & v_1w_2 & v_1w_2 \\ w_1u_2 & w_1v_2 & w_1w_2 \end{bmatrix}
$$

where **ii**, **ij**, **ik** etc. are unit dyads.

The gradient of a vector v, which is a tensor, is given by

$$
\{\nabla\mathbf{v}\} = \left\{\frac{\partial}{\partial x}\mathbf{i}+\frac{\partial}{\partial y}\mathbf{j}+\frac{\partial}{\partial z}\mathbf{k}\right\}(u\mathbf{i}+v\mathbf{j}+w\mathbf{k}) = \begin{bmatrix} \frac{\partial u}{\partial x} & \frac{\partial v}{\partial x} & \frac{\partial w}{\partial x} \\ \frac{\partial u}{\partial y} & \frac{\partial v}{\partial y} & \frac{\partial w}{\partial y} \\ \frac{\partial u}{\partial z} & \frac{\partial v}{\partial z} & \frac{\partial w}{\partial z} \end{bmatrix}
$$

The sum of two tensors $\boldsymbol{\sigma}$ and $\boldsymbol{\tau}$ is given by

$$
\boldsymbol{\sigma}+\boldsymbol{\tau} = \begin{bmatrix} \sigma_{xx}+\tau_{xx} & \sigma_{xy}+\tau_{xy} & \sigma_{xz}+\tau_{xz} \\ \sigma_{yx}+\tau_{yx} & \sigma_{yy}+\tau_{yy} & \sigma_{yz}+\tau_{yz} \\ \sigma_{zx}+\tau_{zx} & \sigma_{zy}+\tau_{zy} & \sigma_{zz}+\tau_{zz} \end{bmatrix}
$$

The multiplication of a tensor $\boldsymbol{\tau}$ by a scalar s is given by

$$
s\boldsymbol{\tau} = \begin{bmatrix} s\tau_{xx} & s\tau_{xy} & s\tau_{xz} \\ s\tau_{yx} & s\tau_{yy} & s\tau_{yz} \\ s\tau_{zx} & s\tau_{zy} & s\tau_{zz} \end{bmatrix}
$$

The dot product of a tensor $\boldsymbol{\tau}$ with a vector **v is** given by

$$
(\boldsymbol{\tau}.\boldsymbol{v}) = \begin{bmatrix} \tau_{xx} & \tau_{xy} & \tau_{xz} \\ \tau_{yx} & \tau_{yy} & \tau_{yz} \\ \tau_{zx} & \tau_{zy} & \tau_{zz} \end{bmatrix} \begin{bmatrix} u \\ v \\ w \end{bmatrix} = \begin{bmatrix} u\tau_{xx}+v\tau_{xy}+w\tau_{xz} \\ u\tau_{yx}+v\tau_{yy}+w\tau_{yz} \\ u\tau_{zx}+v\tau_{zy}+w\tau_{zz} \end{bmatrix} =
$$

$$
(u\tau_{xx}+v\tau_{xy}+w\tau_{xz})\mathbf{i}+(u\tau_{yx}+v\tau_{yy}+w\tau_{yz})\mathbf{j}+(u\tau_{zx}+v\tau_{zy}+w\tau_{zz})k
$$

Similarly, the divergence of a tensor $\boldsymbol{\tau}$ is given by

$$
(\nabla.\boldsymbol{\tau}) = \left(\frac{\partial\tau_{xx}}{\partial x}+\frac{\partial\tau_{yx}}{\partial y}+\frac{\partial\tau_{zx}}{\partial z}\right)\mathbf{i}+\left(\frac{\partial\tau_{xy}}{\partial x}+\frac{\partial\tau_{yy}}{\partial y}+\frac{\partial\tau_{zy}}{\partial z}\right)\mathbf{j}+\left(\frac{\partial\tau_{xz}}{\partial x}+\frac{\partial\tau_{yz}}{\partial y}+\frac{\partial\tau_{zz}}{\partial z}\right)\mathbf{k}
$$

Appendix 4

Case Studies

A.4.1 CASE STUDIES FOR LAMINAR FLOW

CASE STUDY 1 LID-DRIVEN CAVITY FLOW PROBLEM.

Problem definition: The lid-driven cavity flow problem with boundary conditions is shown in Figure A.4.1. The cavity is filled with a fluid, which is initially stationary. The upper plate is moving with a velocity $u = 1$ m/s while the other walls are stationary. This induces fluid to move inside the cavity. The skewness of the cavity side walls is α.

Given:

i) Assume steady flow.
ii) Consider three cases of skew angle: a) $\alpha = 30^0$; b) $\alpha = 45^0$; and c) $\alpha = 90^0$.
iii) For square cavity $(\alpha = 90^0)$, take the Reynolds number $Re = {}^{\rho u H}/_{\mu} =$ 100, 400 and 1000, and for shewed cavity $\alpha = 30^0$ and 45^0; take $Re = 100$ and 1000.
iv) Take $\rho = 1.2$ kg/m^3 while μ can be calculated from the Reynolds number.

Run the cases for $41 \times 41; 61 \times 61$ and129×129 grids with UDS, SOU and QUICK scheme.

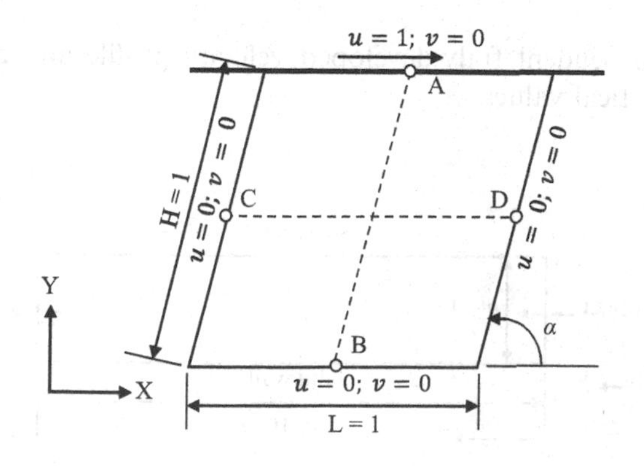

FIGURE A.4.1 Lid-Driven Cavity Problem

Find the following for an error criterion of $\leq 10^{-6}$:

i) Draw u-velocity profile along mid-plane AB and v-velocity along mid-plane CD. For square cavity, compare the results with Ghia et al. [1982] and for skewed cavity, compare the results with Demirdzic et al. [1992].
ii) Plot velocity vector streamline and vorticity contour for all the cases.

Prepare a report on this.

Case Study 2 Poiseuille Flow between Two Parallel Plates.

Problem Definition:
A laminar flow between two infinite parallel plates is known as a plane Poiseuille flow, which is shown in Figure A.4.2. The analytical fully developed velocity profile is given by

$$u = \frac{1}{2\mu}\left(-\frac{dp}{dx}\right)\left[yh - y^2\right].$$

Given:

i) Assume steady flow.
ii) Boundary conditions:
Inlet: Uniform velocity $u = 1$ m/s; $v = 0$; Solid wall: $u = 0$; $v = 0$;
Outlet: $\partial u/\partial x = \partial v/\partial x = 0$; $p = 0$ Pa.
iii) Distance between the plates h = 1m.
iv) Reynolds number = 100, 400 and 1000 where Reynolds number is defined as $Re = \rho u h/\mu$
v) Take $\rho = 1.2 \text{kg/m}^3$ while μ can be calculated from Reynolds number.

Find the following for an error criterion of $\leq 10^{-6}$:

i) Grid independent fully-developed velocity profile and compare with the analytical values.

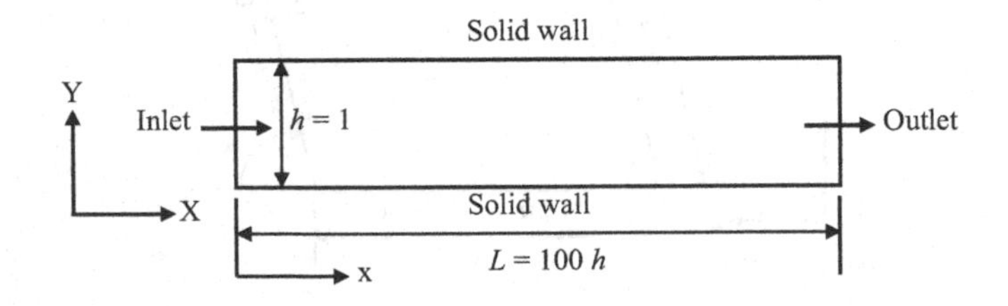

FIGURE A.4.2 Plane Poiseuille Flow

ii) Velocity and pressure profile at x/h = 10, 20, 50.
iii) Fully developed length.

Prepare a report.

Case Study 3 Laminar Flow over Backward facing Step.

Problem definition:

E The flow problem for backward-facing step is shown schematically in Figure A.4.3.

The coordinate of point A is (0,0).

Given:

i) Assume steady flow.
ii) Boundary conditions:
Inlet: Fully developed Plane Poiseuille flow between parallel plates (parabolic profile).
Solid wall: $u = 0$; $v = 0$;
Outlet: $\partial u/\partial x = \partial v/\partial x = 0$; $p = 0$ Pa.
iii) $h = 1$m, $H = 2$m, $L = 300$ h. Inlet at $x/h = -20$
iv) Reynolds number = 100, 400, 800 and 1000 where Reynolds number is defined as $Re = \rho u h/\mu$
v) Take $\rho = 1.2$kg/m^3 while μ can be calculated from Reynolds number.

Find the following for an error criterion of $\leq 10^{-6}$:

i) Carry out grid independence study.
ii) Variation of u and v velocity along Y coordinate at different x/h along the length.
iii) Values of x_1, x_2, x_3, x_4 and x_5.
iv) Stream function contours.

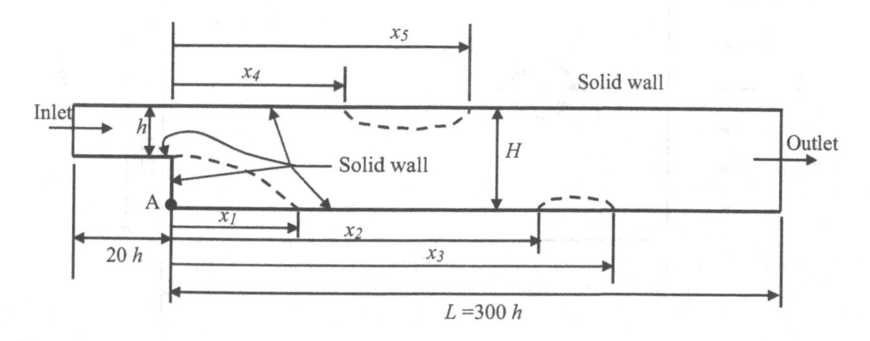

FIGURE A.4.3 Backward Facing Step

Compare your results with Armly et al. [1978] and Erturk [2008]. Prepare a report.

CASE STUDY 4 LAMINAR FLOW PAST CIRCULAR CYLINDER

The flow past a stationary circular geometry is an interesting problem for researchers. Although the geometry is simple, the flow field becomes increasingly complex even with a relatively low Reynolds number. For Reynolds numbers between 5 and 40, the flow is steady with a symmetric pair of vortices formed behind the cylinder. However, as the Reynolds number increases further, for 40 < Re < 200, the vortex shedding occurs behind the cylinder, which is an unsteady phenomenon. Hence, this problem has been selected as a case study.

Problem Definition:

A circular cylinder is subjected to an external uniform flow of U m/s. The cylinder with computational domain is shown in Figure A.4.4. Some of the definitions are

$$\text{Drag coefficient} = C_D = \frac{F_D}{0.5\rho U^2 D}$$

$$\text{Lift coefficient} = C_L = \frac{F_L}{0.5\rho U^2 D}$$

$$\text{Pressure coefficient} = C_p = \frac{(p - p_\infty)}{0.5\rho U^2 D}$$

$$\text{Strouhal number} = St = \frac{fD}{U}; \text{Reynolds number} = Re_D = \frac{\rho U D}{\mu}$$

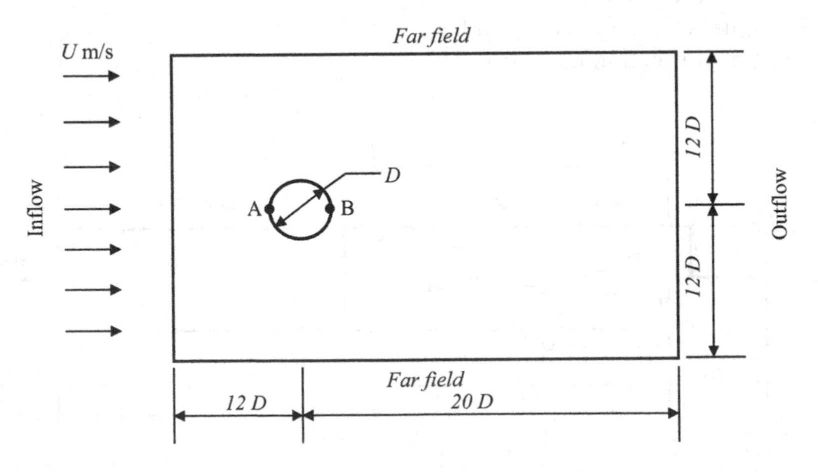

FIGURE A.4.4 Computational Domain for Flow Past Cylinder

where, U = Far stream velocity; p_∞ = Far stream pressure, p = local static pressure over cylinder; F_D = Drag force; F_L = Lift force; θ_s separation angle; f = vortex shedding frequency = $1/T$; T is the time-period of one vortex shedding cycle; A is stagnation point and B is base point.

Given:

i) Boundary conditions:
Inflow: $U = 1\text{m/s}$
Solid wall: $u = 0$; $v = 0$;
Outlet: ${\partial u}/_{\partial x} = {\partial v}/_{\partial x} = 0$; p_∞= atmospheric pressure
ii) $D = 1\text{m}$
iii) Reynolds number = 10, 40, 100.
iv) Take $\rho = 1.0\text{kg/m}^3$ while μ can be calculated from Reynolds number.

Carry out grid independence study and find the following for an error criterion of $\leq 10^{-6}$:

a) Re = 10 and 40

- Assume steady flow.
- Plot C_p vs. θ.
- Calculate drag coefficient, mean separation angle.
- Compare the results with Grove et al. [1964] and Dennis and Chang [1970].

b) Re = 100

- Carry out unsteady analysis.
- Plot C_p vs. θ.
- Calculate drag coefficient, mean separation angle and Strouhal number, and compare the results with Tritton [1959].
- Plot instantaneous stream function, velocity vectors and pressure contours for t/T = 0, 0.25, 0.5, 0.75 and 1.0 where T is the time period of vortex shedding.

Prepare a report.

A.4.2 CASE STUDIES FOR TURBULENT FLOW

CASE STUDY 5 TURBULENT FLOW BETWEEN TWO PARALLEL PLATES

Refer to Figure A.4.2 for the problem definition. You are required to compute fully developed channel flow for friction Reynolds number $Re_\tau = {u_\tau \delta}/_{\nu} = 180$, *where* $u_\tau = \sqrt{{\tau_{wall}}/_{\rho}}$ and $\delta = {h}/_{2}$. This corresponds to the flow Reynolds number $Re_h = 5600$.

Given:

i) Assume steady flow.
ii) Boundary conditions:
  a. Inlet: Uniform velocity $u = 1$ m/s; $v = 0$; $k = i^2u^2$; $i = 0.07$; $\varepsilon = {c_\mu k^{1.5}}/{(0.03h)}$; $\omega = {\varepsilon}/{k}$.
  b. Solid wall: $u = 0$; $v = 0$;
  c. Outlet: ${\partial u}/{\partial x} = {\partial v}/{\partial x} = 0$; $p = 0$ Pa.
iii) Distance between the plates h = 1m.
iv) Reynolds number = 5600 where Reynolds number is defined as $Re = {uh}/{\nu}$.
v) Take $\rho = 1\text{kg/m}^3$ while ν can be calculated from Reynolds number.

Carry out grid independence study and find the following for an error criterion of $\leq 10^{-6}$:

i) Grid independent fully-developed velocity profile
ii) Plot u^+vs. y^+
iii) Velocity and pressure profile at x/h = 10, 20, 50.
iv) Fully developed length.

Compare your results with Kreplin and Eckelmann [1979] and Kim et al. [1987]. Prepare a report.

CASE STUDY 6 TURBULENT FLOW OVER BACKWARD FACING STEP

Refer to Figure A.4.3 for the problem definition. You are required to compute for friction Reynolds number $Re_\tau = {u_\tau \delta}/{\nu} = 550$, where $u_\tau = \sqrt{{\tau_{wall}}/{\rho}}$ and $\delta = {h}/{2}$. This corresponds to the flow Reynolds number $Re_h = 9000$.

Given:

i) Assume steady flow.
ii) Boundary conditions:
  a. Inlet: fully developed turbulent inflow based on channel flow $k = i^2u^2$; $i = 0.07$; $\varepsilon = {c_\mu k^{1.5}}/{(0.03h)}$; $\omega = {\varepsilon}/{k}$.
  b. Solid wall: $u = 0$; $v = 0$.
  c. Outlet: ${\partial u}/{\partial x} = {\partial v}/{\partial x} = 0$; $p = 0$ Pa.
iii) Step height, h = 1m.
iv) Reynolds number = 9000 where the Reynolds number is defined as $Re = {u_{av}h}/{\nu}$.
v) Take $u_{av} = 2.2m/s$ and $\rho = 1\text{kg/m}^3$ while ν can be calculated from the Reynolds number.

Do the following for an error criterion of $\leq 10^{-6}$ and various turbulence models.

i) Carry out grid independence study.

ii) Find reattachment length x_r.
iii) Plot C_f = Coefficient of friction $= \tau_w/(0.5\rho u_{av}^2)$ as a function of distance from the step.
iv) Normalized velocity profile, u/u_{av} and v/u_{av}, at x = 0.5 h, 4h, 8h and 20h.
v) Static pressure profile $(p - p_w)/(0.5\rho u_{av}^2)$ and coefficient of pressure, $C_P = (p - p_0)/(0.5\rho u_{av}^2)$ at x = 0.5 h, 4h, 8h and 20h, where p_0 is the reference pressure taken at x = -4h, y = 1.5h.

Compare your results with Kopera [2011]. Prepare a report.

CASE STUDY 7 TURBULENT FLOW PAST CIRCULAR CYLINDER

Refer to Figure A.4.4 for the problem definition.

Given:

i) Boundary conditions:
   a. Inlet:

   $$U = 1\text{m/s};\ k = i^2u^2;\ i = 0.16Re^{-1/8};\ l = 0.07D;\ \varepsilon = c_\mu^{0.75}k^{1.5}/l;$$
   $$\omega = k^{0.5}/\left(c_\mu^{0.5}l\right).$$

   b. Solid wall: u = 0; v = 0.
   c. Outlet: $\partial u/\partial x = \partial v/\partial x = 0$; p_∞= atmospheric pressure.
ii) D = 1m
iii) Reynolds number = 3900.
iv) Take $\rho = 1.0\text{kg/m}^3$ while μ can be calculated from the Reynolds number.

Carry out a grid independence study and find the following for an error criterion of $\leq 10^{-6}$:

- Carry out unsteady analysis.
- Plot C_p vs. θ and wake centre line velocity u/U vs. x/D
- Calculate the drag coefficient, mean separation angle and Strouhal number.
- Compare the results with Norberg [1987] and Ong and Wallace [1996].
- Plot the instantaneous stream function, velocity vectors and pressure contours for t/T = 0, 0.25, 0.5, 0.75 and 1.0 where T is the time period of vortex shedding.

Prepare a report.

A.4.3 CASE STUDIES FOR NATURAL CONVECTION

In free convection, the fluid flow is induced by buoyancy forces that arise due to density differences caused by temperature gradients. The flow caused by buoyancy forces is dependent on gravitation. In forced convection, it is not essential to solve a separate energy equation for obtaining the velocities. However, for natural convection, since the flow is induced by temperature differences, momentum and energy equations are coupled. Hence, for two dimensions, the governing equations in the absence of other body forces become

Continuity:

$$\frac{\partial u}{\partial x}+\frac{\partial v}{\partial y}=0$$

X-Momentum:

$$\frac{\partial}{\partial t}(\rho u)+\frac{\partial}{\partial x}(\rho uu)+\frac{\partial}{\partial y}(\rho uv)=-\frac{\partial p}{\partial x}+\frac{\partial}{\partial x}\left(\mu\frac{\partial u}{\partial x}\right)+\frac{\partial}{\partial y}\left(\mu\frac{\partial u}{\partial y}\right)$$

Y-Momentum:

$$\frac{\partial}{\partial t}(\rho v)+\frac{\partial}{\partial x}(\rho uv)+\frac{\partial}{\partial y}(\rho vv)=-\frac{\partial p}{\partial y}+\frac{\partial}{\partial x}\left(\mu\frac{\partial v}{\partial x}\right)+\frac{\partial}{\partial y}\left(\mu\frac{\partial v}{\partial y}\right)+\rho_0\beta g(T-T_0)$$

Energy:

$$C_P\left[\frac{\partial}{\partial t}(\rho T)+\frac{\partial}{\partial x}(\rho uT)+\frac{\partial}{\partial y}(\rho vT)\right]=k\left(\frac{\partial^2 T}{\partial x^2}+\frac{\partial^2 T}{\partial y^2}\right)$$

Where, ρ_0 and T_0 are reference density and temperature, respectively, β = coefficient of volumetric expansion $\approx {}^{1}/_{T}$.

CASE STUDY 8 NATURAL CONVECTION IN A SQUARE CAVITY

Problem Definition: Natural convection in a square cavity; insulated top, bottom and side walls are maintained at constant but different temperatures. Refer to Figure A.4.5. The cavity is filled with air.

Given:

i Assume steady flow.

ii Rayleigh number $Ra=\rho^2 g\beta\Delta T L_{ref}^3 Pr \big/ \mu^2 = 10^4$, 10^5 and 10^6 where $Pr = {}^{C_P\mu}/_{k}$. Take ρ = 1.2 kg/m^3, C_p = 1000 J/kg K, k = 0.02514 W/mK, μ = 1.785×10^{-5} kg/ms, β = 0034/K, g = 9.81 m/s^2, $\Delta T = T_H - T_C$ = 1, Pr = 0.71. $L_{ref} = L = H$ can be calculated from Ra.

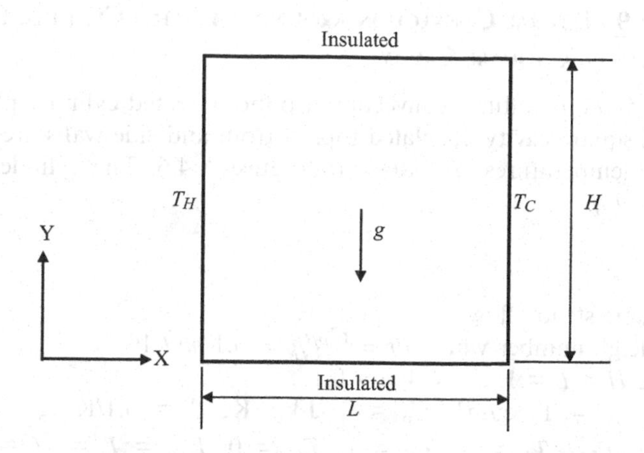

FIGURE A.4.5 Schematic Diagram of Problem of Natural Convection in a Square Cavity.

Carry out grid independence study.

Do the following for an error criterion of $\leq 10^{-6}$:

i) Draw the u-velocity profile along the cavity height at $x = L/2$ and v-velocity along cavity length at $Y = H/2$.
ii) Draw the temperature profile at the horizontal mid-plane.
iii) Find

- u_{max} at on the vertical mid-plane of the cavity and its location;
- v_{max} at on the horizontal mid-plane of the cavity and its location;
- Nu_{av}= average Nusselt number throughout the cavity;
- $Nu_{1/2}$ = average Nusselt number on the vertical mid-plane of the cavity;
- Nu_0= average Nusselt number on the vertical boundary of the cavity at x = 0;
- Nu_{max}= maximum value of the local Nusselt number on the boundary at x = 0 and its location;
- Nu_{min}= minimum value of the local Nusselt number on the boundary at x = 0 and its location.

Compare the results with de Vahl Davis [1983].

iv) Plot isotherm, streamline and velocity vectors for all cases.

Prepare a report on this.

Case Study 9 Natural Convection around a Heated Cylinder Placed Eccentrically in a Square Cavity

Problem Definition: Natural convection around a heated cylinder placed eccentrically in a square cavity; insulated top, bottom and side walls are maintained at constant temperatures, T_C. Refer to Figure A.4.6. The cylinder is kept at temperature T_H.

Given:

i) Assume steady flow.
ii) Rayleigh number where $Pr = {C_P \mu}/{k} = 0.1$ *and* 10.
iii) Take $H = L = 1$; $d = 0.4$; $e = 0.1$.
iv) Take $\rho = 1$ kg/m^3, $C_p = 1$ J/kg K, $\beta = 0.1$/K, $g = 1$ m/s^2, $\Delta T = T_H - T_C = 1$, $T_H = 1$, $T_C = 0$; $L_{ref} = L = H = 1$m. So, $Ra = {\rho^2 g \beta \Delta T L_{ref}^3 Pr}/{\mu^2} = {0.1\ Pr}/{\mu^2}$ is only a function of μ and Pr.

Carry out a grid independence study.

Do the following for an error criterion of $\leq 10^{-6}$:

i) Find the variation of the Nusselt Number along cold wall and cylinder wall.
ii) Find
   - Nu_{av} = average Nusselt number throughout the cavity;
   - Nu_{max} = maximum value of the local Nusselt number both on cold wall and Cylinder surface and their locations;

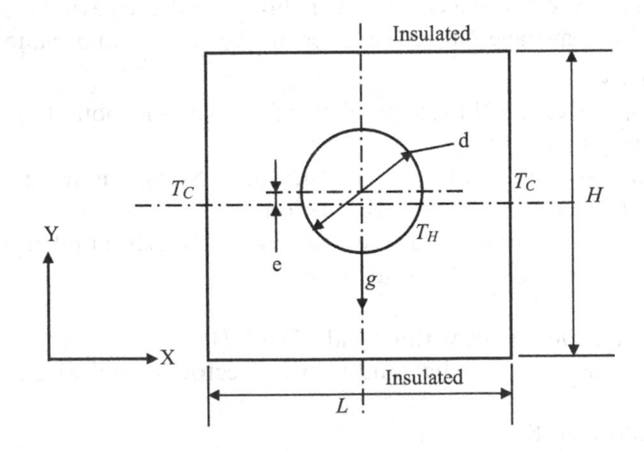

FIGURE A.4.6 Schematic Diagram of Problem of Natural Convection around a Heated Cylinder Placed in a Square Cavity

- Nu_{min} = minimum value of the local Nusselt number both on cold wall and Cylinder surface and their locations.

Compare the results with Demirdzic et al. [1992].

iii) Plot isotherm, streamline and velocity vector for all cases.

Prepare a report on this.

Case Study 10 Natural Convection around a Heated Cylinder Placed at the Centre of a Square Cavity

Refer to Figure A.4.5 for problem definition. Here, $e = 0.0$.

Take the dimensions and properties as mentioned in Case Study 9.

Take $Ra = 10^4$; 10^5; and 10^6 and $Pr = 0.7$ and 10.

Do the following for an error criterion of $\leq 10^{-6}$:

i) Find the variation of the Nusselt number along the cold wall and the cylinder wall.

ii) Find

- Nu_{av} = average Nusselt number throughout the cavity;
- Nu_{max} = maximum value of the local Nusselt number both on cold wall and the cylinder surface;
- Nu_{min} = minimum value of the local Nusselt number both on the cold wall and cylinder surface.

iii) Plot the isotherm, streamline and velocity vector for all cases.

Prepare a report on this.

References

American Institute of Aeronautics and Astronautics (AIAA), 1998. AIAA Guide for the Verification and Validation of Computational Fluid Dynamics Simulations, *AIAA G-077-1998*.

American Society of Mechanical Engineers (ASME), 2009. Standard for Verification and Validation in Computational Fluid Dynamics and Heat Transfer, *V&V 20-2009*.

Armly, B.F., Durst F., Pereira J.C.F. and Schönung, B. 1983. Experimental and theoretical investigation of backward-facing step flow, *J. Fluid Mech.*, **127**: 473–96.

Baldwin, W.S. and Lomax, H. 1978. Thin-Layer Approximation and Algebraic Model for Separated Turbulent Flows, *AIAA Paper* 78–257.

Baker, T.J. 1991. Unstructured meshes and surface fidelity for complex shapes, *10th Proc. AIAA CFD Conf., Honolulu*, 714–25, AIAA Paper 91-1591-CP.

Beam, R.M. and Warming, R.F. 1976. An implicit finite-difference algorithm for hyperbolic systems in conservation law form, *J. Computational Physics*, **22**: 87–100.

Boussinesq, T. V. 1877. Essai Sur La Theorie Des Eaux Courantes, Mem. Presentes Acad. Sci., Paris, **23**: 46.

Bowyer, A. 1981. Computing Dirichlet tessalations, *Comput. J.*, **24**: 162–66.

Brandt, A. 1977. Multilevel Adaptive Solutions to Boundary Value Problems, *Math. Comput.*, **31**: 333–90.

Cebeci T., Shao J. P., Kafyeke F. and Laurendeau E. 2005. *Computational Fluid Dynamics for Engineers*, Springer.

Chew, L.P. 1989. Constrained Delaunay triangulations, *Algorithmica*, **4**: 97–108.

Chorin, A.J. 1967. The numerical solution of the Navier-Stokes equation for an incompressible fluid, *J. Computational Physics*, **2**: 12–26.

Chorin, A.J. 1968. Numerical solution of the Navier-Stokes equation, *Math. Comput.*, **22**: 745–62.

Chung, T. J. 2002. *Computational Fluid Dynamics*, New York: Cambridge University Press.

Crandall, S.H. 1956. *Engineering Analysis*, McGraw Hill Publishing, AMR 12(1959), Rev. 1122.

Crank, J. and Nicolson, P. 1947. A Practical Method for Numerical Evaluation of Solutions of Partial Differential Equations of Heat Conduction Type, *Proc. Cambridge Philos. Soc.*, **43**: 50–67.

Daly, B.J., Harlow, F.H. 1970. Transport equations in turbulence, *Phys. Fluids*, **13**:2634–49.

de Vahl Davis, G. 1983. Natural convection of air in a square cavity: A bench mark numerical solution, *Int. J. Numer. Methods in Fluids*, **3 (3)**: 249–264.

Demirdzic, I., Lilek, Z. and Peric, M. 1992. Fluid flow and heat transfer test problems for non-orthogonal grids: Bench-Mark solutions, *Int. J. Numer. Methods in Fluids*, **15**: 329–54.

Dennis, S.C.R. and Chang, Gau-Zu. 1970. Numerical solutions for steady flow past a circular cylinder at Reynolds numbers upto 100, *J. Fluid Mech.*, **42**: 471–89.

DuFort, E. 1970.C. and Frankel, S.P. 1953. Stability Conditions in Numerical Treatment of Parabolic Differential Equations, *Math. Tables Other Aids Comput.*, **7**: 135–52.

Erturk, E. 2008. Numerical solutions of 2-D steady incompressible flow over a backward-facing step, Part I: High Reynolds number solutions, *Computers & Fluids*, **37**: 633–55.

Ferziger, J. H. and Peric, M. 2002. *Computational Methods for Fluid Dynamics*, Springer, NY, 3rd edition.

Fletcher, A. J. 1988. *Computational Techniques for Fluid Dynamics, Volume I & II*, Springer-Verlag, Berlin.

Franke, J., Hellsten, A., Schlünzen, H. and Carissimo, B. 2007. *Best Practice Guideline for the CFD Simulation of Flows in the Urban Environment*, European Cooperation in Science and Technology (COST) Action 732, Belgium.

Fromm, E.A. 1968. A method for reducing dispersion in Convective Difference Scheme, *J. Computational Physics*, **3**: 176–89.

Gaskell, P.H. and Lau, A.K.C. 1988. Curvature Compensated Convective Transport: SMART, a new Boundedness Preserving Transport Algorithm, *Int. J. Numer. Meth. Fluids*, **8**: 617–41.

Ghia, U., Ghia, K.N. and Shin, C.T. 1982. High Re solution for incompressible flow using Navier-Stokes equation and a multigrid method, *J. Computational Physics*, **48**: 387–411.

Gibson, M. and Launder, B. 1976. On the Calculation of Horizontal, Turbulent, Free Shear Flows under Gravitational Influence, *Journal of Heat Transfer*, **98**: 81–7.

Godunov, S.K. 1959. Finite-Difference Method for Numerical Computation of Discontinuous Solutions of Equations of Fluid Dynamics, *Math. Sbornik*, **47**: 271–306.

Grove, A.S., Shair, F.H., Petersen, E.E. and Andreas, A. 1964. An experimental investigation of steady flow past a circular cylinder, *J. Fluid Mech.*, **19**: 60–80.

Harlow, F.H., and Welch, J.E. 1965. Numerical Calculation of Time-Dependent Viscous Incompressible Flow of Fluids with Free Surface, *Phys. Fluids*, **8**: 2182–89.

Harten, A. and Hyman, J.M. 1983. Self-Adjusting Grid Methods for One-Dimensional Hyperbolic Conservation Laws, *J. Computational Physics*, **50**: 23–69.

Hayase T., Humphery J. A. C. and Greif R. 1992. A consistently formulated QUICK scheme for fast and stable convergence using finite-volume iterative calculation procedures, *J. Computational Physics*, **98**: 108–18.

Hildebrand, F.B. 1956. *Introduction to Numerical Analysis*, McGraw-Hill Book Co., Inc., New York, Toronto, London.

Hinze, J. O. 1975. *Turbulence*, New York: McGraw-Hill.

Hirsch, C. 1990. *Numerical Computation of Internal and External Flows, Computational Methods for Inviscid and Viscous Flows*, Vol. 2, Wiley Interscience.

Hoffmann, Klaus A. and Chiang Steve, T. 1993. *Computational Fluid Dynamics for Engineers, Volume I & II*, Engineering Education System, Wichita, Kansas, 67208–1078, USA.

Issa, R.I., Gosman, A.D. and Watkins, A.P. 1986. The computation of compressible and incompressible recirculating flows by a noniterative implicit scheme, *J. Computational Physics*, **62**: 66–82.

Joly P. and Eymard R. 1990. Preconditioned biconjugate methods for numerical reservoir simulation, *J. Computational Physics*, **91**: 298–309.

Jones W.P. and Launder B. E. 1972. The prediction of laminarization with a two-equation model, *Int. J. Heat Mass Transfer*, **15**: 301–14.

Kim, J. and Moin, P. 1985. Application of the Fractional-Step Method to Incompressible Navier-Stokes Equations. *J. Computational Physics*, **59**: 308–23.

Kim, J., Moin, P., Moser, R. 1987. Turbulence statistics in fully developed channel flows at low Reynolds number, *J. Fluid Mech.*, **177**: 133–66.

Kobayashi, M. H. and Pereira, C. F. 1991. Calculation of Incompressible laminar flows on a non-staggered, non-orthogonal grid, *Numerical Heat Transfer, Part B*, **19**: 243–62.

Könözsy, L., Drikakis, D. 2014. A Unified fractional-step, artificial compressibility and pressure-projection formulation for solving the incompressible Navier-stokes equations, *Communications in Computational Physics*, **16 (5)**: 1135–18.

Kopera, M. 2011. Direct Numerical Simulation of Turbulent flow over a backward-facing step, PhD diss., University of Warwick.

Kreplin, H. and Eckelmann, H. 1979. Behaviour of the three fluctuating components in the wall region of a turbulent channel flow, *Phy. Fluids*, **22**: 1233.

Launder, B.E. and Spalding, D.B., 1974, The numerical computation of turbulent flows, *Computer Meth. Appl. Mech. Eng.*, **3**: 269–89.

Launder, B., Reece, G., and Rodi, W. 1975. Progress in the development of a Reynolds-Stress Turbulence Closure, *J. Fluid Mech.*, **68 (3)**: 537–66.

Lax, P.D. 1954. Weak Solutions of Nonlinear Hyperbolic Equations and their Computation, *Commun. Pure Appl. Math.*, **7**: 159–93.

Lax, P.D. and Wendroff, B. 1960. Systems of Conservative Laws, *Commun. Pure Appl. Math.*, **13**: 217–37.

Leonard, B. P. 1979. A stable and accurate convective modelling procedure based on quadratic upstream interpolation, *Computer Meth. Appl. Mech. Eng.*, **19**: 59–98.

Leonard, B. P. 1987. *SHARP Simulation of Discontinuities in Highly Convective Steady Flow*, NASA TM–100240.

Leonard, B. P. 1988. Simple High-Accuracy Resolution Program for Convective Modelling of Discontinuities, *Int. J. Numer. Meth. Eng.*, **8**:1291–318.

MacCormack, R.W. 1969. The Effect of Viscosity in Hypervelocity Impact Cratering, AIAA Paper 69–354, Cincinnati, Ohio.

Menter, F.R., 1994, Two-equation eddy-viscosity turbulence models for engineering applications, *AIAA J.*, **32**: 1598–605.

Menter, F.R., Kuntz, M. and Langtry, R. 2003. Ten Years of Industrial Experience with SST Turbulence Model, *Turbulence, Heat and Mass Transfer 4*, Ed. Hinjalic, K., Nagono, Y. and Tummers, M., Begell Houce, Inc.

Moukalled, F., Mangani, L. and Darwish, M. 2015. *The Finite Volume Method in Computational Fluid Dynamics- An Advanced Introduction with OpenFOAM and Matlab*, Springer.

Muralidhar, K. and Sundararajan, T. 2008. *Computational Fluid Flow and Heat Transfer*, Narosa Publishing House, New Delhi, 2nd edition.

Norberg, C. 1987. *Effect of Reynolds number and low intensity free stream turbulence in the flow around a circular cylinder*, Report No. 87/2, Dept. of Applied Thermoscience and Fluid Mechanics, Chalmer University of Technology, Gothenurg, Sweden.

Oberkampf, W.L. and Trucano, T.G. 2002. Verification and Validation in Computational Fluid Dynamics, *Progress in Aerospace Sciences*, **38**: 209–72.

O'Brien, G. G., Hyman, M. A. and Kaplan, S. 1950. A Study of the Numerical Solution of Partial Differential Equations, *J. Math. Phys.*, **29**: 223–251.

Ong L., and Wallace J. 1996. The velocity field of the turbulent very near wake of a circular cylinder, *Exp. In Fluids*, 20: 441–53.

Patankar, S.V. 1980. *Numerical Heat Transfer and Fluid Flow*, Series of computational methods in mechanics and thermal sciences, Hemisphere Publication Corporation, McGraw-Hill Book Company.

Patankar, S.V. and Spalding, D.B. 1972. A Calculation Procedure for Heat Mass and Momentum Transfer in Three Dimensional Parabolic Flows, *Int. J. Heat Mass Transfer*, **15**: 1787–1806.

Patankar S. V. 1981. A Calculation Procedure for Two-Dimensional Elliptic Situations, *Numerical Heat Transfer*, 4: 409–425.

Pletcher, R.H., Tannehil, J.C. and Anderson, D.A. 2013. *Computational Fluid Mechanics and Heat Transfer*, 3rd Edition, CRC Press.

Pope, S. 2000. *Turbulent Flows*, Cambridge: Cambridge University Press.

Rhie, C. M. and Chow, W. L. 1983. Numerical study of the turbulent flow past an aerofoil with trailing edge separation, *AIAA J.*, **21**: 1525–32.

Richardson, L.F. 1910. The Approximate Arithmetical Solution by Finite Differences of Physical Problems involving Differential Equations, with an Application to the Stresses in a Masonry Dam, *Philos. Trans. R. Soc. London, Ser. A*, **210**: 307–57.

Roache, P.J. 1972. *Computational Fluid Dynamics*, Hermosa, Albuquerque, New Mexico.

Roache, P.J. 1998. *Verification and Validation in Computational Science and Engineering*, Hermosa Publishers, Albuquerque, NM.

Rodi, W. 2017. *Turbulence Models and Their applications in Hydraulics*, CRC Press, London.

Roe, P.L. 1980. The Use of the Riemann Problem in Finite-Difference Schemes, *Lecture Notes in Physics, Springer-Verlag, New York*, **141**: 354–9.

Roe, P.L. 1981. Approximate Riemann Solvers, Parameter Vectors and Difference Schemes, *J. Computational Physics*, **43**: 357–72.

Roe, P.L. 1986. Characteristic-based Schemes for Euler Equations, *Ann. Rev. Fluid Mech.*, **18**: 337–65.

Roychowdhury, D.G., Das, S.K. and Sundararajan, T. 1999. An Efficient Solution Method for Incompressible N-S Equations using Non-orthogonal Collocated Grid, *Int. J. Numer. Meth. Eng.*, **45**: 741–63.

Spalding, D.B. (1972). A Novel Finite-difference Formulation for Differential Expression Involving Both First and Second Derivatives, *Int. J. Numer. Methods Eng.*, **4**: 551–559.

Speziale, C., Sarker, S., and Gibson, T. 1991. Modeling the Pressure-Strain Correlation of Turbulence: An Invariant Dynamical Systems Approach, *J. Fluid Mech.*, **227**: 245–72.

Stone, H.L. 1968. Iterative solution of implicit approximations of multidimensional partial equations, *SIAM J. Numer. Anal.*, **5**: 530–558.

Sweby, P.K. 1984. High Resolution Schemes using Flux Limiters for Hyperbolic Conservation Laws, *SIAM J. Numer. Anal.*, **21(5)**: 995–1011.

Temam, R. 1969. Sur l'approximation de la solution des equations de Navier-Stokes par la methode des pas fractionnaires (i), *Arch. Rat. Mech. Anal.*, **32**: 377–85.

Tennekes, H. and Lumley, J. L. 1972. *A First Course in Turbulence*, Cambridge, MA: MIT Press.

Thomman, H.U. 1966. Numerical Integration of Navier-Stokes Equations, *Z. Angew. Phys.*, **17**: 369–84.

Thompson, J.F., Warsi, Z.U.A. and Mastin, C.W. 1982. Boundary-fitted Coordinate Systems for Numerical Solution of Partial Differential Equations- A review, *J. Computational Physics*, **47**: 1–108.

Thompson, J.F., Warsi, Z.U.A. and Mastin, C.W. 1985. *Numerical Grid Generation, Foundations and Applications*, North Holland.

Tritton, D.J. 1959. Experiments on flow around a circular cylinder at low Reynolds number, *J. Fluid Mech.*, **6**: 547–67.

Van Doormal J. P. and Raithby, G.D. 1984. Enhancements of the SIMPLE method for predicting incompressible fluid flows, *Numerical Heat Transfer*, **7**: 147–63.

Van Leer, B. 1979. Towards the Ultimate Conservative Difference Scheme V, *J. Computational Physics*, **32**: 101–36.

Versteeg, H.K. and Malalasekera. W. 1998. *An Introduction to Computational Fluid Dynamics: The Finite Volume Method*, Longman.

Warming, R.F. and Beam, R.M. 1975. Upwind Second-Order Difference Schemes and Applications in Unsteady Aerodynamic Flows, *Proc. AIAA 2nd Computational Fluid Dynamics Conference, Hartford, Connecticut*, 17–28.

Watson, D.F. 1981. Computing n-dimensional Delaunay tessalation with application to Voronoi polytopes, *Comput. J.*, **24(2)**: 167–71.

Wilcox, D. C. 1993. *Turbulence Modeling for CFD*, La Canada, CA: DCW Industries.

Index

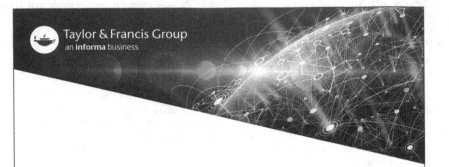

Printed in the United States
by Baker & Taylor Publisher Services

Printed in the United States
by Baker & Taylor Publisher Services